HUMAN ORIGINS & EVOLUTION

EVIDENCE & EXPLANATION

Behavior, bipedal locomotion, the hand, robusticity, throwing and striking, effects on the brain, and other human traits explained by a new theory.

By Richard W. Young, Ph.D.

Los Angeles, California

Copyright © 2013 by Richard W. Young
Los Angeles, California

All rights reserved. No part of the book including text and illustrations may be reproduced without permission in writing from the author, except by reviewers, who may quote brief passages, and by educational institutions who may copy or reprint portions for noncommercial use. Requests should be directed to rwyoung@ucla.edu

Printed and bound in the United States of America.
First Edition.

First printing 2013.

Includes bibliographical references and index.

ISBN-13: 978-1494367152

ISBN-10: 1494367157

Library of Congress Control Number: 2013922768

CreateSpace Independent Publishing Platform, North Charleston, SC

Library of Congress subject headings:
Human origins
Human evolution
Paleoanthropology

This book is dedicated to my wife, *Joyce C. Hagen, Ph.D.* in recognition and appreciation of her love, patience, and understanding of how important this project was for me during the many years it has taken to complete it.

TABLE OF CONTENTS

Author's Preface..vi-x

Acknowledgments...xi-xvi

CHAPTER 1. Science, Evolution, and the Human Lineage...1-10

CHAPTER 2. Our Last Common Ancestor with the Apes...11-17

CHAPTER 3. Human Origins...19-27

CHAPTER 4. Early Hominin Weapons............................29-37

CHAPTER 5. Gender Size Differences............................39-51

CHAPTER 6. The Transition to *Homo erectus*: Part I,
  The Evidence..53-72

CHAPTER 7. The Transition to *Homo erectus*: Part II,
  The Explanations...73-96

CHAPTER 8. Weapons and the Acquisition of Meat.........97-108

CHAPTER 9. The Human Hand....................................109-137

CHAPTER 10. Human Handedness...............................139-151

CHAPTER 11. Diminution of the Canine Teeth..............153-163

CHAPTER 12. Robusticity of Bone and Muscle..............165-177

CHAPTER 13. Development and Maturation of
  Throwing and Striking....................................179-202

CHAPTER 14. Bipedal Locomotion..............................203-227

CHAPTER 15. The Central Nervous System and
    Bipedal Use of Weapons..................................229-238

CHAPTER 16. The Structure of Evolutionary
    Explanations...239-253

CHAPTER 17. Summary and Conclusions....................255-274

APPENDIX. Predictions from the Theory....................275-282

GLOSSARY..283-293

REFERENCES CITED..295-360

INDEX...361-383

AUTHOR'S PREFACE

My theme in this book is "Where did we come from? How did we become the way we are today?" I will present evidence showing that we came from a family of tree-dwelling apes who exploited an innovative behavior that allowed them to proliferate more successfully than their fellow apes. They reproduced, natural selection acted, and we evolved. At first there were very few of us, but now our single, interbreeding species includes more than 7 billion members and is still expanding. In contrast, the dwindling populations of our nearest relatives, the apes, are headed down a path that could lead to extinction. I shall explore how we diverged from a 50-million-year history of adaptation to life in the trees and were transformed into a new kind of creature with new behaviors which enhanced our reproductive success and led to our becoming the greatest former ape the world has ever known. Something very remarkable is involved here!

The title for this book announces that its subject is human origins and evolution, and discloses my dual intention to present the evidence for these events *and* to explain them. These two categories of science—evidence and explanation—are classically separated for good reason: Evidence is the essential ingredient of science, but without an explanation of its significance, it is simply a body of accumulated facts. Alone, they fail to enhance our understanding. We want to know what the evidence *means.*

The facts to be described are in the public record, published in scientific books and articles. My primary contribution, as I see it, is in the *explanation* I provide—an explanation of what the evidence *signifies*—an explanation that aids understanding by revealing how a wide array of information related to human evolution can be made readily comprehensible in a simple manner: a new way of thinking about it. Previous explanations will be surveyed, summarized and cited so they can be compared with this new perspective.

The book is organized as a series of discussions about current issues in the study of human evolution. Because each chapter can be read as a separate essay, readers with special interests can jump right into their favorite topic. For those who wish to examine the original source of the evidence, it is fully documented so that it can be located and examined first-hand. I hope scientists whose specialty area is represented among these chapters will find that I have addressed the salient features of their topic. If they have published research in this field I trust they will find their name in the References Cited section.

I did not start out as an expert on all the issues listed in the Table of Contents—far from it. In most cases I was a novice, but over many years I did my homework. When I reached the stage at which I thought I had clarified my ideas and expressed them adequately, I sent excerpts from the book to established authorities on a particular issue, asking if they would help me improve the presentation of material in their area of expertise. Many responded and took time out from their own work to help me with mine. I gratefully identify these generous scientists in the acknowledgments a few paragraphs below. With this sort of support, and the long years I spent examining and studying these matters, I feel I am on solid, scientific ground, confident the evidence has been fairly stated and my explanations are ready to be evaluated.

The topics I have chosen vary in significance, but all concern evidence that has yet to be satisfactorily explained. Why did I pick these particular subjects? I followed my eureka idea, which took place years ago. I was reading a stirring critique of current explanations for human upright walking which revealed that no proposed benefit of such an unstable mode of travel seemed to exceed its inherent disadvantages. Suddenly, the thought came to mind, *"what about throwing?"* (It turned out that my mind had retrieved a notion I had once read in Darwin's' great book on human evolution [1871, p. 142], but at the time it felt new).

Next, I thought, if throwing was an ancient source of bipedalism (two-legged behavior), then the human hand should be adapted for throwing. It was! Two grips, one for throwing, one for clubbing! Now I definitely had the sensation of having hit upon something worth pursuing. I began to search for other features of human evolution that might yield to this simple throwing-and-clubbing explanation. There proved to be several. Some turned out to be huge (Human origins! The transition from *Australopithecus* to *Homo*! The genetic basis of throwing behavior! Its manifestation in the brain!) Some were topics in which evolutionary aspects had yet to be explored (the ontogeny, biomechanics and neurophysiology of throwing). All were satisfying insofar as they showed the utility of this new point of view to promote understanding. Many important evolutionary issues, such as modern human creativity and spoken language were not elucidated, but the breadth of its explanatory scope seems noteworthy.

Although the chapters on different topics can stand alone and be read independently, the essay format requires repetition of a few elements that are crucial to my explanation of the evidence, which is explicitly Darwinian. It is also based on *the bipedal use of hand-held weapons*. This behavior is identified as the source of the reproductive advantages that drove natural selection in a manner which accounts for hominin origins and major events in the subsequent evolution of our lineage until the emergence of *Homo*. Natural selection creates evolutionary change through differential reproduction. (I believe all evolutionary biologists accept this concept). Consequently, the question, "*what were the reproductive advantages*?" of a particular evolutionary change is repeatedly raised and examined.

Towards the end of the book, I cautiously admit that I think I have crafted a theory. Some long-time scholars in the field of paleoanthropology may find this outrageous and may even think that to mention such a thing is in bad taste. There has never been a theory of human evolution. (There is certainly a grand *general theory of evolution*, but none specifically designed for the uniquely human odyssey). This specialized field of science also is devoid of a discussion of what the structure of such a theory should be. This, in turn, could be linked to the absence of prior examination of the nature of e*xplanation* in the study of human evolution. I investigate this issue in Chapter 16, then formalize my "bipedal use of weapons" theory in what may be called the "classical" format of a scientific theory, which emphasizes simplicity of the explanation and the magnitude of the evidence it explains.

One element philosophers of science expect from a proper theory is that it should serve as a platform from which predictions can be launched. Without a theory of human evolution, predictions seem unattainable. However, with a theory in hand, predications can be made to test it. Accordingly, after Chapter 17 (a condensed summary of the book's contents), an appendix is provided containing twenty predictions suitable for testing the theory.

The book is admittedly not an "easy read" because the flow of the prose is interrupted by nearly 900 references to scientific publications. This format is designed for scientists who want to look at the original sources I am citing. Apart from these interjected names and dates, I have tried to write in a simple style with clarity as my goal, avoiding jargon where possible, defining terms when it seemed necessary, and providing a glossary which includes abbreviations (the first entry) and definitions of terms so it will be clear what I mean when I use them.

About the Author

Richard W. Young is Professor emeritus from the University of California, Los Angeles, Medical School.

After four years in the Marine Corps, he obtained a BA degree in biology from Antioch college (1956), where he was a research assistant at the Fels Research Institute. He next was a Predoctoral Fellow of the US Public Health Service at Columbia University, where he completed a Ph.D program in human anatomy (1959), followed by a National Science Foundation Postdoctoral Fellowship spent at the University of Bari (Italy) and the Karolinska Institute (Stockholm, Sweden). He subsequently received training in the use of radioisotopes at the Oak Ridge Institute of Nuclear Studies (1963) and in electron microscopic autoradiography at the Centre d'Études Nucléaires (Saclay, France, 1966-1967). In 1977 he was a Visiting Professor at Wolfson College, Cambridge University (England).

Dr. Young joined the faculty of the Department of Anatomy, UCLA Medical School in 1960 as an Assistant Professor, attaining the rank of Professor in 1968. His primary teaching assignment was in microscopic anatomy, a course which he chaired for several years. Laboratory research, funded by the US Public Health Service and primarily based upon autoradiography, was applied to questions of cell biology in bones, teeth, the ocular lens, and particularly the visual cells of the retina, in which he discovered the continual renewal of the outer segments of the rods and cones.

He received several awards recognizing his research, including an honorary Doctor of Science from the University of Chicago in 1980.

Dr. Young retired from his academic duties at UCLA in 1991, but continued to follow various aspects of science actively, becoming dedicated to the topic of human evolution in 1998. His motto is "*Ancora imparo*"

Richard W. Young, Ph.D., c. 1980

ACKNOWLEDGMENTS

Randall L. Susman and Gary P. Chimes commented on a very early version of my ideas about the role of throwing and clubbing in human evolution. Their detailed criticisms helped me get started in a better direction.

John Fleagle kindly read and commented on Chapter 2, dealing with the hominin/panin ancestor, and recommended two additional references that were pertinent. Bernard Wood also reviewed this chapter and provided some useful entries on a related topic (the hand) from an in-press publication. John Mitani provided a much appreciated comment that suggested I was on the right track. Bridget Senut also graciously examined Chapter 2 and sent me some valuable suggestions, references and advice regarding *Orrorin*, bipedalism and Miocene apes.

Chapter 4, concerning early hominin weapons, was generously reviewed by Pamela Willoughby. Among several comments that assisted me in improving the presentation, she confirmed that many stone spheroids were the result of use as hammerstones or were naturally shaped cobbles, some of which were manuports. Nicholas Toth concurred that most spheroids were battered hammerstones or exhausted cores. He supplied a crucial reference I had missed, and said the use of stones as weapons could be tested. Mohamed Sahnouni also examined Chapter 4, sent me additional references, a correction on the date for the Ain Hanech site, and pointed out that some faceted spheroids were cores. Barbara Isaac graciously read this chapter and sent me several suggestions which improved the clarity and accuracy of the exposition.

David Lordkipanidze and Bernard Wood kindly read the opening section of Chapter 6 that addresses the question of which hominins may represent the intermediary between *Australopithecus* and *Homo*. Their positive comments were very encouraging. G. Philip Rightmire also took the time to assess this section. He identified an important reference I had overlooked and commented on the relationship between the Dmanisi hominins and *H. erectus*. Henry McHenry read an excerpt from this chapter that concerns the increase in body size that accompanied the emergence of *Homo* and accepted my summary and interpretation of his valuable work on this topic.

Richard Wrangham approved my abbreviated summary of his cooking hypothesis in Chapter 7, and kindly forwarded to me three useful references concerning early hominin control of fire. Karen Steudel thoughtfully commented on the section of Chapter 7 that includes a discussion of the relationship between leg length and human bipedalism. She did not complain about my description of her cited research, but did point out a seriously malformed paragraph that enabled me to avoid an egregious error.

James O'Connell had a significant influence on Chapter 8 (the acquisition of meat). He pointed out several places where I had painted with too broad a brush and neglected the historical developments in archaeology, particularly concerning the hunting-or-scavenging controversy, and stressed that reading the early sites behaviorally was trickier than I had made it seem. These general comments were supplemented by detailed recommendations which I used to make several revisions in the manuscript. Nick Blurton Jones responded to my request for assistance on this chapter with several pertinent references and commentary on them. Pat Shipman's remarks that plausibility is insufficient in science, most hunts are unsuccessful, and scavenging (when it works) is an efficient way of getting meat all aided my efforts to get it right. Manuel Domínguez-Rodrigo approved my summary of his work and offered useful comments.

Major thanks are due to Mary Marzke for sharing her expertise on the evolution of the hominin hand (Chapter 9), beginning in 2000 when I visited her laboratory, and continuing on several occasions in subsequent years when I solicited her advice. This she gave generously and it had a profound effect on my thinking about this topic. Jonathon Hore was a source of valuable assistance in the interpretation of his neurophysiological studies of hand movement during throwing, suggesting deletions, additions, new citations, corrections and other improvements which I gladly accepted. Steven Churchill helpfully shared some thoughts on this chapter, including two important references and his support for the notion that significant hand evolution appears to have preceded stone tool manufacture. Marvin Shrewsbury sent me a detailed and provocative analysis of Chapter 9, with emphasis on his concern about the terminology of grips and their limitation to the two categories identified by Napier. Zeresenay Alemseged suggested a rewording of the description of the Dikika *A. afarensis* hand specimen.

Michel Raymond reviewed Chapter 10 on the evolution of human handedness. He offered me advice, encouragement and additional references, all of which helped me improve this part of the book. William Hopkins gave this chapter a through going-over. His incisive comments forced me to clarify my thinking about pre-hominin handedness, led me to broaden my reading on this topic, and precipitated some substantial rewriting of the manuscript.

I thank Tim White who read the section on canine diminution (Chapter 11) and provided useful suggestions that helped me avoid errors in revising it. Leonard Greenfield also took the time to read this chapter. His valuable assistance is gratefully acknowledged.

Chapter 12, concerning hominin robusticity, was enhanced by communications from Bridget Senut, Martin Pickford and William Jungers, who affirmed the robusticity of the *Orrorin tugenensis* femora. Steven Churchill informed me that the summary of the ongoing analysis regarding diaphyseal strength in the *A. sediba* hominins was accurate. Tim White's cautionary comments regarding non-quantitative statements on the topic of robusticity and the limited evidence on this topic in *Ar. ramidus* led me to make revisions in Chapter 12.

My understanding of the ontogeny of throwing in Chapter 13 was greatly assisted by detailed reviews and copious important suggestions provided by both Stephen Langendorfer and Mary Ann Roberton. Laura Petranek cordially answered my questions about her research on gender differences in throwing among children. Steve Butterfield and Jerry Thomas also examined this section. Both provided information and encouragement. Glenn Fleisig has been a major source of inspiration and has provided noteworthy assistance in my attempt to understand the biomechanics of throwing. He read multiple drafts of Chapter 13, suggested changes in wording and format, recommended alterations to aid clarity, provided important references, advocated more illustrations, and overall exerted a large impact on the final draft. Rafael Escamilla gave the biomechanical section a thorough inspection that was very helpful. Tomoyuki Matsuo also read this segment and offered a welcome suggestion for its improvement.

Several scholars diverted attention from their own work to assist me with Chapter 14, which deals with bipedal locomotion. Kevin Hunt provided valuable comments which led to my changing a number of statements I had made. He addressed several existing hypotheses for bipedal gait, critiquing some and supporting his own view of a postural origin associated with harvesting fruit from small trees by a quadrupedal ancestor. Carston Niemitz helped me craft the section of this chapter which deals with his multidisciplinary "shore-dweller" hypothesis. Robin Crompton pointed out a serious omission which, when remedied, enhanced the presentation. Carol Ward noted that carrying food may require upright walking, but carrying a pointed stick had its own advantages. Craig Stanford sent me several helpful suggestions regarding specific references, interpretations of certain fossils and the important distinction between bipedal posture and locomotion which led me to change parts of this chapter. Herman Pontzer also took the time to send me comments, which he characterized as "hard-nosed and critical". They included three missing hypotheses, the lack of archaeological evidence of early tool use, early diminution of canine size, and the use of sticks by chimpanzees, all of which have been incorporated into the book. In his view, there are many good hypotheses for bipedal locomotion, but a lack of any that can be tested. (In the appendix I offer some).

Jonathon Hore's neurophysiological expertise on the topic of neural control of throwing, a field he and his colleagues essentially created, came to my rescue on several occasions regarding the material in Chapter 15. He helped me navigate through the complex tangle of torques and forces as different interacting joints in the arm and hand are coordinated to produce and control the acceleration and release of a missile toward a target. The research of Jonathon Hore and his colleagues has revealed that natural selection for throwing prowess has measurably affected the evolution of the brain. Insofar as I have been able to understand and describe this concept, it was largely due to his assistance.

Adrienne Zihlman took the time to read and critique the draft of a chapter on scenarios of human evolution which helped me improve it. Ultimately, this discussion was condensed and incorporated into Chapter 16. Sue Parker also contributed to my preparations of this chapter, calling attention to missing references, and emphasizing both the importance of sexual selection and the difference between hypotheses and theories.

Eduard Kirschmann has been a steady source of encouragement and counsel since we first discovered our mutual interest in the role of throwing in human evolution over 12 years ago. His influence on my thinking is acknowledged in the following pages by many references to his pioneering book, *The Age of the Thrower* [1999]. I was able to enlist the skills of Susan Way, who produced an excellent English translation of this publication and was honored to write the Forward for it. Eduard is the only scholar who has used his insights to formulate predictions on this topic. His guidance and wisdom have helped me to persevere as I developed my own version of the view that early hominins became adapted to the bipedal use of weapons. Thank you, Eduard, for your creative thinking and your steady support.

These scholars represent the high levels of altruism and generosity that can be found in the scientific community, where a willingness to give up time that could be spent on one's own research is instead devoted to sharing personal expertise with a scientific colleague. I place great value on this aspect of science, and especially the assistance rendered me by these particular individuals, all of them authorities in their respective fields. They provided me with new perspectives on my work, made me rethink topics, realize weaknesses in my presentation and caused me to alter them accordingly. I hope all of these fellow scientists will understand how much I appreciate their assistance and will find that the changes they suggested to me have been made.

Sharon Belkin took my sketches for Figures 4, 8, 9, 10, 11, 13, and 14, and converted them into marvelous watercolor illustrations. (For examples in color see Figs. 1-4, Young, 2010). Some of these illustrations have been slightly modified to improve the greyscale printing in this book. I produced the other figures by ink, stippling or pencil.

All of these illustrations were skillfully scanned and wondrously enhanced by James B. Young, photoshopper extraordinaire, who also designed, produced and rendered the digital files for the cover.

Finally, I want to emphasize my debt to the University ofCalifornia for funding the pension which has sustained me during the many years I have devoted to this book, and to the University's outstanding library system, which enabled me to gain access to the hundreds of scientific references I have cited (as well as others that did not make the final cut). As time passed and technology improved, the library system kept pace. Working through the University of California, Los Angeles Library and its Collections, Search and Request services, in recent years I have been able to locate, obtain, print, and read anything and everything it seemed to me I required to pursue the several different scientific aspects touched upon in this book, without having to leave the comfortable sanctity of my home office.

Considering all the advantages and assistance I have had in this project, I certainly hope the reader will find something of value in the book that emerged from it. I will respond to comments, plaudits or criticisms concerning its contents. Send them to: rwyoung@ucla.edu

CHAPTER 1

Science, Evolution, and the Human Lineage

"Nothing in biology makes sense except in the light of evolution". [Dobzhansky, 1973, title].

Science and the search for our place in nature. Where did we come from? How did we become the way we are today? These are among the most profound of human questions—common, it seems, to all cultures. Before the rise of science, answers to such queries could only be imagined, resulting in a dazzling display of creative myth-making which often invoked the activities of one or more supernatural beings. After the emergence of biological science about three centuries ago, a new way of thinking about these questions began to gain currency. It too involved creativity, but in a manner that was constrained by an emphasis on evidence and verifiability.

Science is a method for learning about the world and our place in the universe. It yields explanations and answers from reliable evidence, rather than assertions from ancient books or opinions of modern authority figures. (Miracles and superpowers are excluded from its deliberations). Instead, by careful observation and experiment, scientists have discovered that underlying the bewildering heterogeneity of the natural world there are a number of general operating principles, such as the laws of physics and chemistry. Science is humanity's most dependable, cumulative method for comprehending reality. It gives a valid description of the way objects interact with each other and how they change from one form to another [Koch, 2012].

Biological laws are required to account for the workings of living creatures, which are much too complicated to be explained by physics and chemistry alone. The unfathomable complexity of living systems (millions of molecules of thousands of different kinds in every living cell) delayed the discovery of biological laws. The fact that living organisms are either cells or made of cells was not revealed until 1839. Twenty years later the greatest of all biological laws was identified: all life on earth—past, present and future—stems from a natural process of evolution [Darwin, 1859]. Now upgraded to a theory, Darwin's concept of evolution is the foundation of biological science. Modern biologists heartily agree with Dobzhansky [1973], whose famous observation is stated in the epigraph above. It is largely within the domain of biology that answers to the age-old queries about human origins and evolution have been profitably examined, with major advances having been made in the last few decades.

Origins and evolution. This book addresses two linked issues: How did the human lineage get started and what happened next? I approach these topics using a new explanatory principle (based on Darwin's theory) which accounts for human origins, covers millions of years of human evolution, sheds light on the numerous ways our bodies changed during that transformation, and accounts for many features of our current anatomy and its related behavioral innovations.

Some important aspects of our evolution remain outside its scope: the rapid brain enlargement that took place late in the human saga, spoken language, social behaviors, and the cultural florescence in *H. sapiens* are left in the dark. However, its explanatory range is considerable, extending from rotation of the pelvis to the opposable thumb, from bipedalism to the orientation of the shoulder joint, from the ontogeny of throwing to the redesign of canine teeth, from unique handgrips to more massive legs, from diminution of the forearm to unprecedented motor patterns stored in the brain. It explains how and why our ancestors diverged from the apes, why they abandoned the trees, how they obtained meat, and provides a rich source of predictions. In addition, this new explanation is explicitly Darwinian. I have labored relentlessly throughout the book to construct an explanation of the evidence that is based on modern Darwinism.

Evidence and explanation. Scientific evidence consists of entities or processes that can be verified by other scientists. Such confirmable information is also called empirical or objective evidence. It is objective (not subjective) because the scientist's wishes and opinions are eliminated insofar as is possible by use of measuring devices and analytical techniques to examine the item of interest. Objective evidence represents the foundation of science. In the case of early human evolution it consists primarily of fossilized bones and teeth.

A particularly interesting question regards the *significance* of the evidence. What does the accumulated evidence *mean*? Scientists begin the quest for understanding by searching through the facts (verified units of evidence) for regularities. This can be done objectively (for example by statistical analysis), but eventually the only way to comprehend the meaning of evidence is through human intervention: the creativity and insight of the human mind must be applied to make sense of it. The ultimate goal of science is not simply more evidence, but *explaining* the evidence as completely as possible in the simplest manner. Only a very parsimonious explanation of a large body of evidence is worthy of the term "theory". A theory represents the highest level of scientific explanation attainable. In contrast, a "hypothesis" is commonly an impromptu explanation for a particular observation. Sometimes these terms are misused, or used interchangeably, even by scientists—but not by this one in this book!

In the following chapters I will review the evidence for human origins and evolution, examine earlier explanations of the evidence, and then present a new explanatory structure which (I assert and demonstrate) explains the largest body of evidence in the simplest manner—and as an added bonus can be *tested*. The nature of explanation, hypotheses and theories is treated at some length in Chapter 16. Because my explanation is explicitly Darwinian, before plunging into our evolutionary history, I will describe what I take to be the essence of modern

Darwinism.

Darwinian evolution and genetics. Today, 150 years after the publication of Darwin's masterwork, *The Origin of Species* [1859], his bold and fertile theory has been so thoroughly documented, verified and confirmed that the notion it might some day be replaced is no longer seriously entertained. That is why some scientists say "it's not just a theory, it's a *fact!*" (By this they mean it is irrevocably established. Facts and theories are not synonyms).

Darwin's theory has been fine-tuned in the decades since 1859, although its basic structure remains intact. Not only has accumulated evidence indicated that Darwin's insight was correct, it has also provided enhanced understanding of how evolution works, enriching the substance of his insight. Today the theory is expanded, improved and more intellectually satisfying than the original. The science of genetics, missing back then, has become a central element of modern Darwinism. It identifies the units of inherited information, the *genes*, which embody sets of instructions coded in DNA molecules. The genes provide continuity from one generation to the next, as individuals are born, reproduce and die. In Dawkins' classic metaphor [1976], each individual is a transitory vehicle, transporting its genes for a while, then passing them on to the next generation.

In this book I employ the word "genes" as shorthand for the inherited elements of living beings, including gene alleles (different versions of the same gene). Sometimes genes control other genes. There are complex networks of regulatory regions in the DNA which turn genes on and off or alter their activity. Rearranging the position of genes may also modify their expression patterns. Alternative splicing may lead to multiple proteins being coded by a single gene. Some genetic differences are due to extra copies of a gene, others to gene loss. New genes are occasionally created, often by copying errors, but these mutations occur by chance, not design. They bear no relation to what might be useful to the organism then or in future generations. While in the long run evolution depends on mutations, it can also draw on novelties provided by recombination, chromosomal rearrangement and other DNA copying errors.

In 1974 Hull wrote that the effort to reduce Mendelian genetics to molecular genetics had not been very instructive. Today, the concept of the gene at work in evolution still cannot be systematically related to what the molecular biologist studying nucleic acid structure calls a gene. I use the term to indicate the hereditary units whose changing frequencies underlie evolutionary change. The *genotype* comprises the inherited instructions in an individual's genetic code, the *genome* is the total of all genetic material in a species, and the *phenotype* encompasses the traits that emerge during development as the information in the genes interacts with environmental influences such as nutrition.

Natural selection. Darwin's great discovery, natural selection, is the essential causal process of evolution. Natural selection usually acts at the level of individual organisms, based on heritable phenotypic properties that are linked to reproductive success. Selection occurs when organisms reproduce. Those who produce more offspring transmit more copies of their genes to the next generation: their genes are naturally selected compared to the genes of those who produce few descendants or none at all. In this way a population, over the generations, comes more and more to resemble the individuals who reproduce most frequently. The population evolves as gene frequencies change due to differential breeding of individuals.

Phenotypic variation due to underlying genetic variation is a prerequisite for evolution by natural selection. (If every individual had exactly the same genes, no evolution would occur). A casual look at our own species shows that there is plenty of inherited variability to work with: height, weight, muscularity, athleticism, skin and hair color, and many other traits "run in families". Some of these attributes provide an advantage to those who inherit them, yielding enhanced likelihood of survival to reproductive age and a greater chance of reproducing successfully and often.

The effect of environment. The "environment" is the context in which development or evolution occurs. It encompasses any factor outside the individual that may affect its well-being. Nutrition is an important aspect of the environment that influences growth, development, and reproductive success.

Climate is another environmental factor. Local environments change constantly (for example by growing colder or hotter, wetter or drier). Natural selection may track these changes to favor organisms whose genotypes are better designed to live in the changed habitats. In regard to *human* evolution, because we are preeminently social animals, and almost certainly have been since our lineage began, heritable variations that affected our relationships with each other, and with other species of animals (such as prey and predators) are environmental factors that were probably even more significant than climate or weather. Individuals of the same species compete among themselves for access to scarce resources, such as food, mating opportunities and enriched habitats. Nevertheless, a change in climate does seem to have had a major effect on the trajectory of our evolution, about half-way through our lineage, when our ancestors abandoned life in the trees.

Natural selection of behavior. Natural selection resulting from reproductive success due to a particular heritable trait can act continuously and cumulatively to produce dramatic and complex evolutionary changes called adaptations. The adaptive process is most striking when natural selection acts to improve *behaviors* which enhance reproduction, such as developing a new way to win fights over food and mating opportunities. As long as such a behavior provides reproductive advantages, it will continue to be improved because those who are born with inherited variations that enhance the behavior will be naturally selected and their genes will increase in frequency in the population's gene pool. This can go on for *millions of years*, yielding a behavior that reaches unprecedented levels of prowess and results in a major transformation of an animal's body—as it does in the centerpiece of this book.

We are a special case of a general phenomenon. Each of the millions of species on the earth presents a specific case stemming from its unique genome, evolutionary history, and the nature of the environments in which it evolved. There are many provocative, unresolved questions concerning the singular narrative of our own lineage (the *hominin* lineage) which began about seven million years ago (7 Mya). I will use the term hominin (rather than the earlier "hominid") to refer to both modern humans and our extinct relatives back to the time when our lineage diverged from that of the chimpanzees.

Scientific research has revealed an outline of the way in which our earliest hominin ancestors were gradually transformed from tree-dwelling apes into the way we are today. Eventually, they abandoned the trees and, traveling on two legs, spread throughout the world, reproducing their kind, ineluctably molded by natural selection to create a singular, peculiar, former ape, *Homo sapiens,* in which no bone, muscle, nerve or organ remained unchanged and various new behaviors abound. *Homo* is our genus and *sapiens* is our species (optimistically, it means "wise").

The origin of species. How do new species arise, and how can they be identified? The short but incomplete answer is that species arise through natural selection and different species are distinguished by their inability to breed with each other. This clear and satisfying concept (the "biological species") is unfortunately difficult to demonstrate, even in living populations. It is useful when it can be applied, but the more distant two populations are in space and time, the more difficult it becomes to test their species status relative to each other [Mayr, 1963].

The essential element for producing two biological species from one breeding population is that part of the population loses the ability to breed with the other part. How can something like that be lost? The reproductive isolation of one part of a formerly intact breeding population can occur as a result of geographic isolation: two parts of the original population become physically separated and subsequently evolve separately, each being naturally selected for optimal success in its own niche. In time, this can lead to divergence that results in sterile or nonviable hybrids when they reencounter each other and attempt to mate. Another way that new species can arise is when a subgroup comes to use resources in a way that sets them apart from other members of their species [NAS, 2008]. I will propose that something like this happened when the hominin and chimpanzee lineages diverged. Hominins initiated a distinctive behavior involving natural resources which eventually led to their reproductive isolation from other members of the breeding population.

The problem of how to define an *extinct* species is especially contentious. At the present time there are 23 species and seven genera on the hominin family tree. How can one determine interbreeding behavior in animals known only from fossilized bones and teeth? The species and genus names assigned to fossil hominins help us to keep track of which ones we are discussing, but the assignment of species names is necessarily based on *phenotypic* differences, sometimes as unimposing as tooth structure. The useful features that can be discerned in scarce fossil relics are inevitably few in number and individual variability is unknown, so the result is a tentative hypothesis [Wood and Lonergan, 2008]. A genus is a category whose borders are even less biological; generally it represents a collection of species grouped together because they appear to have had a common ancestry [Beckner, 1959]. Wood and Collard [1999] similarly defined a genus is a group of species of common ancestry, but added the idea that a genus is also adapted to a different ecological situation from that of other genera. Fortunately, the vexatious questions of fossil family trees are largely avoidable in discussion of the *process* of hominin evolution I shall describe.

The hominin lineage. A species comes to an end when it becomes extinct. If it first gave rise to another species, that part of the genome survives and may also split to form additional species. The result is a branching structure, with some branches dying without descendants. The fossil record shows that most branches end in extinction and this is true in our own ancestral branching bush. We are the last remaining hominins, the only survivors of a complex family tree of ancient ancestors (but reproduction is one of our strong points!) Somewhere in the maze of expired relatives there existed a continuous lineage—a single, surviving, often-changing packet of genes of which we are now the sole vehicles.

A few authors have expressed the view that there is a disjunction between pre-*Homo* and post-*Homo* hominins (although they do not question the genetic continuity) leading them to conclude that whatever explains hominin origins cannot also explain human origins because not all hominins became humans. I prefer to emphasize the continuity of the hominin lineage during which the genome persisted, while undergoing modification by natural selection, even though the path through the hominin family tree cannot yet be precisely identified. The transition to *Homo* was certainly dramatic, but there is no reason to doubt it was due to the same process of natural selection that had previously acted on hominin genes for several million years and will continue to act in the future. A new genus designation may be appropriate for the first hominins to abandon the trees, particularly if one accepts the importance of adaptation to different ecological situations [Wood and Collard, 1999], but there is nothing biologically aberrant in the way it happened.

Human uniqueness. The genes we share today bear the instructions for an animal remarkably different from our earliest ancestors, who were arboreal apes. Why was our special collection of genes naturally selected? What were the reproductive advantages bestowed by our astounding mix of inherited traits? In this book I will develop the proposal that many of them can be explained by natural selection acting to improve an innovative *behavior* initiated by the first hominins— a behavior that provided reproductive benefits driving a process of evolutionary adaptation that persisted for millions of years.

If true, our species should be particularly skilled at this behavior, it should be inherited (developing in every child without teaching), numerous features of our body structure should be adapted for it, and it should have characteristics that reasonably could have provided enhanced survival and breeding opportunities throughout most of hominin history.

In what follows I will show that natural selection to improve this behavior can account for many alterations that occurred as our forbears were transformed from arboreal specialists to dedicated ground-dwellers. Modifications in body size, gender size differences, proportional changes in the size and mass of the arms and legs, transmutations that produced a waist, independent rotation of the pelvis and thorax, increased mobility of the shoulder joint and changes in its orientation, remodeling of the hand, the onset of handedness, smaller canine teeth, robusticity of bone and muscle, the innate development of throwing and club-swinging, a crucial contribution to the onset of bipedal locomotion, and modifications of the central nervous system in controlling the behavior in question are all accounted for. Each of these topics can be clarified by the proposition that our ancestors underwent a prolonged selection for improved *use of hand-held weapons from a bipedal stance*. The first use of weapons signaled human origins. This behavior continued to yield reproductive benefits in competition for enriched habitats, food and mates for millions of years.

What were we like at the beginning of the hominin odyssey? To identify the starting point of our lineage, we must look millions of years into the past for a glimpse of our earliest ancestors, the first hominins. This is the topic of the next chapter.

CHAPTER 2

Our Last Common Ancestor With The Apes

"While phylogenetic evidence points toward chimpanzees, and fossil evidence remains ambiguous, experimental studies of humans and other primates point squarely toward an arboreal, climbing ancestor of hominids" [Schmitt, 2003, p. 1441].

"The last common ancestor of chimpanzees and hominins was a predominantly arboreal, orthograde, short-legged, long-armed great ape with long, curved fingers and toes "[Crompton, et al., 2008, p. 533].

When and where did our lineage begin? For well over a century evolutionary scientists have known that we are descended from an ancient primate ancestor that lived mainly in the trees, an ancestor that also gave rise to the surviving great apes. Man's place in nature is with the arboreal apes, Huxley asserted in 1863, based on his studies of comparative anatomy. Darwin [1871] thought it was likely that the human lineage had its origins in Africa, the homeland of chimpanzees and gorillas. The discovery there by Dart [1925] of an ancient, ape-like creature that walked upright (*Australopithecus africanus*) has focused the search for our earliest ancestors on that continent ever since. Numerous potential human predecessors have subsequently been found in Africa, some younger and more human-like than *A. africanus,* others older and more ape-like.

The oldest have features that make it difficult to decide if they are ancestors of hominins, ancestors of apes, or simply ancient apes who left no descendants and became extinct. The *where* question seems settled: our lineage arose in Africa.

When did the African apes arise? Our arboreal ancestry goes back well beyond the time hominins first appeared. The last common ancestor of all the primates arose about 77 Mya [Steiper and Young, 2006]. Subsequently, ape and monkey lineages split around 24 to 35 Mya [Kumar, et al., 2005; Bradley, 2008; Harrison, 2010]. The orangutan branched off about 18 Mya in Asia, and the first of the African apes (the gorilla) diverged approximately 9 Mya by molecular analysis [Steiper and Young, 2006] or 10-12 Mya, according to fossil evidence [Suwa, et al., 2007b]. This left a separate line of descent, the last common ancestor of chimpanzees (panins) and hominins.

When did the first hominins appear? The answer to this question is unknown, but estimates derived from fossil and molecular evidence are gradually narrowing the range of possibilities. The classical scientific plea that *more evidence is needed* applies here, because fossils from the relevant time range are very scarce and not readily identified as hominin, panin or something else. They are few in number, no knowledge is available about individual variability in apes or hominins from that era and the similarities between the earliest hominins, panins, and their immediate ancestors may have persisted for many generations until natural selection produced changes that allow paleontologists to distinguish them. Interbreeding could have occurred for a long time before they became reproductively isolated from each other [Patterson, et al., 2006]. A precise date for human origins will be very difficult to determine.

Estimates of the time of the hominin/panin split are based on a combination of fossil evidence (dated by geological techniques) and molecular genetic evidence. Branch points derived from analyses of DNA changes have to be calibrated by a divergence event documented in the paleontological record. For example, analysis of protein-coding genes suggests that the divergence of humans and chimpanzees took place about 20% of the time back to the divergence of apes and Old World monkeys, which is taken (from fossil evidence) to be 24 to 35 Mya. When the minimum of this range is used, it yields a hominin-panin split at 5-7 Mya [Kumar, et al., 2005]. If instead the African ape-orangutan divergence is taken as 12 to 16 Mya, this leads to an estimate that hominins and panins diverged 4.6 to 6.2 Mya [Chen and Li, 2001].

Currently, the earliest undisputed hominin is *Australopithecus anamensis* at 4.2 Mya [Kumar, et al., 2005; Wood and Harrison, 2011]. This provides a *minimum* constraint; the divergence from the common ancestor must have been earlier than that. However, there are three older candidate hominins not yet certified by a consensus of expert opinion. They are *Sahelanthropus*, dated at 6-7 Mya, *Orrorin* at 5.8-6.1 Mya, and *Ardipithecus,* known from 4.4-5.8 Mya. If these are hominins, the first members of our lineage must have arisen before then.

Paleontologists use two markers of hominin status: reduced canine tooth size and skeletal evidence indicating some sort of bipedal behavior. Wood and Harrison [2011] argue that in these three candidate hominins, smaller canines and bony evidence for bipedality may not signify hominin status because they also occur in Miocene apes which are not in the hominin lineage—apes such as *Oreopithecus*, dated at about 8 Mya [Rook, et al., 1999; Begun, 2007]. The broad ilium in *Oreopithecus* tends towards the hominin condition [Harrison, 1991]. So does stance, because the associated iliofemoral ligament stabilizes the knee and hip joints and increases the mechanical advantage of thigh muscles in upright postures [Straus, 1962; Harrison, 1991]. Wood and Harrison [2011, p. 348] note that several aspects of human anatomy (such as the foramen magnum, pelvis and proximal femur) "typically identified as being uniquely related to bipedalism, are, in fact, also found in non-hominoid primates associated with quite different locomotor behaviors" and may not signal "human-like" bipedalism. Similar traits observed in the three candidate hominins could be due to parallel evolution in similar environments, rather than common descent [Wood and Harrison, 2011].

Fossil evidence puts the time of hominin origins before 7

Mya if the three oldest candidates are validated. Results of genetic analysis fall within the same general time frame. Bradley [2008] reported that estimated dates for the divergence of chimpanzees and humans ranged from 4-8 Mya. Benton and Donoghue [2007] calculated the range to be 5 to 7 Mya. Based on the combined evidence, a conclusion that the hominin/panin divergence occurred at "about 7 Mya" seems a reasonable estimate, sufficient for the purposes of this book. This means "about seven million years" was how long it took for natural selection to convert arboreal apes into modern humans. If the hominin generation time was 16 years (roughly halfway between the minimums in chimpanzees and humans), during 7,000,000 years there would be 437,500 generations and a like number of opportunities for natural selection to retain the genes of those individuals whose heritable variations promoted reproductive success.

What did the first hominins look like and how did they behave? Answers to these questions must be speculative due to the paucity of appropriate fossil specimens from this time period. Nevertheless, the speculation concerns profound and classic issues in the study of paleoanthropology. What caused human origins? What was distinctive about the first hominins? What was the innovative behavior that drove natural selection to change their bodies, bones and teeth? What differentiated them from their conspecifics who continued to follow the ancestral lifestyle? *What were the reproductive advantages?*

Studies of living African apes, supplemented by the evidence of early *Australopithecus* fossils (from about 4 Mya), provide dual support for a model of the first hominins. Just before the gorillas branched off around 9-12 Mya, Miocene apes were agile arboreal primates, adapted to vertical climbing and arm-hanging from branches. They had mobile wrist, shoulder, hip, knee and ankle joints, grasping hands and feet. Thumbs were short, fingers were long, slender and ventrally curved.

Males were roughly the size of modern chimpanzee males. Females were about half as large [Harrison, 1991]. During the past seven million years of evolution, it is likely that the chimpanzee lineage *(Pan troglodytes)* changed in various ways. About 1 Mya, as an example, it gave rise to another species, *Pan paniscus*—the bonobos, now isolated in central equatorial Africa [Hey, 2010]. Both species of *Pan* remained in the woodlands, retaining or enhancing their arboreal adaptations.

Most researchers accept that the common ancestor of hominins and chimpanzees was a forest-dwelling ape that vaguely resembled modern chimpanzees, but differed from living species because after the divergence, natural selection continued to improve their adaptation to arborealism [Schmitt, 2003; Wood and Longergan, 2008].

There are different viewpoints about this. McGrew [2010] assumes that anything a chimpanzee can do today, the common hominin/panin ancestor could have done 6-7 Mya. Chimpanzees have been used as a model for the last common ancestor in studies of bipedal locomotion [Pontzer, et al., 2009]. On the other hand, Lovejoy and McCallum [2010] take the position that extant apes are irrelevant to the reconstruction of the earliest hominins, as far as their locomotion was concerned. Some believe the last common ancestor of chimpanzees and hominins may have been more like a Miocene ape such as *Pierolapithecus* or *Oreopithecus* [Ward, 2007].

Nevertheless, it is generally agreed that the first hominins, descendants of ancestors whose adaptation to a woodland habitat had endured for more than 70 million years, were arboreal apes.

They had elongated jaws and prominent canine teeth that were larger in males [Wood and Lonergan, 2008]. Their hands had small thumbs and long, curved fingers [Tocheri, et al., 2008]. Their feet, like their hands, were adapted for climbing trees and moving among the branches, being mobile and prehensile, with long, curved toes and a large, movable big toe. They used their quadrupedal gripping ability to move along the upper surfaces of branches. For below-branch travel, their hands gripped tree limbs overhead, the full body weight borne by the arms [Fleagle, et al., 1981; Harrison, 1991; Crompton, et al., 2008]. Their shoulders were mobile and muscular; their arms, long and strong. In contrast, their legs were short and less muscular, reducing the mass that needed to be transported during climbing and moving around in the trees, but retaining sufficient strength to contribute upward thrust during vertical climbing. The lumbar spine was short, tightly linked to the thorax and relatively resistant to rotation. Their body orientation in trees was commonly upright (orthograde) [Rose, 1991]. Sometimes they may have used hand-assisted bipedality when walking on top of branches, gripping the limbs against the soles of their feet in an upright mode while hanging on to an overhead or lateral branch with one or both hands [Thorpe, et al., 2007a; Crompton, et al., 2008]. Possibly they could have fully extended their knee joints when they did this [Tuttle, 1981; Thorpe, et al., 2007a]. On the ground they moved with the long axis of the body parallel to the ground (pronograde), as other quadrupeds do. They may have walked on the palms of their hands [Crompton, et al., 2008; Lovejoy, et al., 2009d], although chimpanzees and gorillas use their knuckles.

Like modern chimpanzees, early hominin females probably gave birth every 5 or 6 years, followed by a short period of sexual receptivity, so that at any time the number of reproductively active males was much greater than that of females available for mating. This sets the stage for male-male competition for reproductive opportunities [Mitani, 2009] and promotes sexual size dimorphism (males are larger than females). Larger, stronger males win more fights, but females do not have to fight to become inseminated. Lovejoy [2009a, b] has proposed an alternative view, that hominin males hardly ever fought with each other, although this runs counter to fossil evidence that sexual size dimorphism was high in *Australopithecus*. The issue is potentially important because gender size differences may provide an insight into the behavioral proclivities of our ancient ancestors (Chapter 5).

Chimpanzee males form aggressive coalitions to improve their own dominance status within their community and for killing chimpanzee males in other communities to expand their feeding territories and mating opportunities [Mitani, 2009]. This sort of "team aggression" can be identified in modern humans [Wrangham and Peterson, 1996; Potts and Hayden, 2008] and was likely a feature of behavior among the earliest hominins, as described in Chapter 8.

A noteworthy signal that this depiction contains a major element of truth is found in the much richer fossil record of later hominins. Roughly halfway along the timeline from the hominin/panin ancestor to modern humans, hominins were

CHAPTER 2: OUR LAST COMMON ANCESTOR WITH THE APES

walking upright on two legs, their jaws jutting forward in an ape-like manner, with arms that were long compared to their legs. They were the size of chimpanzees, males were larger than females, the fingers were curved towards the palm [Johanson and White, 1979] and, despite their two-legged walking on the ground, many skeletal signs show they were still adapted to an arboreal lifestyle.

In the next chapter, I will begin to develop the proposal that the bipedal use of hand-held weapons was the innovative hominin behavior that marked the onset of the human lineage. Hominins thereafter became adapted to this behavior by natural selection during millions of years because of the reproductive benefits it provided.

CHAPTER 3

Human Origins

"I can see no reason why it should not have been advantageous to the progenitors of man to have become more and more erect or bipedal. They would thus have been better able to have defended themselves with stones or clubs, or to have attacked their prey or otherwise obtained food. The best constructed individuals would in the long run have succeeded best, and have survived in larger numbers"
[Darwin, 1871, p. 142].

What caused our lineage to branch off from the apes? About seven million years go, some individuals from a community of arboreal apes began behaving in an unusual manner, marking the onset of a new lineage—the hominin lineage—which diverged from the ancestral ape ancestry. There is no accepted explanation for how and why this happened. It remains "the enduring mystery of human origins" [Berger, 2000].

There has long been widespread agreement that the first hominins were distinguished by their practice of a bipedal behavior of some kind, commonly presumed to be bipedal walking. Nevertheless, how bipedal walking itself could have provided reproductive benefits is a mystery.

This obstacle is surmounted if the innovative behavior was not bipedal walking, but bipedal use of hand-held weapons [Young, 2003, 2009, 2010, Chapter 14]. In the present chapter I will show how this activity, performed in a bipedal stance, would have provided immediate and continuing reproductive benefits to our earliest ancestors that promoted natural selection to improve this behavior for millions of years. In later chapters I will describe how this turned an arboreal ape into a hominin with modern body proportions who walked bipedally.

Fighting and reproductive success in chimpanzees. Although we are not descended from chimpanzees, we share with them a common ancestor (Chapter 2). As our nearest living relative, and one which retains its arboreal adaptation, the common chimpanzee (*Pan troglodytes*) provides a model for understanding how use of weapons could have arisen, and why it would have yielded reproductive benefits.

Chimpanzee males are larger, more aggressive, and fight more often than females, using physical aggression to establish and maintain a dominance hierarchy. High rank is frequently gained by direct confrontation, although more disputes are settled by display and threat than by fighting [Goodall, 1986, 1990a]. Displays may include dragging or flailing branches, throwing rocks or other material in random directions and slapping the ground or a tree. When attacks occur, they do not involve weapons, but may include hitting, kicking, stamping, dragging, scratching, grabbing and biting [Goodall, 1986, 1992]. High rank yields reproductive benefits. The future genetic composition of a chimpanzee community is strongly influenced by the breeding privileges of the alpha male. Although females also copulate with lower-ranking males, this may be thwarted by the alpha male, who is especially possessive of mature females during the ovulation period [Goodall, 1986, 1988, 1990a; deWaal, 1989; Nishida, 1990]. The male hierarchy also regulates access to food [Boesch and Boesch-Acherman, 2000].

Chimpanzee behavior is strikingly different toward familiar and unrecognized conspecifics. Wild chimpanzee males display a hostile and aggressive attitude towards strangers (except females in estrus). Outgroup members may be attacked, killed and even eaten [Goodall, 1986, 1990a, 1992; Nishida, 1990]. Territories are patrolled and defended by groups of males [Manson and Wrangham, 1991; Wrangham, 1999]. When patrol groups encounter trespassers, the group with the fewest males withdraws. Consequently, the more males, the more easily a neighbor's territory can be invaded and occupied, yielding more food and females for raising more offspring [Goodall, 1986; Boesch and Boesch, 1989; Wrangham and Peterson, 1996].

The chimpanzee diet is predominantly vegetarian, but small mammals acquired by hunting (usually a male group activity) are also eaten. Prey animals such as Colobus monkeys, are taken by grabbing, then killed by biting, battering or eating [Teleki, 1973; van Lawick-Goodall, 1975; Goodall, 1986; Boesch and Boesch, 1989; Stanford, et al., 1994]. A male who eats the meat he has procured increases his intake of a nutritionally valuable food. When a hunt is successful, however, the carcass is commonly shared, especially with females in estrus, members of the hunt and dominant males [Teleki, 1973; Goodall, 1986; Boesch and Boesch, 1989; Boesch and Boesch-Acherman, 2000]. Females copulate more frequently with males who have shared meat with them. Thus, males increase their mating success by sharing meat with females and females increase their caloric intake by mating more frequently with males who share meat with them [Gomes and Boesch, 2009]. (Gilby, et al. [2010] are not convinced that chimpanzee males gain more copulations by sharing meat with sexually receptive females).

Chimpanzee throwing and striking. Throwing objects and flailing with sticks and branches commonly occurs on the ground from a bipedal stance and involves only the arm (or arms). Throwing may be overarm, sidearm, underarm or back-handed [Goodall, 1964,1986; van Lawick-Goodall, 1968, 1975; Kortlandt, 1972; Brewer, 1978; deWaal, 1989; Marzke and Wullstein, 1996]. When it seems to be aimed, throwing is highly inaccurate [Goodall, 1964, 1986, 1988]. If the throw is underarm, the object sometimes lands behind the thrower [van Lawick-Goodall, 1968]. In their natural habitat they throw sticks, stones, leaves, twigs, nuts and branches. During conflicts, they also may throw grass, leaves, or sand, suggesting that throwing is a threat display, rather than an intentional use of weapons to cause injury. They also grasp sticks or tree branches and swing them like clubs [Köhler, 1927; Kortlandt, 1972, 1986; Goodall, 1964, 1986]. Flailing with sticks from close enough to hit a target is seldom successful [Goodall, 1986]. Throwing and striking appear to be unaimed in most instances. Both are almost entirely male behaviors [Goodall, 1986], although with large individual differences. About half the males do not throw at all [van Lawick-Goodall, 1968; Goodall, 1986, 1990a]. Chimpanzees have never been observed to use weapons for predation [Kortlandt, 1986; McGrew, 1992]. They seem to lack the "concept" of weaponry [Köhler, 1927; Wrangham and Peterson, 1996].

Hominin throwing, striking, and reproductive benefits. Darwin [1871] was the first to suggest that throwing and clubbing would have yielded advantages to early hominins. He thought it would have enabled them to defend themselves with stones or clubs, attack their prey, obtain food, and fight with their enemies. Those best constructed for this behavior, in his view, would have proliferated. Since then, many authors have expressed the opinion that ancient hominins used weapons, such as "stones thrown as missiles, or stones, bones, or wood used as clubs" [Schick and Toth, 1993, p. 183].

Supporters of this view include Dart [1949a, b, 1953; Dart and Craig 1959], Broom [1951], Oakley [1964], Howell [1965], Laughlin [1968], Washburn and Lancaster [1968], M. Leakey [1971], Washburn and Moore [1974], Darlington [1975], Wolpoff [1976], G. Isaac [1977], Lee [1979], Kortlandt [1980], Calvin [1983], B. Isaac [1987], Fifer [1987], Marzke [1983, 1992b], Thomas and Marzke [1992], Knüsel [1992], Straus [1993], Bunn and Ezzo [1993], Bingham [1999], Kirschmann [1999], Van Valkenburgh [2001], Dunsworth, et al. [2003], Schmitt, et al. [2003], Kingdon [2003], Carrier [2004], Young [2003, 2009, 2010], Tattersall [2012], and Stringer [2012].

In chimpanzee communities, as just described, the males fight, compete for dominance, and do most of the throwing and branch-swinging. Females avoid these aggressive interactions and nurture the children. This conception of gender differences in behavior is fully consistent with what is observed in modern humans. In humans as in chimpanzees, males are actively involved in between-group fighting whereas females have less interest in combat and do not defend resources aggressively [Manson and Wrangham, 1991; Kaplan and Hill, 1992]. Fighting is more important for males than females because it generates reproductive opportunities for those who are victorious [Symons, 1979]. Males also predominate as hunters, as shown in ethnographic studies [Watanabe, 1968; Coon, 1971; Stanford, 1999; Hawkes, 2001]. Women probably avoid hunting because of the reproductive costs associated with pursuing dangerous, mobile prey [Tooby and DeVore, 1986; Kaplan and Hill, 1992]. Hominin males are more likely to have been the hunters and warriors, resulting in greater reproductive selection for those who threw effectively, in contrast to females, whose greater investment in gestation, lactation and child care may have inhibited their throwing action [Kirschmann, 1999; J. Thomas, 2000; K. Thomas, et al., 2001]. This innate gender difference forms the backdrop for an explanation of the origin of the hominin lineage.

In a population of the last common ancestor of chimpanzees and hominins it is proposed that a young adult male with an innate drive to dominate stood up on his hindlimbs, threw a rock during a threat display that hit an adversary, prompting him to flee. Something as simple as that could have had major consequences, if the thrower noted the effect and repeated it in subsequent confrontations. A similar event was observed among wild chimpanzees and recorded by Goodall. A chimpanzee male noted the response he elicited from his unprecedented behavior of kicking kerosene cans (stolen from his observers). He then used this behavior repeatedly to aid his rise to alpha status. An adolescent male was subsequently observed practicing this behavior alone [Goodall, 1986]. This confirms other reports which indicate that behavior of the highest ranking male is subject to imitation by immature males [Teleki, 1973; Goodall, 1986]. Imitative stone-throwing has also been observed [Köhler, 1927; Goodall, 1988].

Use of hand-held weapons from an erect stance would have provided benefits to its initial practitioners. The first ape males to rise up bipedally to hurl rocks or swing sticks with an intent to injure an opponent would have *immediately* gained an advantage. The belligerent thrower and clubber could still grab and bite; however, he could also injure his opponent without approaching closer than the distance he could throw or the length of his stick. Males who used this behavior effectively would have risen in the dominance hierarchy [Kirschmann, 1999]. Other males would have mimicked their actions, expanding the use of aggressive throwing and clubbing in agonistic situations. Such behavior would have escalated in subsequent generations due to the reproductive advantages it yielded in competition for scarce resources. Natural selection must have become involved at an early stage among these first hominins. If continued for an extended period, this process would have led to a suite of inherited attributes that increased the effectiveness of the behavior, raising the frequencies of the selected genes. It would also pave the way to the eventual formation of a new species, which used resources in a way that set them apart from the members of their ancestral species.

Higher dominance rank would increase access to breeding females, allowing males to sire more offspring. A rise in dominance status also would improve access to food, a significant component of reproductive success. Competing with conspecifics who shared the same biological needs for access to limited natural resources was likely the first major application of this novel behavior. Furthermore, use of weapons to deter predators would raise the likelihood of survival to reproductive age and thus the chance of leaving more offspring [Kirschmann, 1999]. Serious injury to predators might not have been necessary, particularly if several hominins participated. Lacking instinctive reactions to creatures who threw rocks or swung clubs, some threatening carnivores may have retreated from surprise, pain or fear.

When outgroup males were encountered by male coalitions patrolling the group's territory, those with weapons would have an advantage. During invasion of the territory of another community, in the conflict between males that ensued, the group which used weapons skillfully would be more likely to vanquish those who lacked weapons or who used them less adeptly when attempting to protect themselves, their females and their territory. Communities of armed apes with greater throwing and striking prowess would have improved access to females, food and other territorial resources.

Because males who were better throwers and clubbers rose to higher ranks, wielded more power, dominated other males, commandeered the best feeding sites, were better hunters and confrontational scavengers, and could protect women and children more effectively, females were more likely to select such males for mating, when the opportunity arose.

Aggressive use of weapons enhanced male access to nutritious food and fertile mates, but the nutritional aspect is more important for females. Access to food resources that promote good nutrition improves reproductive capacity in female chimpanzees. For example, consumption of nutritious fruit raises estrogen levels and increases the frequency of sexual swellings, a predictor of female attractiveness, copulation rate and associations with males. When average fruit consumption remains high for several weeks, females conceive more quickly [Thompson and Wrangham, 2008].

Evidence from modern humans. Successful acquisition of food resources is especially vital to the reproductive capacities of present-day human females. Because they carry the energetic burden of reproduction, they are constrained by the availability of energy and the time consumed by gestation and lactation. Male reproductive success, by contrast, is primarily limited by mating opportunities. It does not require the same investment of time and energy that female reproduction does. Fecundity in human males is normally continuous, sustained into the late stages of life and insensitive to energetic constraints [Ellison, 2001, 2003].

Positive energy balance is an important predictor of a woman's reproductive success. The ability to divert metabolic energy towards reproduction in the early gestation period significantly affects the success of that pregnancy. Ovarian function improves in the well-nourished female before and at the time of conception. Fat stored during this period is subsequently drawn upon to meet the costs of gestation and lactation [Ellison, 2001, 2003]. In contrast, chronic undernutrition is associated with lower levels of ovarian function, which may result in low fecundity. In addition to undermining the mother's health, depleted maternal nutritional reserves reduce the probability of conception and jeopardize the health and survivability of her offspring, leading to low birth weight, premature birth, and increased risk of infant mortality. Malnourished mothers may not produce sufficient milk. An inadequate diet also increases the interval between births. Underfed children are smaller and have delayed skeletal maturation leading to smaller adult size, including reduced height, weight, muscle, skeletal mass, head circumference, greater deficits in cognitive development and susceptibility to infectious disease [McFarland, 1997; Bogin, 1999; Ellison, 2001, 2003].

It is clear that reproductive advantages are associated with improved access to food. Modern humans fight over scarce resources, and during much of the past the most frequently scarce and valuable resource was food. Success in fighting would have been an important factor in determining which groups of early humans survived and proliferated [LeBlanc, 2003]. Early hominin females gained advantages from male use of weapons when it led to the expansion of territory that opened access to new sources of nutritious food. Females do not need to fight for mating opportunities. (There will always be male volunteers). However, living in a community that controls nutritious food resources is beneficial to females. Furthermore, they and their dependent children would have profited from increased protection by males using weapons against outgroup marauders and carnivorous predators.

Nutrition and natural selection. Living organisms can be thought of as systems that capture energy from the environment for the purpose of replicating themselves. This pithy and fertile concept, due to Ellison [2003], supports the idea I have emphasized in these first three chapters and will refer to in those that follow. Natural selection is a result of differential reproduction that is linked to the production of vigorous offspring by well-nourished parents. The hominin innovation of hand-held weapons opened a new way of fulfilling these fundamental biological needs from the moment of its inception. Eventually, this enabled hominins who were proficient in such behavior to occupy safer habitats with more sources of shelter, water and food, thereby promoting population growth and increasing the proportion of their genes in the breeding population. This open-ended selective process would have continued as long as it enhanced reproductive success.

Human origins and human evolution. Once natural selection began to preserve preferentially the genes of individuals whose heritable traits facilitated their throwing and club-swinging prowess, the issue of human origins becomes one of human evolution. As evolution continued, the reproductive advantages of this unprecedented behavior were extended in numerous ways as hominin use of weapons became more effective due to a selective process that endured for millions of years and progressively adapted them to this behavior.

The utility of weapons during conflicts within a hominin community would soon have led to its use by coalitions of males in conflicts with adjacent communities over control of breeding females and territories with rich natural resources. As skill with weapons improved, opportunities for aggressive scavenging of carcasses would have increased when predators could be driven from prey they had killed. Greater proficiency with weapons would also have made hunting more efficient, eventually providing access to larger prey animals augmenting the availability of nutritious food, thereby yielding reproductive benefits. Meat acquired by males by aggressive scavenging or by hunting could be traded for sex with females, increasing the propagation of their genes.

The use of weapons by ancient hominins has sometimes been challenged by the assertion that there is no evidence of weapons. There is wholehearted support that weapons would have been useful, but frequently iterated doubts that they are associated with hominin fossils. This topic will be addressed in the next chapter.

CHAPTER 4

Early Hominin Weapons

"The apparently simple ability to throw overarm with force and accuracy is a skill uniquely developed in the human animal and one which was probably practised in deepest antiquity. Yet the lack of any evidence convincing to archaeologists results in the human ability to throw being rarely discussed or even referred to in most accounts of human evolution" [B. Isaac, 1987, p. 3].

Introduction. In Chapter 3 it was proposed that hominin origins coincided with the first use of sticks and stones as weapons and that the reproductive benefits obtained by use of this behavior led to natural selection that increased the prowess of this activity during millions of years of evolution. Some have opposed this idea by asserting that our early ancestors could not have used weapons, because there are no weapons that archeologists can identify as *made by hominins* [Lewin, 1987]. I shall refer to this as "the no hand-made weapons = no weapons argument" and refute it in this chapter by demonstrating that hominin weapons were available in their environment.

The earliest weapons. The earliest weapons were unmodified natural objects like sticks, stones and bones that lay ready at hand [Oakley, 1959]. Hominins did not *make* these weapons, they picked them up and used them. Because they were unmodified natural objects, they would have left no archeological trace. The first weapons would have been relatively simple, such as stones and tree branches [Young, 2003; Carrier, 2004]. Stones would serve as missiles and "percussive instruments" [Burton, 1884]. When choosing a rock for throwing, hominins would likely have selected those with a spheroidal form, because this shape has the best aerodynamic properties [Washburn and Lancaster, 1968; Coon, 1971; Adair, 1990; Kirschmann, 1999]. Ethnographic studies indicate that hunters who carry stones "as premeditated missiles either select water-rounded pebbles or peck rough stones into a globular shape" [Coon, 1971, p. 96].

Rocks of a size that could be held in a firm grip would have been sought. Based on estimates from available hominin hand fossils (Chapter 9), stones suitable for throwing would range from about the size of a golf ball to that of a softball (4.2-9.0 cm diameter). Those of high density would have been most useful. Potential clubs would have had an overall cylindrical shape, with cylindrical handles of a size suitable for gripping by hominin hands. Smooth surfaces on missiles and clubhandles were likely preferred to limit pain or injury to the hand as force was applied to the weapon. The mass of a missile or club needs to be sufficient to injure an adversary, but not too large to be accelerated to high velocity by an arm motion (later to evolve into a full-body motion). Fossil evidence of hominin hands shows that well before the earliest known archaeological sites with hominin-made tools, spherical and cylindrical tools of some kind were having an impact on hominin reproductive success [Young, 2003, 2009, 2010]. The gradual evolution of two unique hominin handgrips, one for spheres, one for cylinders, began near the time of hominin origins (Chapter 9).

No ancient clubs or spears have been identified. The oldest known spears are from 0.4-0.5 Mya (Chapter 8). However, the long bones of large animals found at hominin sites would have made effective clubs, their smooth-surfaced, cylindrical shafts gripped by muscular, robust hominins (Chapter 12) swinging from a bipedal stance. Nevertheless, the evidence of ancient clubbing is partly indirect, consisting of the structure of the evolving hominin hand and modern club-swinging prowess, but this is supported by the direct evidence that men emerge in late prehistoric times swinging clubs, axes, maces and swords, all with cylindrical handles [Burton 1884; Frayer, 1997; Lambert, 1997].

Unmodified stones. Natural objects found at ancient hominin sites include stones appropriate for throwing located in nearby stream beds or rock piles. Rocks transported to sites from more distant regions are called "manuports" (hand carried) [M. Leakey, 1971; G. Isaac, 1977; Potts, 1994]. Some of these naturally weathered cobbles were spherical in form. Many could have been used for clubs as well as throwing stones. Localities near the Kada Gona river in Ethiopia are presently the earliest known archaeological sites (2.6 Mya) where primitive, hand-chipped, stone tools are found [Semaw, 2000]. The main raw material sources for the Gona tool-making hominins were rocks of volcanic origin from local streams in the form of water-worn, rounded, fist-sized cobbles. The smooth-surfaced stones ranged from 5.8 to 10.6 cm in diameter, with an average of about 8 cm [Hay, 1976; Semaw, 2000].

Water-rounded stones were readily available in local stream beds at Olduvai Gorge (Tanzania) at the early archaeological sites [Jones, 1995]. Olduvai Bed I assemblages (1.9 -1.7 Mya) are notable for huge quantities of manuports. At site HWK E, levels 3-5, there were 1,184 manuports [M. Leakey, 1971]; the MNK site contained over 800 [Potts, 1988]. "For what purpose would such large stockpiles of stone be necessary or even useful, particularly for a technological mode relying upon very simple cores/core tools and flakes?" asked Schick [1987, p. 795]. No hominin would be expected to accumulate rocks unless they were considered crucial for survival, yet they showed no signs of use. Among several possibilities, Potts [1984, 1988] hypothesized that hominins stockpiled rocks in different areas for use while foraging locally. Food that required tool processing was then brought to the rock piles. Predation at Olduvai was a potential threat and hominin strategies of predator avoidance were at a premium [Potts, 1988]. This suggests that the rock collections may also have been used as a source of missiles.

Mary Leakey [1971] was the first to state explicitly that manuports could have served as weapons. At Olduvai Upper Bed I site FLK North, where 83 cobblestone manuports mainly composed of vesicular lava were imported by hominins (mean diameter, 6.4 cm, tennis-ball size), she noted that this material was seldom used for tool-making at that level. "This seems to indicate that the nodules and blocks were not brought in to serve as raw material but for some different purpose. There is no indication of utilisation and they may have been kept in readiness for use as missiles against predators or scavenging animals" [Leakey, 1971, p. 83]. Comparable manuports occur in the oldest sediments of Lower Bed I site FLK NN, where there are water-worn lava cobblestones which also average 6.4 cm in diameter [M. Leakey, 1971]. Manuports occur at all levels at Olduvai [M. Leakey, 1971] and many are of suitable size and shape for throwing [B. Isaac, 1987]. They may have been used for weapons [Lee, 1979; B. Isaac, 1987; Cannell, 2002; LeBlanc, 2003]. Many other assemblages also include unmodified stones that would have been effective, lethal projectiles [G. Isaac, 1984]. Sites at Koobi Fora, Kenya (1.9-1.3 Mya) contained spheroidally shaped stones of a size appropriate for throwing. Local cobble-bearing channels provided plentiful well-rounded, stream-worn cobbles, mainly of fine-grained basalt [G. Isaac, 1984; Toth, 1987, 1997; G. Isaac, et al., 1997]. Toth [1997] refers to the cobbles as "spherical and subspherical." Those between 4 and 10 cm in diameter, commonly chosen for tool-making [G. Isaac, et al., 1997], were also suitable for throwing [B. Isaac, 1987, 1992; Cannell, 2002].

At the Ain Hanech site, dated at 1.8 Mya [Sahnouni, et al., 2002, 2004], water-worn limestone cobbles were obtained locally by hominins for stone tool-making. Their mean maximum diameter was 8.4 cm (range 4.4 to 13.7 cm) [Sahnouni and de Heinzelin, 1998]. Most would have made effective throwing weapons. Some of the cores used as a source of flakes assumed a spheroidal form [Willoughby, 1987; Sahnouni, et al., 1997] and could also have served as missiles.

Dmanisi, in the Republic of Georgia, contains a hominin fossil site (1.8 Mya) where stone tools and cobbles have been recovered. The diameter of the core tools is about 6-7 cm. Some of these are spheroidal; others are more elongated. As of 1995, 995 intact cobbles larger than 4 cm in diameter had been found [Gabunia, et al, 2001].

At the Olorgesailie site in Kenya, dated at 0.8-0.9 Mya [Potts, et al., 2004], abundant stockpiled, unworked stones suitable for throwing were present (as were well-worked spheroids of lava, quartzite and quartz) [G. Isaac, 1977; B. Isaac, 1992; Willoughby, 1987]. Many cobbles of vesicular trachyte carried in from nearby outcrops occur at most sites. Trachyte is generally unsuitable for tool manufacture, yet the hominins carried as much as two tons of stone to this vicinity. The rocks might have been used as hammerstones and anvils, footings for windbreaks, or missiles for hunting or warding off predators [G. Isaac, 1977].

The Cave of Hearths is a South African Upper Acheulean hominin site (~0.2-0.25 Mya), where spheroidal rocks from Bed 3 analyzed by Willoughby [1985, 1987] are mainly unmodified specimens with an even size distribution. Diameters larger than 10 cm are rare. At this site 215 spheres of smooth dolerite, appropriate for throwing, were collected about 1.6 km from the cave.

Hominin-made spheroids. These modified rocks could also have been used for throwing, but were not necessarily fabricated for that purpose. Shape defines a rock as a spheroid, whether it is produced by hominins or by natural forces. The term denotes a class of stones associated with hominin sites that includes partly angular subspheroids with projecting ridges and more spherical and smooth-surfaced spheroids [M. Leakey, 1971; Willoughby, 1987]. The size of spheroids is variable, but "many could quite comfortably be held in the hand for use" [Willoughby, 1985, p. 52].

Beginning about 1.9 Mya and thereafter, hominins produced spheroids from a variety of raw materials, including granite, limestone, quartz, quartzite, gneiss, basalt, other lava, flint, chert, dolerite, ironstone, sandstone, pegmatite and welded tuff. Spheroids follow the general rule that fine-grained, homogeneous, hard and durable rocks were chosen for tool manufacture, although sharp-edged tools from the same site (such as flakes and retouched flakes) were often made from raw materials different from that of some spheroids [Willoughby, 1987]. Hominins produced some of them by flaking a cobble until it assumed an angular, polyhedral form. When this was followed by battering on stone, the flake ridges were reduced, producing a more symmetrical and smooth-surfaced "subspheroid" [Figure 1]. Further reduction of flake ridges by crushing the angular facets then yielded a spheroid [Willoughby, 1985]. The epitome of this class of tools, achieved by careful

CHAPTER 4: EARLY HOMININ WEAPONS

Figure 1. A subspheroid (left) and a spheroid from FLK North, Olduvai Gorge, about 1.9 Mya [After Leakey, 1971, Plate 16]. Spheroids found at hominin sites that show signs of flaking and battering were probably byproducts of stone tool-making. Others were naturally occurring cobbles collected by hominins and transported to the sites (manuports). All would have been effective throwing-stones.

---

pecking away of surface irregularities, was a product of such perfect sphericity and surface smoothness that M. Leakey [1971] named them "stone balls."

Spheroids made by hominins have been found all over the globe, beginning in the African Lower Paleolithic and later extending into Eurasia in the Middle Paleolithic, including African sites at Olduvai Gorge [M. Leakey, 1971; Willoughby, 1987], Ain Hanech, 'Ubeidiya, Sterkfontein, Isimila, Olorgesailie, Guettar and Windhoek [Willoughby, 1987], Swartkrans [M. Leakey, 1970; Brain, et al., 1988] and Makapansgat [Dart, 1925a], in Zambia and Zimbabwe [Oakley, 1964; Willoughby, 1987], and at Xujiayao in China [Willoughby, 1987], at Ngandong in Java [Oakley, 1964], in Europe at Grotte de l'Ours [Dart, 1925a], and La Quina, France, where in a Mousterian cave site more than 100 spherical balls of limestone 3.5 to 9.0 cm in diameter were found [Sollas, 1924; Willoughby, 1987; B. Isaac, 1992].

M. Leakey [1971] was the first to undertake the quantitative analysis of these ancient artifacts. She measured the length, breadth and thickness for many specimens, and reported the range and means for each of these dimensions, then calculated the mean diameter, and presented the range and average of the means. Analysis of these data and additional measurements [B. Isaac, 1987] indicates that in Olduvai Beds I-III (1.9 to 1.7 Mya), the average diameter of spheroids is about 6.0-6.5 cm. The unmodified cobbles at Koobi Fora (n = 110) have a mean diameter of 5.6 cm [G. Isaac, et al., 1997]. Taken together, these results indicate that the average diameters of subspheroids and spheroids in these samples ranges from 5.6 to 6.7 cm and the range for manuports and unmodified cobbles is 5.6 to 6.4 cm, an average about the size of a billiard ball (5.8 cm) to a tennis ball (6.5 cm). These stones could have been used as throwing weapons.

The most comprehensive data source regarding spheroids is the publication by Willoughby [1987, Appendices III, IV], who reports 21 parameters of 1630 spheroids and subspheroids sampled from seven African Early Stone Age assemblages and two Middle Stone Age (MSA) sites in Zambia and Zimbabwe. The ages of the assemblages range from about 2 Mya (Sterkfontein 5, Ain Hanech, and Olduvai Bed I) to 0.3 Mya or younger (Isimila, Cave of Hearths, and the MSA sites). The average mean width of these 1781 chipped and pounded stones is 6.8 cm. Using Willoughby's data, the average spheroid or subspheroid is about the size of a tennis or cricket ball, but heavier. Its density (3.17) is twice that of a billiard ball (1.61) and five times greater than a baseball (0.69). Most would be appropriate for use as missiles.

The function of spheroids. Many spheroids had properties suitable for throwing weapons [Dart, 1925a; Oakley, 1964; M. Leakey, 1971; Willoughby, 1985, 1987; B. Isaac, 1987; G. Isaac, et al., 1997; Kirschmann, 1999], but this frequently suggested explanation of their function is not clearly established [Schick and Toth, 1993]. The conclusion that spheroids could have served as weapons does not mean they were intentionally fabricated for this purpose.

Jelinek [1977] pointed this out when he noted that some

spheroids might be exhausted cores from which additional flakes could not be struck because the surface was no longer sufficiently angular. Sahnouni, Schick and Toth put this concept to an experimental test [Sahnouni, et al., 1997] and found that heavy reduction of limestone cores tended to produce subspheroids and, more rarely, faceted spheroids. Such exhausted cores, if subsequently used as hammerstones, would show signs of battering. These authors concluded that faceted limestone subspheroids and spheroids were end products of flake production, not deliberately produced shapes [Sahnouni, et al., 1997]. If basalt is used as a core until exhausted (when platform angles approach 90°) subspheroids are among the shapes that may be created [Braun, et al., 2005]. Schick and Toth [1993, 1994] revealed by experiment that when quartz chunks are used as cores until they are exhausted, many assume a subspherical shape, becoming more spherical if they are subsequently used as hammerstones. Quartz is more easily pulverized than most other types of stone used in flaking [Schick and Toth, 1993]. When used repeatedly as a hammerstone, a chunk eventually becomes spherical. The tendency to strike the core with the end of the longest axis of the hammer tends to produce a spheroidal form through deterioration from impact [Schick and Toth, 1994]. Jones [1995] reported a similar observation.

The primary goal of early stone tool-making is thought to be the production of sharp and durable edges for cutting, piercing and scraping [Jelinek, 1977; G. Isaac, 1984; Potts, 1988; Schick and Toth, 1993]. It appears that the majority of worked spheroidal tools—the subspheroids, faceted spheroids, battered and pecked spheroids—could have been the result of the production of sharp-edged flakes, rather than objects deliberately fashioned to yield aerodynamically superior missiles.

Were weapons available to early hominins? The evidence shows that natural weapons were present in the surroundings of the earliest hominins and all those who followed them. They were present in their habitats in the form of unmodified sticks, stones, and bones, and they were abundant in stream beds as smooth-surfaced, water-worn cobbles thatcould be gripped in the hand. They also were present in stockpiles of rocks transported to hominin sites from far and near, and they were available in the spheroidal shapes of hammerstones and cores from which flakes had been knapped when hominins began to make sharp-edged stone tools. As Barbara Isaac [1987] phrased it, in an emergency, they would have used any stone that could be thrown.

CHAPTER 5

Gender Size Differences

"Some questions, including those pertaining to...sexual dimorphism...appear to be resistant to solution—or are at least ongoing subjects of polarized debate" [Kimbel and Delezene, 2009, p. 39]

Introduction. A difference in body size between males and females is referred to as sexual (or gender) size dimorphism. In most animals males are larger than females and their physical weapons such as antlers, horns or canine teeth are inevitably larger. Males have been naturally selected in these ways because they fight with each other over access to females. Those who are bigger, stronger and better equipped for combat are more likely to win these battles, giving them an advantage in the competition for mates and increasing the frequency of their genes in the next generation.

Females don't fight to get pregnant because males willing to perform this service are not a scarce quantity. They are not naturally selected for combat because there is no reproductive benefit in it for them. Females of many species are not sexually receptive when they are pregnant or lactating, which is most of the time. Thus receptive females *are* a scarce quantity, males fight over them, and the victors gain the most mating opportunities. This is the prevailing explanation of gender size differences.

In the field of human evolution, sexual size dimorphism looms large. The major evidence consists of fossilized bones and teeth, and the predominant goal of analysis of these precious relics is to learn something about the *behavior* of our ancient ancestors. "Sexual dimorphism is the primary morphological evidence for social behavior in early hominins" [Plavcan, et al., 2005, p. 318]. It is one of the few potential means of inferring behavior in hominin groups [Harman, 2006]. Gender size differences provide "a window onto behavior in earlier hominids and added perspective on the evolution of human social behavior and mating systems" [Larson, 2003, p. 9104].

Were hominin males fighting with each other for mating opportunities? The message in the bones makes it seem likely. *Australpithicus afarensis* and later fossil hominins had greater sexual dimorphism than exists among modern humans [Johanson and White, 1979; Harman, 2006]. In this chapter, the evidence and its interpretation will be summarized. The result proves to fit seamlessly with the proposal that hominins underwent an evolutionary adaptation to the bipedal use of hand-held weapons in which males did most of the fighting.

Sexual selection and sexual size dimorphism. Darwin [1871] placed male against male competition for females in a special subcategory of natural selection which he called "sexual selection" (that is also based on reproductive success). Darwin proposed two forms, one for each gender. Males battled among themselves for direct access to females or for possession of resources like food that females need. Females were the choosy sex, selecting males to breed with based on their evaluation of personal attributes such as physical signs of health and vigor. Darwin stated that the greater size and strength of modern human males compared to females, together with their broader shoulders, more developed muscles, greater courage and pugnacity were due to the success of the strongest and boldest men in the general struggle for life and in their contest for mates. Success would have ensured their leaving more progeny than the males they defeated.

Darwin's view remains in the forefront of current thinking. Fighting typically has been far more important to male than female reproductive success (Chapter 3). Because males can father more offspring than a female can produce, they will compete for access to more potential mothers. Large size and better weaponry therefore will be favored by natural selection

[Gaulin and Sailor, 1984; Foley, 1987; Mitani, et al., 1996]. Selection for fighting presumably explains the marked sexual dimorphism of early hominins [Kirschmann, 1999; Plavcan, 2000; Carrier, 2004; Harman, 2006].

The role of female choice is often difficult to determine. The ability of males to defeat rivals may indicate that they are healthy and vigorous, which might entice a female to choose them for mating. However, sometimes females may have no choice. The male may be capable of injuring her if she tries to refuse his advances [Gould and Gould, 1989].

Polygyny, monogamy, and size dimorphism. When analysis of hominin bones suggests that the population was sexually dimorphic in size, the deduction is usually made that the males were "polygynous." Polygyny is often defined as having more than one "wife" at a time, but since there is little likelihood that early hominin males had wives, in this context it means males mated with more than one partner.

Hominins might have lived in small groups of one male who defended a harem with children (as do gorillas), but this view is not favored because at some hominin sites there is evidence of several individuals of both genders, suggesting that early hominins lived in mixed groups of males and females, like chimpanzees [Johanson and White, 1979], and mated with multiple partners. In chimpanzees and bonobos this is displayed as "promiscuity" with both males and females having sexual relations with different partners. There is no pair-bonding and no paternal support [Ryan and Jethá, 2010].

Polygyny, where males potentially have access to multiple mates, is linked with sexual size dimorphism and inter-male fighting for mating partners. The proposal that early hominin males added the use of hand-held weapons to their fighting behavior, that females remained adapted to the role of gestating and nurturing children, and that the two genders were sexually dimorphic in size is symbolically depicted in Figure 2.

Gender size monomorphism (males and females are the same size) is associated with monogamy. With one exception (cited below), numerous analyses of the hominin fossil record

CHAPTER 5: GENDER SIZE DIFFERENCES

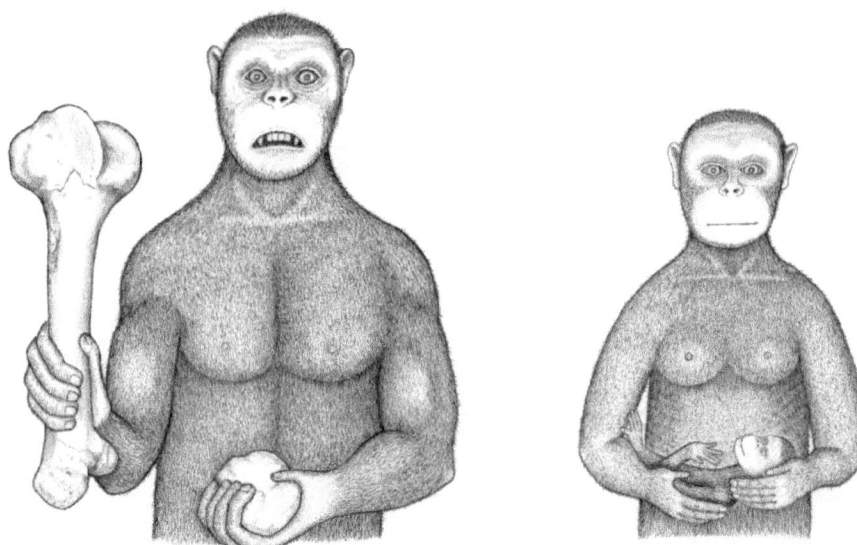

Figure 2. Artist's conception of an early hominin male and female, illustrating sexual size dimorphism, related to the male's use of hand-held weapons (a club for striking and a spheroidal rock for throwing), an addition to that gender's ancestral role of fighting. The smaller female, holding an infant, depicts her major role in producing and raising the next generation. The male is flashing his sharp canines, which in subsequent hominins became obsolete for fighting and were naturally selected for other purposes. The male and female are not married or pair-bonded (although they might be friends) because early hominin mating was presumably promiscuous, with advantages going to the best male fighters because they achieved higher dominance status.

have rejected this form of social organization. The traditional concept of monogamy in primates is a prolonged association and mating relationship between one male and one female. The family unit consists of pair-bonded male and female adults and their children in an exclusive sexual relationship [Fuentes, 1998]. Among the apes, only gibbons (genus *Hylobates*) are considered monogamous, but even here the groupings may be temporary or transitional and the two-adult group may not reflect a monogamous mating strategy. In the small number of other primates classed as monogamous, there is a huge range of variation, not a single type of social organization. In primates as in other mammals, the rate of monogamy is about 3% [Fuentes, 1998]. Among most mammals the basic social unit is the mother and her children.

Nothing in modern human male sexuality hints of an adaptation to monogamy, according to Symons [1979]. Rather, there seems to be a natural and universal tendency of modern males to want a variety of sexual partners. This suggests to Symons that human males, like chimpanzee males, would tend more toward promiscuity if there were no social restrictions to prevent it. Fisher [1992], on the other hand, asserts that modern humans practice "serial monogamy", in which monogamy does not mean fidelity by either gender, but rather a mix of monogamy and adultery.

Modern humans display a wide range of grouping patterns and mating practices, from small, two-adult groups to large multi-male multi-female bands, none of which seem to be inherited [Fuentes, 1998]. Ryan and Jethá [2010] conclude that our hominin ancestors, up until the onset of agriculture, lived in multi-male multi-female bands, like our closest relatives, the two species of *Pan*, who continue to do so today. Their gender size dimorphism is similar to our own, a topic that is described below.

Fossil evidence of mixed-gender hominin groups, such as *Australopithecus afarensis* [Johanson and Taieb, 1976; Johanson and White, 1979] supports this view, which implies that both genders had multiple sexual partners, without pair-bonding, and paternity was not an issue [Ryan and Jethá, 2010].

Besides fighting for sex, other factors that may influence gender dimorphism include feeding ecology, reproductive behavior, social variation, ontogenetic processes, body size, life history strategies, diet, habitat, allometry, phylogenetic inertia or selection for female characters [Ford, 1994; Leigh and Shea, 1995; Plavcan, 2000]. However, there is no current indication that any of these factors are widely applicable to interpreting size dimorphism in the hominin fossil record, where males are always larger and male-male competition holds sway. As we shall see, a relative increase in female size, possibly related to birthing larger babies, reduced gender size dimorphism in *Homo* (Chapter 7).

Modern human gender size dimorphism and the effects of puberty. Modern human male adults range in average weight from 58-65 kg; females average 34-56 kg. The female average is ~86% of that of the male. With regard to stature, the average for males is 175 cm. Females average 161 cm, about ~92% of that of males [McHenry, 1992; Smith and Jungers, 1997; McHenry and Coffing, 2000]. Modern human males are 1.1-1.2 times larger than females.

This difference is related to sex hormones. In males, androgens, particularly testosterone, are largely responsible. Testosterone levels in males are high before birth, biasing the brain in a male direction, but after birth the level falls, remaining low until puberty [Tanner, 1978; Symons, 1979]. The multiple effects of testosterone in males also underlie the greater propensity for aggression in that gender. Before puberty, boys typically are already more aggressive than girls, and this difference continues in adults [Symons, 1979]. Throwing and club-swinging ability is also markedly advanced in boys compared to girls, independent of teaching, from early childhood through puberty into adulthood (Chapter 13).

Before puberty boys are about the same size and strength as girls, and their shoulders and hips are equally broad. An increase in muscle mass and strength after puberty is linked to testosterone production, which is 20 times higher in adult men than in women [Ellison, 2003]. At puberty, shoulder cartilage cells in males respond to rising levels of testosterone resulting in broader shoulders [Tanner, 1978]. The high heritability of this feature indicates that broad shoulders may be important to male reproductive success [Ellison, 2001]. The increase of limb musculature is also much higher in boys [Round, et al., 1999]. Males become stronger than females in both upper and lower limbs, but proportionately greater in the upper extremity [Round, et al., 1999; Ransdall and Wells, 1999]. The strength of internal and external rotation of the shoulder in females is about 60% of that of males [Hughes, et al., 1999]. Women perform just under 50% as well as men at push-ups, and only 15% as well at chin-ups [Wells, 1991].

Testosterone also helps regulate libido and mating effort in males. Rising testosterone levels give young men sex drive and motivation for status and aggression [Ellison, 2003; Potts and Hayden, 2008]. Males also develop larger hearts and lungs, higher systolic blood pressure, lower resting heart-rate, a greater capacity for carrying oxygen in the blood and of neutralizing the chemical products of muscular exercise. Power, athletic skill and physical endurance all increase progressively and rapidly throughout adolescence in males. In short, males display an adaptation for fighting [Tanner, 1978; Wrangham and Peterson, 1996]—a way of fighting that benefits from larger body size and muscular strength, particularly in the upper body. This is all concordant with the proposition that natural selection acted to improve the behavior of use of weapons from a bipedal stance, predominantly in males. Men who won these fights left more offspring.

In girls, subcutaneous fat on the limbs rises at puberty, as does hip width, due to the response of cartilage cells in the hip joint to female sex hormone (estrogen) [Tanner, 1978]. The wider hips in post-pubertal females and their increased stores of fat have long been recognized as part of an evolutionary adaptation to child-bearing.

Gender size dimorphism in primates. In non-human primates, sexual selection has played a role in the evolution of gender size dimorphism, highlighting the relationship between male mating competition and sexual dimorphism. The effect may be even greater in large primates (such as African apes and hominins) because large size requires longer gestation, longer birth intervals and therefore a reduction in the availability of fertile females which leads to greater male competition for access to them [Mitani, et al., 1996].

Table I shows that in our living great-ape relatives, sexual size dimorphism in body mass is significantly greater than among modern humans (where females average about 86% of the weight of males). Males prove to be the much more massive gender. Sexual size dimorphism in mass is lower in chimpanzees compared to gorillas and orangutans. In *Pan*, female average

TABLE I

Gender Size Dimorphism in Apes

Body Weight (kg)	Male	Female	F/M	Reference
Gorilla (*Gorilla gorilla*)	170.4	71.5	42%	Fleagle [1999]
	165.2	81.6	49%	Leigh & Shea [1995]
	169.4	80.0	47%	Smith & Jungers [1997]
Orangutan (*Pongo pygmaeus*)	78.2	35.7	46%	Fleagle [1999]
	98.8	46.0	47%	Leigh & Shea [1995]
Chimpanzee (*Pan troglodytes*)	49.6	41.4	83%	Fleagle [1999]
	42.0	35.2	84%	Mitani, et al. [1996]
	58.6	46.2	79%	Leigh & Shea [1995]
	42.7	33.2	78%	Smith & Jungers [1997]
	49.6	41.4	83%	Fleagle [1999]
	58.6	46.2	79%	Leigh & Shea [1995]
	42.7	33.2	78%	Smith & Jungers [1997]
	54.0	40.0	74%	McHenry [1992]
	49.0	41.0	84%	McHenry, Coffing [2000]
Bonobo (*Pan paniscus*)	45.0	33.2	74%	Fleagle [1999]

weight is about 80% as much as in males. The gorilla and orangutan dimorphism might relate to a higher degree of inter-male competition, whereas the chimpanzee and modern human differences could reflect a lower degree of this behavior [Leigh and Shea, 1995].

Gender size dimorphism in fossil hominins. Fossil evidence of hominins is relatively scanty, although it has been markedly expanded in recent years due to the hard-earned discoveries of fossil-hunters. Teeth, due to their composition, are more easily preserved and therefore more plentiful than bones.

There are few examples of entire bones, or even parts of the same bones that match others, so it is difficult to get a clear picture of gender body size dimorphism. Assessing size dimorphism in canine teeth is less problematic, but it is not necessarily linked to body size. A second difficulty, besides scarcity, concerns the assignment of gender to teeth and to bone fragments. It is nearly impossible to determine sex from small bone fragments, so that when different specimens of the same tooth or bone are compared, the larger ones are simply assigned to males and the smaller ones are declared to be from females. This seems like circular reasoning, but it has some predictive value in modern apes and humans.

The role of calculations. Stature and body weight in fossil hominins are derived from calculations based on measurements of bone fragments. Particular emphasis is given to the dimensions of the articular surfaces of the lower limb [McHenry, 1994; Ruff, et al., 1997], especially the proximal femur, since in bipedal stance it is related to the mass of the body it supports. Such measurements are then entered into formulae, some based on measurements of modern apes, others based on data from modern humans [McHenry 1991]. The results are estimates which become more reliable in fossil species where more skeletal remains are available. At the present time, *Australopithecus afarensis* provides the largest body of evidence.

Table II is based largely on the work of McHenry [1991, 1994; McHenry and Coffing, 2000]. The weight dimorphism for *Paranthropus robustus* is from Susman, et al. [2001]. The ages of the fossils are in millions of years (Mya).

Hominin body mass estimates remain in the chimpanzee range for several million years, then rise markedly in *Homo*. The gender difference in weight also diminishes at that time. White and coworkers [2009] think sexual size dimorphism in *Ardipithecus ramidus* (4.4 Mya) may be lower than in *Australopithecus*.

Table II reveals that sexual size dimorphism in body weight in hominins between *A. anamensis* and *Homo* is higher than in chimpanzees and modern humans but lower than in orangutans and gorillas. Females are about 68% as heavy as males, compared to about 86% in chimpanzees and just under

TABLE II

Gender Size Dimorphism in Hominins

		Body Weight (kg)			Stature (cm)		
	Age (Mya)	Male	Female	F/M	Male	Female	F/M
A. anamensis	4.2-3.9	51	33	65%	----	----	---
A. afarensis	3.7-2.9	45	29	64%	151	105	70%
A. africanus	3.0-2.4	41	30	73%	138	115	83%
P. robustus	1.8-1.4	40	30	75%	132	110	83%
P. boisei	2.3-1.4	49	34	69%	137	124	90%
A./H. habilis	1.9-1.6	43	32	74%	157	125	80%
H. erectus	1.8-0.7	66	56	85%	180	160	89%

50% in the gorilla and orangutan. The origin of *H. erectus* marked a dramatic increase in body size, especially in the female [McHenry, 1994] (Chapter 6).

Gender size dimorphism in *A. afarensis* has been examined by several researchers. Johanson and White [1979] proposed that certain hominin remains found in Ethiopia and Tanzania were members of this species (which they named). The fossils included both large and small specimens, which they attributed to a gender disparity. The small size of the individual dubbed "Lucy" (A.L. 288-1) (about 105 cm in stature) is matched by other small postcranial remains which can be contrasted with other larger but morphologically identical individuals. "We consider that much of this body size difference reflects sexual dimorphism" [Johanson and White, 1979, p. 324]. The difference was greater than in chimpanzees and modern humans, and near that of gorillas.

Following the reasoning developed above, this should mean that male-male fighting was *elevated* in these early hominins. McHenry [1994] asserted that the relatively high level of body size sexual dimorphism in early hominins implies a polygynous mating system, not monogamy. Klein [1999, p. 192], concurred: "strong sexual dimorphism in the australopithecines implies an apelike social organization in which males competed vigorously for females and the sexes did not cooperate economically."

Based on an expanded sample of specimens assigned to *A. afarensis*, Richmond and Jungers [1995] concluded that the size extremes displayed sexual dimorphism as great as it is in gorillas and orangutans. A subsequent analysis which included intermediate values [Lockwood, et al., 1996] found the dimorphism in *A. afarensis* to be intermediate between gorillas and orangutans. Harman [2006] studied seven well-preserved *A. afarensis* proximal femora. She found that size variation was close to that of orangutans and gorillas and concluded that male-male competition for mates might best characterize this hominin's behavior. Gordon and coworkers [2008] then used eight measurements on bone articular surfaces and cross-sections of the femur, tibia, humerus and radius. They also found that in *A. afarensis* sexual dimorphism was similar to that of gorillas and orangutans, suggesting a moderate to high degree of competition among males for mating opportunities.

The disparity between body size and canine size dimorphism. Johanson and White [1979] had observed that *A. afarensis* hominins showed marked body size dimorphism, but the metric and morphological dimorphism of the canine teeth was not as pronounced as in most other extant ground-dwelling primates, which perhaps had behavioral implications. Variation in canine size predicted a male/female index of 1.2, similar to chimpanzees and modern humans, whereas body size dimorphism appeared to be much higher than that [McHenry, 1991]. Lockwood and coworkers [1996] found that size variation in the femur and humerus was greater than for the mandible and thought this might be due to reduced canine dimorphism.

Plavcan and van Schaik [1997a,b] then opened a discussion of this issue. Body weight dimorphism and canine size (crown height) dimorphism in monkeys and apes have both been linked to sexual selection resulting from males fighting over access to females. In *A. africanus* and *A. robustus* this connection is broken. Large male body size is associated with reduced canines in both genders. Thus, the two variables offer ambiguous evidence about the mating system of these animals [Plavcan and van Schaik, 1997a]. Rather, they provide contradictory evidence [Plavcan and van Schaik, 1997b; Plavcan, 2000]. Harmon [2006] reiterated this puzzle, as did Kimbel and Delezene [2009], who concluded that the issue of sexual dimorphism in *A. afarensis* is paradoxical.

What does it mean when the two predictors give mixed signals, one pointing towards combat among males, the other pointing away from it? This vexing problem is solved by the proposition that hominin males underwent an evolutionary adaptation to the bipedal use of weapons. First, before elaborating that point, an alternative explanation will be considered which is based on the argument that early hominins were *not* more sexually dimorphic in body size than modern humans.

Solutions to the gender size dimorphism puzzle.
In 2003 Reno and coauthors reported analysis of a sample of *A. afarensis* fossils that led them to conclude that skeletal dimorphism was "human-like". This, they maintained, indicated that *A. afarensis* had a pair-bonded, monogamous reproductive strategy, with little competition between males (a concept originated by one of the coauthors [Lovejoy, 1981] based on male provisioning driven by female choice, to be examined in Chapter 16). In short, there was no problem because there was little sexual size dimorphism in either the skeleton or the canine teeth.

This was contested by Plavcan, et al. [2005]. When they added a small specimen to the Reno team's [2003] sample, it moved the *A. afarensis* estimate to a gorilla level of sexual size dimorphism. This corroborated earlier reports that supported the "widely held [view] that early hominins were strongly sexually dimorphic in body mass, but showed little canine dimorphism—a unique pattern among primates that makes inferring behavior difficult" [Plavcan, et al., 2005, p. 31]. Reno and coworkers [2005] acknowledged the widespread view, but maintained that skeletal dimorphism is not a surrogate for body mass and appears to be a poor predictor of polygyny or monogamy. Two subsequent reports [Harman, 2006 and Gordon, et al., 2008] rejected the Reno group's proposal and reaffirmed earlier conclusions about male-male competition.

Ward [2002] proposed another way of thinking about the

CHAPTER 5: GENDER SIZE DIFFERENCES

so-called paradox. Since all extant anthropoids (monkeys and apes) with an *A. afarensis* level of body size sexual dimorphism are polygynous with fairly intense levels of male-male competition, body size dimorphism must have been valuable in these hominins. The solution? Relatively large canines had ceased to confer reproductive advantages to male hominins.

This conclusion is fully consistent with the bipedal use of weapons proposal. The explanation for the high level of sexual size dimorphism in the bodies of early hominins coupled with a low level of dimorphism in more diminutive canine teeth is as follows: Males were larger in body mass, stature and muscularity because they had been naturally selected for prowess in fighting during millions of years, using weapons consisting of stones thrown as missiles and sticks or bones used as clubs. During this time span, the canine teeth became obsolete as weapons for combat. Consequently, they decreased in size in males, reducing the sexual dimorphism, as they underwent natural selection that improved their function for *chewing* (Chapter 11).

The search for evolutionary explanations hinges upon the question "where are the reproductive advantages?" In a contentious group of males vying for mating opportunities by fighting with each other, the advantage will often go to the bigger and stronger combatant. Hominin males were doing something involving "brute force" that increased their opportunities to father more offspring bearing their genes. Natural selection maintained or increased testosterone levels, which helped promote aggressiveness, large body size, muscle mass, shoulder width, upper body strength and other features valuable during combat. Females did not have to fight to become inseminated when they came into estrus. Engaging in dangerous combat provided no reproductive advantages for them. Instead, natural selection acted to produce higher levels of estrogen, which fostered less aggressiveness, broader pelves, more fat storage, less upper body musculature and other effects that facilitated the gestation and birth of healthy children and the disposition to nurture them after birth.

Hominins remained about the same size for millions of years. This changed during the transition from *Australopithicus* to *Homo*, as described in the next chapter.

CHAPTER 6

The Transition to *Homo Erectus*

PART I. THE EVIDENCE

"Only with Homo erectus might body size, culture and other factors have combined to 'release' hominids from their dependence on trees" [Susman, et al., 1984, p. 113].

"The arrival of Homo… marks the single most important transformation below the neck in the entire course of human evolution" [Tattersall, 1995, p. 238].

Introduction. The transition from *Australopithecus* to *Homo erectus* in Africa involved the most profound metamorphosis in the history of the hominin lineage. The appearance of *Homo erectus* announced an abrupt evolutionary divergence from an ancestral anatomy that had been stable for millions of years. This took place during a brief interval around 1.9 Mya [Asfaw, et al., 2002]. In the process, a group of small, partly arboreal, partly bipedal, long-armed, short-legged hominins abandoned the trees and were transformed into much larger hominins with a body form similar to our own, fully dedicated to bipedal behavior on the ground. If we seek to understand how we came to be the way we are today, this transition looms large. What caused this abrupt alteration? *What were the reproductive advantages?*

The australopithecine stage of hominin evolution, from

about 4 to 2 Mya, was a relatively stable period. Body size and physical form, brain size and limb proportions were not conspicuously modified, an arboreal adaptation was evident, as was bipedal behavior on the ground. About 2.6 to 1.8 Mya a group branched off, distinguished by their massive and powerful jaws and teeth, becoming a different species (*Australopithecus robustus,* a.k.a *Paranthropus*). Apparently they were dietary specialists. Their postcranial bones (the skeleton below the neck) remained similar to those of other australopithecines.

Another divergence from structural conservatism that continued in these *Australopithecus* species concerns the canine teeth. Long and sharp at the start of the lineage, they gradually diminished in size and became more incisor-like. The hand also changed: its long, curved fingers became shorter and straighter, the diminutive thumb became larger, stronger and more mobile. Two new handgrips emerged. These modifications in the hand (Chapter 9) and canine teeth (Chapter 11) were largely completed before the transition to *Homo erectus*. While retaining their adaptation to life in the trees, the australopithecine were evidently practicing a bipedal behavior involving the hand and one type of tooth—a behavior that was yielding reproductive advantages, thereby causing natural selection of inherited traits that changed these body structures.

Otherwise, during and after *A. anamensis* (~4.2-3.9 Mya), the general body plan of the australopithecines remained intact for two million years. If the small body, long arms and short legs had been disadvantageous, natural selection would have had ample time to alter them [Ward, 2002]. Why did they persist for so long, then suddenly change?

Perhaps climate played a role. From 8 to 3 Mya, parts of Africa were warmer and more humid than at present, nurturing wet, lowland, rainforest vegetation in regions that today support seasonally dry, savannah grasses and shrubs [deMenocal, 1995]. Cooling after 3.0 Mya in Africa seems tied to a shift from relatively stable global warmth to high-amplitude cooling and warming cycles [Marlow, et al., 2000]. The emergence of *Homo* might have been due to "climate forcing" as fluctuating cooler, drier conditions promoted more open habitats in tropical Africa [Behrensmeyer, et al., 1997]. East Africa experienced an increase in wooded grasslands between 2.5 and 1.5 Mya [Cerling, et al.,

2011]. The appearance of some new species at this time may have been "climatically mediated" by ecologic fragmentation and genetic isolation that favored adaptations to aridity [deMenocal, 1995]. Possibly an irregular but overall expansion of open habitats combined with persisting woodlands and forests to provide increased opportunities for mammals and a rise in diversity from 3.0 to 2.0 Mya [Behrensmeyer, et al., 1997]. It may have provided new opportunities for hominins.

The evidence of the anatomical changes that took place during the transition from *Australopithecus* to *Homo erectus* will be examined below, after a brief look at hominins that may have participated in that event. In Chapter 7, explanations that have been proposed to account for the new hominins will be presented, including the idea than an adaptation to bipedal use of hand-held weapons seems to have been a significant factor.

When did *Homo erectus* appear? An occipital bone fragment from the Koobi Fora region of the Lake Turkana basin in Kenya (KNM-ER 2598, dated at 2.0-1.9 Mya) is considered by some to be the earliest evidence of *H. erectus* [Antón, 2003; Ferring, et al., 2011; Wood, 2011]. However, Suwa, et al. [2007a] note that the fossil was found on a surface that may sample overlying deposits. Other fossils from that area include postcranial remains from large-bodied hominins that may be *H. erectus* (KNM-ER 3228, 2.0 Mya and KNM-ER 1481, 1.9 Mya), and a partial skeleton (KNM-ER 1808) from 1.7-1.8 Mya [Klein, 1999; Antón, 2003]. The earliest definitive *H. erectus* cranium is KNM-ER 3733 at 1.8 Mya [Walker and Shipman, 1997; Antón, 2003; Suwa, et al., 2007a; Wood, 2011]. KNM-WT 15000 (Nariokotome boy), found on the west side of the Turkana basin, dates to 1.5 Mya [Brown and McDougall, 1993]. The term, *Homo ergaster,* has sometimes been used to designate early African *Homo erectus*, but some find this "doubtfully necessary or useful" [Asfaw, et al., 2002].

Who were the transitional hominins? Most authorities currently agree that the lineage to *Homo* passed through *A. afarensis,* which was present in East Africa until ~3.0 Mya. Since *H. erectus* is first documented in that same region about 2 Mya, the transition from *Australopithecus* to *H.*

erectus may have occurred there before that date. Several possible transitional hominins have been reported.

Hominin remains from Ethiopia's Middle Awash region, dated at 2.5 Mya and assigned to the species *Australopithecus garhi*, might be an ancestor for early *Homo* [Asfaw, et al., 1999]. A femur and arm elements suggest that the humero/femoral index would have been "humanlike," possibly marking the earliest sign of the femoral elongation that characterizes later hominins. *A. garhi* is in the right place at the right time to be the ancestor of early *Homo* [Asfaw, et al., 1999], insofar as all the earliest *H. erectus* specimens are from East Africa.

Australopithecus sediba is represented by the remains of four individuals dated at 2.0 Mya [Woodhead, et al., 2011]. They were encased in cave deposits at Malapa in South Africa, near sites bearing *A. africanus,* their possible ancestor [Berger, et al., 2010a]. The *A. sediba* hominins comprise a paleodeme: individuals of similar geological age from a small area. They were diminutive with small brains, long upper limbs, relatively long forearms with large joint surfaces, and primitive traits in both upper and lower limbs, but they had some *Homo*-like features in the pelvis and lower limb. The shoulder joint was oriented upward and the manual phalanges were curved (as in other australopithecines). Features of the hip, knee and ankle indicate they were habitual bipeds, but the foot was primitive [Berger, et al., 2010a]. *A. sediba* appears to be too recent and possibly in the wrong place (South rather than East Africa) to be ancestral to *Homo erectus*. Perhaps they arose earlier, gave rise to *H. erectus*, then survived as a sister species.

Specimens attributed to *Homo habilis* have been identified that are 300,000 years older than *A. sediba*, and in the putative right place, East Africa. *Homo habilis* from Olduvai Gorge, Tanzania was first named and described by L. Leakey and coworkers [1964]. The authors considered the fossils to be relics of a hominin that was transitional between *Australopithecus* and *Homo*. The cranial capacity, form of the chin and the molar teeth were intermediate between these two hominins; the clavicle, hand and foot specimens were within the modern human range. Stone tools were linked to *H. habilis* [L. Leakey, et al., 1964; Tobias, 1964]. Stone artifacts are now known from australopithecine sites a half-million years older.

Homo habilis fossils proved to be widespread, in Kenya, Ethiopia, and Malawa. Most are from ~1.9-1.8 Mya, but the oldest is dated at 2.3 Mya [Kimbel, et al., 1997] and the youngest, at 1.4 Mya [Spoor, et al., 2007] (both from Koobi Fora). Survival until 1.4 Mya may make *H. habilis* a less likely ancestor *of H. erectus* [Spoor, et al., 2007], but *erectus* might have branched off earlier. The 2.3 Mya specimen is the right age for an *H. erectus* predecessor and the East African setting is appropriate. Why two hominin species inhabited the same area for nearly 500,000 years is unknown [Spoor, et al., 2007]

Specimens attributed to *H. habilis* are not only widely dispersed and temporally extended, they encompass a diverse group of specimens with teeth of different sizes and shapes, different palates, different faces, cranial capacities ranging from about 500 cc to 800 cc and large differences in body size. Some individuals at Koobi Fora had relatively large skulls and large australopithecine-sized teeth whereas others had small australopithecine-sized skulls combined with small erectus-sized teeth [Klein, 1999].Variability within *H. habilis* is so great it might confound two species. The large individuals might have been ancestral to *H. erectus*, but the small-brained specimens with long arms and short legs are "false pretenders" [Walker and Shipman, 1997, p. 259].

Wood has expressed doubts whether fossils attributed to *H. habilis* should be assigned to *Homo* [1992, 2010, 2011; Wood and Lonergan, 2008]. (Susman [2008] would retain *habilis* in *Homo*). Wood and Collard [1999], however, defined a genus as groups of species of common ancestry that are adapted to different ecological situations, then decided *Homo habilis* is adapted more like *Australopithecus* than *Homo*. Their small body size and their teeth resemble those of the australopithecine, as does their body form and their capacity for proficient climbing (the long arms, short legs, curved fingers and toes, for example). Although bipedal on the ground, they display a mixed locomotive strategy, whereas the *Homo* group is fully committed to terrestrial bipedalism [Wood, 2011]. In this view, the genus *Homo* begins with *H. erectus*, who abandoned the trees. In the spirit of ambivalence, I shall hereafter refer to this group of hominins as "*A./H. habilis*".

The Dmanisi hominins are the earliest known to have

emerged from Africa. Remains of four individuals, numerous manuports and stone artifacts were found in a layer dated at 1.77 Mya at Dmanisi, Republic of Georgia, in the southern Caucasus. The fossils are all from this layer, but there are stone artifacts documenting use of the site beginning ~1.85 Mya. Dmanisi hominins had primitive (Oldowan-type) tools. More advanced (Acheulian) artifacts, mixed with Oldowan artifacts, occur at about the same time in West Turkana, Kenya (1.76 Mya) [Lepre, et al., 2011]. Dmanisi might have been a way station on the route from Africa to China and Java (1.7-1.6 Mya) [Ferring, et al., 2011].

The Georgia hominins have been assigned to the primitive end of the *H. erectus* spectrum [Lordkipanidze, et al., 2007, 2013]. Like most *A./H. habilis* specimens they have small brains and small bodies. Their brains are small even relative to their diminutive body size [Rightmire, et al., 2006; Lordkipanidze, et al., 2007]. Some aspects of skull morphology resemble *A./H. habilis*; others are like those of early *H. erectus* as are the teeth [Rightmire, et al., 2006]. Their limb proportions (femoral/tibial and humero/femoral ratios) are also similar to those of modern humans [Lordkipanidze, et al., 2007]. Although they were small in stature, their leg length relative to body weight was at the low end of the modern human range and beyond the upper end of the chimpanzee range [Pontzer, et al., 2010]. The lower limb is longer than in chimpanzees, *A. afarensis* and *A./H. habilis*, but much shorter than in *H. erectus*. The foot resembles that of the *A./habilis* hominin from Olduvai Gorge (O.H. 8). It seems to have had transverse and longitudinal arches, an adducted hallux (big toe), and a human-like ankle joint [Pontzer, et al., 2010] signifying a full commitment to bipedal locomotion but some locomotor traits found in *H. erectus* are lacking [Lordkipanidze, et al., 2007; Pontzer, et al., 2010]. The four Dmanisi hominins constitute a paleodeme. There is variation within it, but they share a common bauplan [Rightmire, et al., 2006].

Perhaps 1.85 Mya is too late for small hominins with tiny brains living in the southern Caucasus to be direct ancestors of the large *H. erectus* individuals in East Africa by ~1.9 Mya. An African hominin more primitive than *H. erectus* could have arrived in Dmanisi before 1.85 Mya, evolved there, then returned to Africa as the stem hominin that completed the transition to *H. erectus* [Rightmire, et al., 2006; Ferring, et al., 2011]. If specimen KNM-ER 3733 (1.8 Mya) proves to be the earliest African *Homo erectus*, this would raise the possibility that Dmanisi hominins may have been its precursors, a transformation that would have required a significant increase in body and brain size.

Which of these hominins gave rise to *Homo erectus*? Each has a mixture of physical traits, some which resemble *Australopithecus* and others which signal *Homo*. This is the typical "mosaic" aspect of evolution. A combination of ancestral ("primitive") and new ("derived") traits is observed in virtually all early hominins [Susman and de Ruiter, 2004]. Some that overlapped in time were "mosaic in different ways" [Harcourt-Smith and Aiello, 2004]. The transition to a larger-bodied, full-striding terrestrial biped like *H. erectus* occurred in a "mosaic fashion" [Berger, et al., 2010a]. The entire skeleton of the Dmanisi hominins had a mosaic of primitive and derived features [Lordkipanidze, et al., 2007; Pontzer, et al., 2010]. The *Ardipithecus* pelvis was an "odd mosaic" [Lovejoy, et al., 2009i], its foot an "amalgam" of retained primitive and derived traits [White, et al., 2009], i. e., a "a mosaic foot" [Haile-Selassie, et al., 2012]. This is what we *expect* in evolving hominins, making it difficult to decide in this case which of the possible transitional hominins bore the genome that was naturally selected to produce *H. erectus*.

The intermediary stage is hazy, but major changes that took place during the transition from *Australopithecus* to *Homo erectus* are clear, despite the scarcity and incompleteness of paleontological evidence, in which all fossil skeletons and most individual bones are incomplete, and the dimensions of missing parts are calculated judgments.

Body size increased. *Homo* erectus was larger than its ancestors. No one has charted this increase more diligently than McHenry. In 1992 he stated unequivocally that "all *Australopithecus* and *Homo habilis* seem to have been relatively small, whereas later hominids from *Homo erectus* to *Homo sapiens*, have been generally similar in size to modern people" [McHenry,

CHAPTER 6: THE TRANSITION TO *HOMO ERECTUS*, PART I

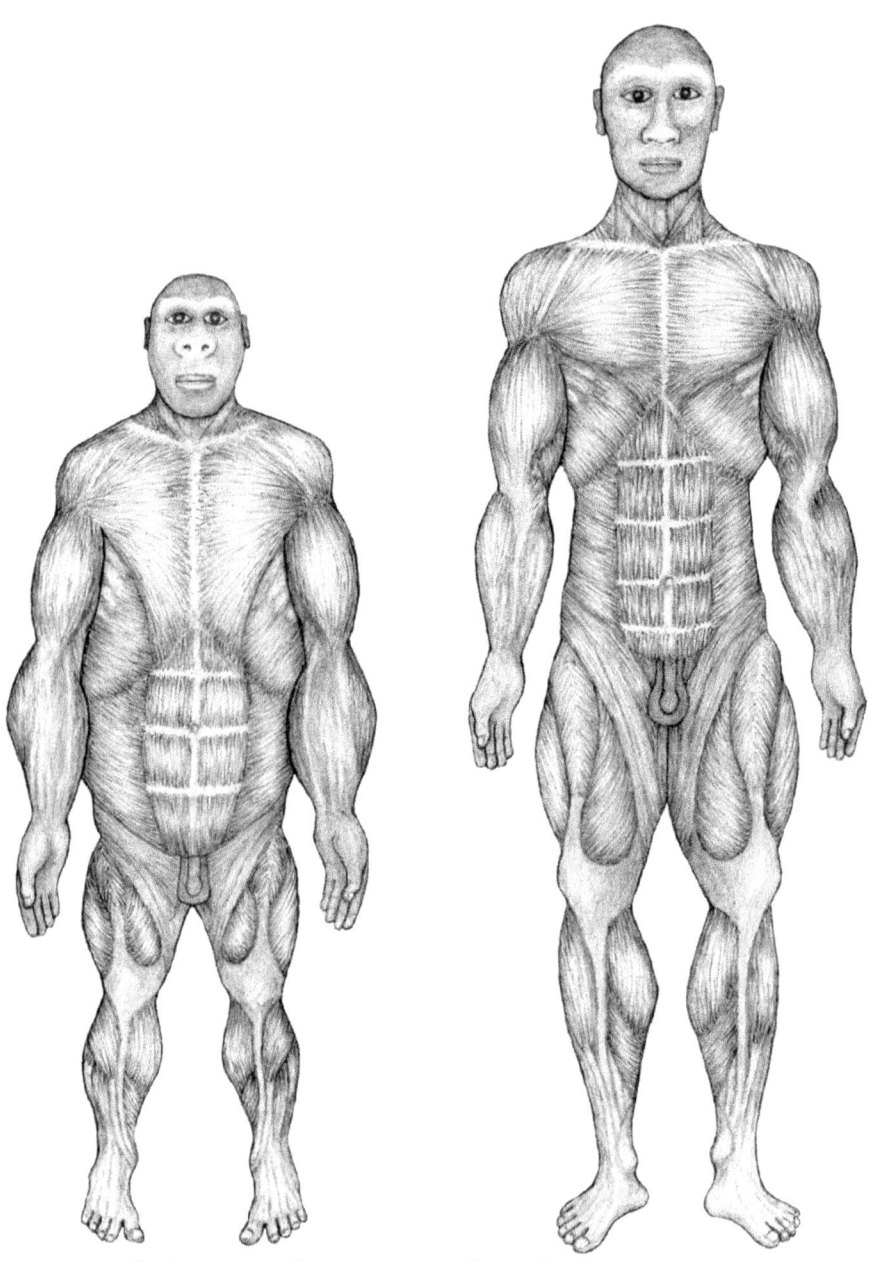

1992, p. 19]. In 1994 he presented evidence that *H. erectus*
Figure 3. Major anatomical changes in males during the transition from *Australopithecus* (left) to *Homo erectus* (right), based on fossil evidence. The differences in stature (143 cm and 180 cm) are derived from Table II, Chapter 5. Changes in

CHAPTER 6: THE TRANSITION TO *HOMO ERECTUS*, PART I

the hand are given in Chapter 9. (*Caption continued on next page*) Musculoskeletal robusticity characterized pre-*sapiens* hominins (Chapter 12). Body proportions are based on information given in this Chapter (6). In *Homo erectus* body size increased (mass increased from 43 kg to 66 kg: Table II). The length of the upper limbs changed little, but in *Homo* the distal elements became relatively less massive, the forearms diminished in size relative to the upper arm, and the hand became smaller, except for the thumb. In contrast, the lower limbs became longer and more massive in *Homo erectus* while the thorax became more cylindrical. This was associated with the appearance of a waist and the ability to rotate the thorax and pelvis independently. These several changes can be explained by an accelerated adaptation to bipedal throwing and club-swinging when hominins abandoned the trees.

males were 30%-40% more massive than pre-erectus hominins [McHenry 1994] (Figure 3).

When additional fossils became available, McHenry showed that from *A. anamensis* (4.2 Mya) to *A./H. habilis* the estimated average body weight of hominin males was 43.8 kg, whereas in early *Homo erectus* it was 66 kg, an increase of about 50% [McHenry and Coffing, 2000]. These are estimates, obtained indirectly from small samples, but they indicate that a major augmentation of body mass occurred in the transition to *Homo erectus* [Antón, 2003].

In the period immediately preceding *H. erectus* (1.9 to 1.8 Mya) there are many tiny individuals represented by postcranial bones such as the *A./H. habilis* partial skeleton (O.H. 62). By 1.7 Mya, however, the East African hominin fossils are all large-bodied. Adults with small bodies disappear and sexual size dimorphism is reduced [McHenry, 1994]. There is an increase of about 16% in estimated stature and 26% in body weight between pre- and post-1.7 Mya hominins specimens. The mass estimates of six *H. erectus* individuals from 1.9-1.6 Mya average 52 kg. The three oldest, dated at 1.9 Mya, average 46 kg [Ruff and Walker, 1993, Table 11.15], suggesting that body size was still increasing in early *Homo*. Soon, *H. erectus* size reached approximately 170 cm and 58 kg, with no subsequent increase [Ruff and Walker,

TABLE III

Cranial capacity in hominins

Hominins	Age (Mya)	Cranial capacity (cc)*	Reference
Australopithecus			
A. *afarensis*	3.7-2.9	400-550	Wood [2010]
A. *africanus*	3.0-2.4	457	Wood & Collard [1999]
		460	Wood & Lonergan [2008]
A. *robustus*	2.3-1.4	410-513	Wood and Collard [1999]
		480	Wood and Lonergan [2008]
Transitional hominins			
A./H. *habilis*	1.9-1.6	552	Wood & Collard [1999]
		500-600	Wood and Lonergan [2008]
		601-730	Lordkipanidze, et al. 2007]
H. *erectus* †	1.77	775	Rightmire [2004]
H. *erectus* †	1.77	800	Gabunia, et al. [2001]
H. *erectus* †	1.77	600	Vekua, et al. [2002]
Homo			
H. *erectus*	2.4-1.5	725**	Wood [2010]
H. *erectus*	1.8-1.5	838	Rightmire [2004].
H. *erectus*	1.8-1.5	790	ibid. + Spoor, et al.,2007]#
H. *erectus*		904	Lordkipanidze, et al. [2007]
H. *erectus.*	1.55	691	Spoor, et al. [2007]
H. *erectus*	1.5	760	Wood and Lonergan [2008]
H. *heidelbergensis*	0.6-0.2	1,198	Wood & Collard [1999]
H. *heidelbergensis*		1,206	Rightmire [2004]
H. *sapiens* (Modern)		1,355	Wood and Collard [1999]

*Cranial capacity (in cubic centimeters) measured on skulls, an estimate of brain volume, is often reconstructed from fragments.
†Hominins from Dmanisi.
** KNM-ER 1470 from Koobi Fora.
# Includes 600 cc. cranium (D2700) from Dmansi and 691 cc. cranium (KNM-ER 42700) from Ileret [Spoor, et al., 2007].

1993]. Increase in body mass was greater in females, reducing the gender size dimorphism that had previously characterized the hominin lineage [McHenry, 1994]. Males were still larger than females, but females were now about 85% of the weight of males, similar to chimpanzees (74-84%) and much less than the gender disparity of ~68% in hominins during the 2.5 million years since *A. anamensis* (Chapter 5).

The relationship between brain and body size did not change. The brain, like most other parts of the body, participated in the overall enlargement that took place during the transitional period, but there is scant evidence that there was a *relative* increase in brain size. Growth in brain volume was not more pronounced than the overall growth of the body.

Table III presents the evidence concerning changes in estimated cranial capacity in the hominin lineage, with special reference to the transition from *Australopithecus* to *Homo*. The estimated range of brain size in *Australopithecus* overlaps the range for *A./H. habilis* and the smallest skull from Dmanisi. Another small skull (691 cc) has also been attributed to African *H. erectus* [Spoor, et al., 2007].

The cranial capacity of early African *H. erectus* exceeds that of *Australopithecus*. The australopithecine brain was about 500 cc and the early *H. erectus* brain in Africa ranged from 600 to 800 cc. An increase from 500 to 800 cc would be 60%. During the same period, the estimated body weight in *Australopithecus* males including *A./H. habilis* was 43.8 kg; in early *Homo* it was 66 kg. This represents an increase of about 50% [McHenry and Coffing, 2000]. Simultaneously, the increase in female weight was about 32 kg to 56 kg (Table II, Chapter 5), an increase of 75%. It therefore seems that the increase in brain size that accompanied the transition from *Australopithecus* to *Homo erectus* was of the same magnitude as the increase in body size. The increase in body size could account for the increase in brain size.

It once was thought that in *Homo* the brain became relatively larger [McHenry, 1994]. However, recent authorities have shied away from this conclusion, citing the rise in body weight. According to Wrangham [2009], the large increase in cranial capacity in *H. erectus* compared to *A/H. habilis* was accompanied by a parallel growth in body weight, so that whether the brains were relatively larger is uncertain. In Nariokotome boy (KNM-WT 15000), an example of early *H. erectus* [1.53 Mya: Walker and R. Leakey, 1993], the endocranial capacity (~ 760 cc), when scaled to body mass shows relatively little advance over the levels seen in earlier hominins [Wood and Lonergan, 2008].

Parenthetically, there was a later era in which hominin brain size did undergo an expansion while body size remained stable. This occurred in a branch of the *H. erectus* lineage, *H. heidelburgensis,* about 0.6 to 0.2 Mya (Table III). Ruff and coworkers [1997] estimated body mass from femoral breadth and from reconstructed stature and body breadth, then calculated relative brain mass using a formula that yields the "encephalization quotient." In *Homo erectus* from 1.8 to 0.6 Mya there was no increase, but a major rise took place from 0.6 to 0.2 Mya in *H. heidelbergensis*. Rightmire [2004] compared the cranial capacity of 30 *H. erectus* specimens with that of ten *H. heidelbergensis* crania and found that the increase in estimated brain size exceeded the consequences of larger body size and was the result of an accelerated growth of the brain [Rightmire, 2004]. In the other branch, *H. erectus*, the increase in cranial capacity during its nearly 2 million-year-long existence (1.9 Mya to 0.04 Mya) was very small. Rightmire [2004] calculated that it was about 165 cc *per million years.*

Shoulder breadth, mobility and joint orientation was modified. In *A. afarensis*, the shoulder joint (the humerus articulation with the glenoid cavity of the scapula) was directed upward, as it is in apes, indicating that abducting the arm above the head was a frequent, important movement [Stern and Susman, 1983; Haile-Selassie, et al., 2010]. In arboreal primates the shoulder joint faces upward to aid reaching above the head [Oxnard, 1963; Stern and Susman, 1983; Fleagle, 1999; Larson, 2007]. An *A. afarensis* child's scapula was also found to have a glenoid fossa with an upward orientation [Alemseged, et al., 2006; Green and Alemseged, 2012]. So too did an *A. africanus* scapula, suggesting that the arm was used importantly in an elevated position [Aiello and Dean, 1990]. The clavicle was short and the glenoid cavity in *A. sediba* had an upward orientation [Berger, et al., 2010a; Churchill, et al., 2013], as did the shoulder joint in *A./H. habilis* [Larson, 2007] and the Dmanisi hominins [Lordkipanidze, et al., 2007].

In early *Homo erectus* (Nariokotome Boy, 1.5 Mya), however, the glenoid fossa faced laterally [Larson, 2007], as it does in all subsequent species of *Homo*, including modern humans. This is sometimes said to reflect the typically "lowered" position of the upper limb [Ashton and Oxnard, 1964; Aiello and Dean, 1990; Larson, 2007].

In early *Homo* the clavicle remained relatively short [Larson, 2007]. Elongation came later, but it is difficult to determine when that happened because clavicles are rare in the hominin fossil record. One from a *Homo* species dated about 0.8 Mya (*H. antecessor*) is very long [Larson, 2007]. The Neanderthals also had broad shoulders with long clavicles [Aiello and Dean, 1990]. The shoulders were already broader in Nariokotome boy than in the australopithecine because the thorax was wider at the top, [Ruff and Walker, 1993].

The longer clavicle in *Homo*, evident in our own species, increased the range of motion of the upper limb [Ashton and Oxnard, 1964; Larson, 2007]. Indeed, the modern human shoulder is the most mobile of all primates from prosimians through monkeys and apes, according to Oxnard [1969]. Thus, during hominin evolution, the shoulders became broader, the shoulder joint increased in mobility and the articulation changed from an upward orientation in apes and pre-*Homo* hominins to a horizontal orientation in the transition to *Homo*.

The thorax and waist were modified. In early hominins, as in apes, the thorax was "cone-shaped" [Hunt, 1996], wide at the bottom—a form also called "funnel-shaped" [Schmid, 1991, 2004]—and the lumbar region of the vertebral column was short. The short lumbar spine coupled with cranially elongated iliac blades and strong muscles bridging the small gap between thorax and pelvis is such that there is no waist in an ape's trunk [Schmid, 2004; Walker and Shipman, 2005]. Rigidity of this region due to the binding of thorax and pelvis is thought to facilitate arboreal movement because it reduces bending of the trunk during arm-hanging [Fleagle,

1999; Lovejoy and McCollum, 2010]. Arm-hanging in chimpanzees requires full abduction of the humerus and suspension of the entire body weight from one or both arms. The cone shape of the chimpanzee ribcage more closely conforms to the lines of tensile force during one-arm suspension [Hunt, 1996].

A. afarensis and *A. sediba* exhibit an ape-like anatomical structure of this type [Schmid, 2004; Schmid, et al., 2013] (contra Haile-Selassie, et al., [2010]). In *A. afarensis* the shape of the iliac crest and the broad, lower rib cage indicate that external and internal oblique muscles would have been ineffective in producing torsion in the abdominal region [Schmid, 2004]. There would have been no waist and a restricted ability to rotate the trunk without also rotating the pelvis. However, the muscles which ascend from the vertebral column and ribs to the shoulder and upper arm in *Australopithecus* were large [Schmid, 1991].

Australopithecines have a spectacularly large hip width which seems at odds with an adaptation for efficient bipedal locomotion because it requires greater muscular activity during walking and produces high stress on the femoral neck. It is less disadvantageous for bipedal posture because these stresses are reduced during standing. Wide hips lower the center of gravity (mass), and make it easier to maintain upright balance [Hunt, 1996]. *Australopithecus* had a chimpanzee-like thoracic and lumbar region. Chimpanzees have 13 thoracic and 3-4 lumbar vertebrae, whereas modern humans have 12 thoracic and five lumbar vertebrae (about 4% have six) [Swindler and Wood, 1982; Latimer and Ward, 1993].

Stiffness of the midsection in apes and australopithecines was altered during the transformation to *Homo*, when hominins acquired a waist and the capacity to rotate the upper and lower body independently. Dramatic differences appeared by 1.5 Mya, when the boy from Nariokotome, Kenya (KNM-WT 15000, 1.5 Mya) provided the first opportunity to study the thoracic skeleton in *H. erectus*. The waist was fully formed in this hominin, who had six lumbar vertebrae [Walker and R. Leakey, 1993; Latimer and Ward, 1993]. His rib cage was barrel-shaped rather than conical; it was narrowed at the bottom to conform to the narrower pelvis of more modern form, with reoriented iliac blades that were reduced in height.

The relative lengths of the arms and legs changed. *Australopithecus* species were roughly the size of modern chimpanzees and were built like them too, with large upper bodies, long arms, and relatively short legs. The relative length of the upper arm (humerus) compared to the length of the upper part of the lower limb (femur) is measured by the humero/femoral index (humerus length/femur length x 100). In modern apes, the upper extremity dominates, as shown in Table IV (next page) [data from Aiello and Dean, 1990]. Long forelimbs serve to extend reach during arboreal travel and increase the number of food items accessible to an individual feeding among terminal branches. Attenuated hind limbs lighten the lower body and bring the center of gravity closer to arboreal supports, thereby decreasing the risk of falling [Hunt, 1996].

In Table IV the humerofemoral index in living apes and humans is compared with estimates from hominin fossil material. All of the pre-modern fossil indices are calculations based on incomplete specimens. The estimates reveal a chronological trend: pre-*Homo* hominins, from *Ardipithecus* to *A./H. habilis* have slightly lower ratios (longer legs compared to their arms) than in living apes, but are not markedly different from the two species of *Pan* (chimpanzees and bonobos). The first sign of a change is in the Dmanisi hominins (1.77 Mya), considered an early form of *H. erectus*. but one in which body size was still small. The humerofemoral index in modern humans is nowhere near the upper 90s or lower 100s of modern apes, or the mid 90s to upper 80s of ardipithecines and australopithecines. In modern humans the index is 71.2 [Aiello and Dean, 1990]. The change is largely due to increased length of the lower limbs.

By 1.5 Mya, in Nariokotome Boy (KNM-WT 15000), the first *H. erectus* individual of large size in which the humero/femoral index can be measured, the transformation of the arm and leg structure had been completed. This partial skeleton, found in northern Kenya, was a youth of about 12 years of age who stood approximately 1.6 m tall, and might have reached 1.9 m at maturity. The general postcranial proportions of KNM-WT 15000 are well within the modern human range [Ruff and Walker, 1993]. The maximum length of the femur in

Nariokotome Boy was 43.2 cm [Ruff and Walker, 1993, p. 247].

TABLE IV

Humero/femoral index

Primate Class	Humero/femoral Index*
Modern Apes	
Orangutan	128.8
Gorilla	117.9
Chimpanzee	101.9
Bonobo	97.0
Pre-*Homo erectus* hominins	
Ardipithecus	91.0
Australopithecus afarensis	85.1
A./H. Habilis	95.0
Homo erectus (Dmanisi)	76.4
Early African *Homo erectus*	
Homo erectus	74.0
Modern *Homo*	
Homo sapiens	71.2

*The humero/femoral index indicates the relative lengths of the upper arms and thighs. It is determined by comparing the lengths of the humerus and femur: (humerus length/femur length x 100).

The length of the humerus (allowing for the missing superior epiphysis) was estimated to be 31.9 cm [Walker and R. Leakey, 1993, p. 127]. This represents a humero/femoral index of about 74, as recorded by Richmond and associates [2002]. The transition in the lower limbs, from *Australopithecus* to *Homo* appears to have been very rapid. In the oldest hominin, *Ardipithecus* (4.4 Mya), the humero/femoral index estimated from an artist's drawing of a reconstruction is 91, although no humerus was recovered [J. Matterne in Lovejoy, et al., 2009a, p. 101; Shreeve, 2010, p. 67]. In *A. afarensis* (A. L. 288.1, "Lucy," 3.2 Mya) the humerofemoral index is less (85.1), according to

Jungers [1982] who notes that the arms were in the range of modern humans but the legs were much shorter and most of the mass was in the upper body. Richmond, et al. [2002] give a range of 84.5-85.0 for A. L. 288.1. McHenry and Berger [1998] calculated that in *A. africanus* specimens from Sterkfontein (3.0-2.4 Mya) the arms were relatively large, based on joint size, corresponding to a modern human with a body mass of 53 kg, but the hindlimbs were much smaller, matching a modern human of 33 kg. This is *more* ape-like than in earlier *A. afarensis* and *A. anamensis*, suggesting that there might have been a greater dedication to arboreality in *A. africanus*. The arm-to-leg ratio was even greater in *A./H. habilis*. In Olduvai Hominid OH 62 (~1.8 Mya) from Lower Bed I at Olduvai Gorge, Tanzania, consisting of radial, humeral, femoral and tibial fragments, apparently from one individual, the upper arm was *longer* than that of *A. afarensis* (A.L. 288-1), but the femur was *shorter*. The estimated humerofemoral index is 94-95 [Johanson, et al., 1987; Richmond, et al., 2002]. The body size was very small.

Natural selection is a constant, but mutations and other genetic changes introduce unpredictable chance, genetic recombination among variable genomes produces variability within breeding populations, populations become isolated, the climate changes, new predators and diseases are introduced, innovative behaviors appear, the pace of change quickens and slows and, as noted above, the result is a mosaic of primitive and derived features. Excruciatingly small sample sizes introduce more irregularity. Ancient populations were variable and when a single individual is involved, there is no way of knowing whether it is an outlier or near the population average.

Despite these caveats, the general conclusion seems justified that *Homo* appeared rapidly after a brief transitional phase, larger in size and with a distinctly altered arm/leg ratio. In living apes and humans, the ratio of length of the upper and lower limbs can be measured directly to calculate the *intermembral index* (arm length/leg length x 100). Species with long upper limbs have high indices: In the orangutan, it is about 140; in the gorilla, 117; in the chimpanzee, 103, and in bonobos, 102 [Aiello and Dean, 1990; Fleagle, 1999]. In striking contrast, in modern humans it is about 70 [Aiello and Dean, 1990;

Lovejoy, et al., 2009c]. Because upper limb length in modern *Homo* in relation to trunk length is similar to that of chimpanzees whereas the ratio of lower limb and trunk is much higher in humans, the low intermembral index in *Homo* must be due largely to an increase in length of the lower limbs [Schultz, 1969]. The upper limbs in Nariokotome boy (KNM-WT 15000) were longer than Lucy's (A. L. 288-1), but the lower limbs were *much* longer [Klein, 1999, p. 291]. In *Homo*, the entire body increased in size and the legs enlarged disproportionately compared to the arms.

The relative strengths of the arm and leg bones changed. Upper limb length may have increased somewhat, but there appears to have been a significant diminution of upper limb muscularity and mass with the onset of *Homo*. The long upper limbs of *Australopithecus* were robust and muscular (Chapter 12), presumably related to arboreal locomotion. Chimpanzees have notoriously powerful upper body strength, three to five times greater than modern humans in adult males [Goodall, 1988; Savage-Rumbaugh & Lewin, 1994; Wrangham & Peterson, 1996; Fouts, 1997]. This strength advantage resides in the musculature that controls the upper limbs. In contrast, the lower limb muscles of humans are more powerful than those of chimpanzees. Femoral to humeral bone strength proportions in *A./H. habilis* are similar to those of chimpanzees, whereas in *H. erectus* specimens the proportions fall within or above those of modern humans. "The most likely interpretation is that *H. habilis*, although bipedal when terrestrial, still engaged in frequent arboreal behavior, while *H. erectus* was a completely committed terrestrial biped" [Ruff, 2009, p. 90].

Forearm length diminished compared to upper arm length. During the transition from *Australopithicus* to *Homo erectus*, the forearm became reduced in length compared to the upper arm. This is measured by the *brachial index*—the ratio between radius length and humerus length (radius length/humerus length X 100). In chimpanzees, the brachial index is 92.2, in the gorilla it is 81; in orangutans it is 101 [Reno, et al., 2005]. Among fossil hominins, *A. afarensis* (3.2 Mya) had a brachial index of 92 [Lovejoy, et al., 2009c]. *A. garhi* (2.5 Mya) also had an ape-like upper/lower arm ratio [Asfaw, et al., 1999]. *A. sediba*'s index was 84.3 (2.0 Mya) [Berger, et al., 2010b; Churchill, et al., 2013]. The median of various estimates for *A./H. habilis* is ~87. In *H. erectus* (KNM-WT-15000), it drops to ~80; in modern humans it is 77 [Richmond, et al., 2002]. Thus, *H. erectus* already had a shortened forearm compared to the upper arm, close to its modern proportions [Aiello and Dean, 1990; Asfaw, et al., 1999; Richmond, et al., 2002; Reno et al., 2005; Lovejoy, et al., 2009b; Berger, et al., 2010b].

The evidence supports the conclusion that during human evolution early hominins—up to and including *A./H. habilis*— had ape-like, arboreally-adapted limb proportions, with large upper bodies, long and powerful arms, long forearms, relatively small legs, and a body size significantly smaller than *Homo*. In later pre-*Homo* hominins (*A. africanus, A. garhi. A./H. habilis, A. sediba,*) these proportions show no tendency to decline and may even have been exaggerated. However, in early *Homo,* modern limb proportions suddenly appeared and remained that way thereafter (Figure 3).

The chewing apparatus diminished in size. Early hominins had ape-like faces with projecting upper and lower jaws [Johanson and White, 1979]. The canine teeth, which lie at the junction of the front cutting teeth (the incisors) and the rear chewing and grinding teeth (the premolars and molars) are featured in Chapter 11; here the emphasis is on the chewing or "cheek" teeth. The molars are relatively small, with thick enamel in *Orrorin tugenensis* (~6 Mya) [Senut, et al., 2001]. *Ardipithecus ramidus* (4.4 Mya) also has small molars, but they have thinner enamel than those of *Orrorin* or the australopithecine [Suwa, et al., 2009a]. A shift towards thicker tooth enamel appears first in *A. anamensis* (~4 Mya) [Ward, et al., 1999b]. A trend to larger molars may have begun later, in *A. afarensis* [Ward, et al., 2001].

Incorporation of the premolars into the grinding mechanism had its onset in *Australopithecus* and continued until *H. erectus* (~1.8 Mya) as part of a highly derived masticatory complex [Kimbel, et al., 2006]. *Australopithecus* was associated with enlargement in the basal dimension of both premolars and molars ("megadontia") [Teaford and Ungar, 2000; Suwa, et al.,

2009b], a trend carried to extremes in the robust australopithecines, whose craniofacial and dental features apparently reflect an adaptation to a diet requiring heavy mastication [Johanson and White, 1979]. (*A. robustus* became extinct about 1.4 Mya). Tobias [1983] provided a quantitative assessment of these trends by measuring the summed crown areas of mandibular teeth. If *A. africanus* is used as a baseline, beginning with 861 (units not given) there was a reduction to 787 in *A./H. habilis*, 665 in African *H. erectus* and 484 in modern *H. sapiens*. The decrease in *H. erectus* is the largest reduction in tooth size in hominin evolution [Wrangham, 2009].

This concludes an examination of the evidence of the great hominin transition phase. The fossils reveal that a major remodeling of hominin size and form took place in a relatively short time. What does this mean? Why did it happen? What were the reproductive advantages? Explanations will be presented in the next chapter.

CHAPTER 7

The Transition to *Homo Erectus*

PART II. THE EXPLANATIONS

"The reduction in tooth size, the signs of increased energy availability in larger brains and bodies, the indication of smaller guts, and the ability to exploit new kinds of habitat all support the idea that cooking was responsible for the evolution of Homo erectus" [Wrangham, 2009, p. 98].

"In the transition from Australopithecus to Homo, the greatest transformation in the hominin lineage, natural selection for improved throwing and club-swinging prowess played a major role. Most of the anatomical changes preserved in the fossil record can be explained in this manner" [Young, 2013, last paragraph below].

Introduction. How many different explanations are required to account for the evolution of a relatively small, partly arboreal animal into a large, earthbound creature, striding off bipedally toward a new kind of future with a thoroughly remodeled body? What is the simplest explanation of why natural selection acted to elicit the recorded changes?

What are the changes that need to be explained? As described in the preceding chapter, the last common ape ancestor from which the hominin and chimpanzee lineages diverged was adapted to life in the trees, and in hominins prior to *Homo* significant arboreal adaptations were retained. Early hominins were small animals, their arms long and muscular, their fingers and toes curved, their lower limbs short and relatively weak. Arborealism includes hanging beneath limbs for support or locomotion, arms raised above the head, capable of hauling the entire body upward against gravity. Powerful arms, lightweight legs and a stiff midriff promote these actions, coupled with an upward-facing shoulder joint that facilitates reaching above the head and a funnel-shaped thorax and narrow shoulder breadth that reduces stress on the capsule of the shoulder joint during arm-hanging. The persistence of this body form in *Ar. ramidus, A. afarensis, A. africanus* and *A./H. habilis* indicates that pre-*erectus* hominins continued to make use of their arboreal heritage [Stern and Susman, 1983; Hunt, 1991, 1994; Fleagle, 1999; Ward, 2002; Larson, 2007].

In the transition from *Australopithecus* to *Homo* all of this changed. In *Homo erectus* the body was enlarged and the lower limbs were increased in length and strength compared to the upper limbs. Forearm length diminished relative to upper arm length, the hands became relatively smaller, the curved fingers and toes associated with arboreal life were straightened (Chapters 9, 14), the shoulders became broader and more mobile, and the shoulder joint was redirected laterally. In *Homo* a waist appeared and independent rotation of pelvis and thorax became possible (Chapter 6).

Why did hominins retain for millions of years a body structure adapted for life in the trees then lose these arboreal adaptations in *Homo erectus*? Why did body size increase while the chewing apparatus diminished? Why did the thorax change shape in concert with shoulders that become broader and more mobile as the axis of the glenoid fossa become reoriented? Why did *Homo* acquire independent rotation of the thorax and pelvis? Why did the legs become larger and the arms (particularly the forearms) become less massive? *What were the reproductive advantages?*

EXPLANATIONS OFFERED BY PREVIOUS AUTHORS

Early hominins retained a body structure that was adapted for life in the trees. Why did *Homo erectus* lose these features? There is no reason to doubt that the advantages of continuing a behavioral and anatomical adaptation to life in the trees were the same for early hominins as they had been for their ancestors during tens of millions of years. The adaptation was maintained by natural selection because it met their biological needs, such as providing access to food and a place of refuge for sleeping or escape from predators. *Homo erectus* must have abandoned the trees for life on the ground because it provided *increased reproductive advantages*. This concept has had scant prior examination.

However, recently Wrangham [2001, 2009] presented and developed the provocative idea that controlling fire and its use in cooking were the key factors in the transition to *Homo*. These are cultural traits, but Wrangham's argument is that they provided benefits that led to natural selection of heritable variations.

In brief, Wrangham's "cooking hypothesis" proposes that controlling fire began with the "habilines". "According to our hypothesis," he wrote, "hominization happened because a late gracile population of australopith apes—such as *A. habilis*—learned to use fire to improve the digestibility and range of its plant foods. It is possible also that cooking of meat was highly significant" [Wrangham, 2001, p. 237]. After they were able to keep fire going at night, some of them occasionally dropped food morsels into it by accident, ate them after they had been heated, observed that they tasted better, and so deliberate cooking began. This released more calories from their existing diet which enhanced their survival and reproduction. Those who cooked reproduced more copies of their genes because they ingested more calories (rather than gaining access to more food or more nutritious food). Since cooking was a cultural behavior, no evolutionary changes were involved. However, cooked diets provided more energy by increasing the digestibility of starch and protein and fewer calories were expended for digestion, detoxification and defense against pathogens [Carmody and Wrangham, 2009a]. As a result, repetition of this habit led to the evolution of *Homo erectus*.

On the face of it, this major transition may seem an unlikely result of caloric enhancement. Here is Wrangham's explication: Hominin bodies were adapted by natural selection to cooked food. Because it was easier to digest, it resulted in smaller digestive tracts. Body size increased and the shoulder, arm and trunk adaptations that enabled habilines to climb well disappeared [Wrangham, 2009]. The "protection fire provided at night enabled them to sleep on the ground and lose their climbing ability" [Wrangham, 2009, p. 194]. "When they no longer needed to climb trees to find food or sleep safely, natural selection rapidly favored the anatomical changes that facilitated long-distance locomotion and led to living completely on the ground" [Wrangham, 2001, p. 102].

The role of cooking in human evolution has been compellingly documented by Wrangham and coworkers. It seems that modern humans have indeed undergone natural selection for eating food that has been cooked. In mice, the positive energetic effects of cooked meat and tubers are higher than if they are eaten raw [Carmody, et al., 2011]. Modern humans also fare poorly on raw diets. Cooking reduces the structural integrity of most foods, leading to a reduction of chewing time and masticatory stress. It denatures proteins, increasing their susceptibility to proteolytic enzymes. It also lowers the costs of food consumption and immune defense and improves the net energy value of plant and animal foods regularly consumed by humans. The adoption of cooking would have helped ancestral humans thrive [Carmody, et al., 2011].

A problem with this explanation, which Wrangham acknowledges, is that evidence of controlled fire first appears in the archaeological record no earlier than 1 Mya, and probably after 0.5 Mya, well after the transition to *Homo*. If one accepts Wrangham's assumption that only the onset of cooking can explain the major anatomical features of this transition, then the control of fire should predate 2 Mya.

The evidence for controlled fire in association with *H. erectus* is scant and inconclusive, according to Berna and coworkers [2012], who report a possible example of burned vegetation dated at~1 Mya in sediments 30 m from the present-day entrance to a South African cave. Pickering [2012] supports the conclusion that hominins may have created these indications of burning, and perhaps also those of similar age he is studying at the Swartkrans Cave in South Africa.

Nevertheless, in Europe early hominins moved into northern latitudes without the habitual use of fire. They lived there for 800,000 years despite temperatures which at times dropped below freezing. One of the earliest signs of controlled fire comes from the Schönigen, Germany, hominin site (famous for its throwing spears) about 400,000 years ago [Carmody and Wrangham, 2009b; Roebroeks and Villa, 2011].

Because the evidence currently indicates that control of fire began less than 0.5 Mya, too late to account for the transition to *Homo*, it is averred that the bipedal use of weapons provides a more substantial explanation of the changes which accompanied that transition. However, this proposal sheds no light on the question of cooking, which offers a better match with features of the modern human mouth, jaws and digestive tract. Carmody and Wrangham [2009b] point out that humans, compared to chimpanzees have smaller oral cavities, reduced molar dentition, more gracile mandibles, smaller chewing muscles, longer, more massive and sacculated intestinal tracts. If cooking began as recently as 0.25 Mya, even such a late onset would be sufficient for these adaptations to have occurred [Carmody and Wrangham, 2009b].

The alternative explanation of the transition from *Australopithecus* to *Homo* I describe below and have emphasized throughout this book is that the behavior used by *Homo* to gain improved reproductive benefits on the ground was not a new discovery of the transitional hominins, but one that had begun *four million years earlier*, and had been improved by natural selection throughout that long era. Hominins had been undergoing adaptation to bipedal throwing and club-swinging for a very long time, but its full expression had been inhibited by the arboreal adaptation. When the trees were abandoned, a rapid transformation took place, yielding a new body form that enhanced the bipedal use of weapons.

Why was there a significant increase in body size? All preceding hominin species were smaller than *Homo*. *H. erectus* males were nearly twice as heavy as their australopithecine ancestors. By 1.7 Mya in East Africa, small-bodied hominins had disappeared. What were the reproductive advantages?

Large mass and body dimensions constrain arboreal travel. Abandoning the trees eliminated a major restriction on body size, but increased mass also acts negatively on terrestrial locomotion. Whether perambulating on two legs or four, one must expend more calories in the muscles that produce locomotion when the mass being moved is increased. Furthermore, augmenting body size introduces an enhanced *nutritional* requirement to grow and maintain the enlarged bulk. If there were no change in dietary sources, bigger hominins would have had to eat more of the food they were already eating (e.g., fruit from trees) but their larger size would have made them less adept at obtaining it.

For males, increased size is explicable as an adaptation to fighting for scarce resources (Chapters 3, 5), which includes competition for food. Cartmill and Smith [2009] attributed the adaptive advantages to improved hunting and scavenging ability, and the capacity "to put more force behind the use of spears, clubs and other implements for hunting and defensive purposes" [p. 261]. They also support the idea that increased stature would have yielded benefits in locomotion (but see below).

For females, larger bodies meant that daily energy requirements during gestation and lactation would be significantly higher, as would the total energetic cost for each child. *H. erectus* mothers also would have incurred extra costs gathering sufficient food to support and transport their own increased body mass, to maintain themselves through their next pregnancy and lactation, to provide for their young offspring, and for carrying and protecting them [Aiello and Key, 2002; Antón, 2003]. Steudel-Numbers [2006] calculated that the total daily energy expenditure of a female *H. erectus* was 84% greater than that of a female *Australopithecus*.

An increase in stature in a slender body might have provided benefits for thermoregulation and water balance in drier, open environments due to an increased surface-to-volume ratio [Ruff and Walker, 1993; Wheeler, 1993; Antón, et al., 2002; Aiello and Key, 2002]. However, as a rule, *Homo* erectus was not slender. Both genders were robust and muscular (Chapter 12). This would have reduced the effect of stature on temperature regulation. Furthermore, the unequal augmentation in size in males and females suggests that there may have been a gender effect superimposed on the reproductive benefits of increased body size.

The advantage to *Homo* females (who increased in size disproportionately, thereby diminishing sexual dimorphism) may have been due to the additional effects it had on the reproductive process itself. The relative increase in female size was attributed by Antón [2003] to larger mothers being able to deliver more energy to their offspring at lower cost to themselves. Alternative views are that sexual dimorphism declined because males stopped fighting with each other, or that new, unidentified weapons reduced the size advantage in males, or that males and females of similar size yielded more economical distance transport [Carrier, 2004].

There is considerable support for the concept that larger females have increased success in gestating, lactating and nurturing larger, healthier offspring. Larger mothers can command more nutritional resources that can be transferred to their children through the placenta or during lactation. In several mammalian species, including humans, larger females typically have more surviving offspring [Symons, 1979]. Taller human females have lower rates of cephalo-pelvic disproportion, stillbirth and perinatal mortality than short ones [Lovejoy, 2005].

Did *Homo* mothers require help? Some assistance was essential, Aiello and Key [2002] suggested, if *H. erectus* females were to meet the increased energetic costs of their own large body mass and their large-bodied infants. Mothers might have enlisted the help of others; raising children could have been a community affair. Food sharing and other forms of cooperation between the sexes could have reduced the energetic burden of mothers [Aiello and Key, 2002]. Why would males want to do that? (This a difficult problem: see Chapters 16 and 17). Males, by contributing to the care of children, might be investing in their own inclusive fitness, if they knew which children were their own (but they probably didn't). Male provisioning might reduce the interbirth interval, producing a higher incidence of reproductively active females (if males understood such a cause and effect). Males might trade food for sex [Aiello and Key, 2002].

Both genders and all ages of early *Homo*, due to their larger size, needed more nutritious food. It seems possible that in early *Homo* populations where males acquired meat by scavenging or hunting and *ate it all themselves*, the likelihood of extinction was greater than in those where the males shared some of it with females and children, whatever their motivation may have been for doing so.

Wrangham puts a twist to the question of when *Homo* mothers started getting help. They may have received some of the meat that males had procured, but when they started cooking it, they became trapped. Cooking brought nutritional benefits, but mainly to men. It trapped females into a subservient role enforced by a male-dominated culture [Wrangham, 2009].

What nutritional source enabled the enlarged body mass in *Homo*? Improved nutrition was required to grow and maintain a larger body. Eating more fruit was not the solution, because of the lost arboreal adaptation. They must have been eating more food of a different kind [Aiello and Key, 2002; Leonard, et al., 2007]. Some have hypothesized that *Homo* might have taken to digging up "underground storage organs" (tubers) [Aiello and Key, 2002]. However, these plant parts, although a potential source of calories, are otherwise low-quality foods almost devoid of nutrients, and some contain cyanide-yielding toxins [Milton, 1999]. Tubers are unlikely to have provided sufficient quantities of the essential nutritional elements required to grow and renew the enlarged bodies. As is the case with vegetable foods in general, there is no evidence that hominins ate tubers, although their diet almost certainly included plant parts.

There has been widespread speculation and near unanimity that the new, high-quality, source of nutrition exploited by *Homo erectus* was meat. "Meat" in this context means "animal tissues". Carnivores eat muscle tissue and fat, brains, viscera, bone marrow, liver and other organs, which supply many essential minerals, vitamins and other micronutrients [Milton, 1999]. All parts of animals that could be chewed and swallowed would have been eaten. Coupled with ancestral plant foods, meat would satisfy the dietary needs of producing larger hominins. It has often been stressed that a larger brain must be paid for with increased energy in the diet, because it is a relatively high consumer of calories. Aiello and Wheeler [1995] suggested that these costs could be offset by smaller digestive tracts and this could reflect increased carnivory. Indeed, the common conclusion is that the only practical source of augmented calories for the brain was meat (animal tissues) [Rightmire, 2004]. This issue is part of the larger topic of the need for increased food to grow and maintain *all* of the components of a larger body, and the same response seems applicable: eat more meat.

Many scholars have pointed to the advantages of eating more animal tissues as a solution to the problem of supporting the enlarged body of *Homo*. Among these are Milton [1999], Rose [2001], Foley, [2001], Roebroeks [2001], Vasey and Walker [2001], O'Connell, et al. [2002], Aiello and Key [2002], Antón, et al. [2002], Antón [2003], Kingdon [2003], Leonard, et al. [2007], Bennett, et al. [2009], Pontzer, et al. [2010], and Lieberman [2012b]. All conclude that early *Homo erectus,* in addition to having a larger body, was characterized by increased carnivory. The evidence from paleontology and archaeology also indicates that enhanced access to meat was coincident with the emergence of *Homo* (Chapter 8). Nevertheless, something is missing in this argument. How did hominins *take possession* of meat from animals presumably reluctant to be eaten? It has been variously asserted that they sought, detected, stalked, pursued, targeted, caught, took, captured, subdued, and dispatched large animals. Meat was acquired, obtained, and procured. But how? (See Chapter 8).

Wrangham's cooking hypothesis emphasizes the role of meat in brain expansion which he believes occurred in the habilines and may have been due to meat-processing. Dietary shifts toward roots, meat-eating, and meat-processing can explain the growth in brains from a chimpanzee-like ancestor to the habilines [Wrangham, 2009]. He notes the problem of *acquiring* the meat and speculates that hominins might have stolen carcasses from cheetahs or jackals, but not from lions or saber-tooth cats who were too dangerous. However, "it is unclear how australopithecines obtained access to the meat of antelope and other game animals. Maybe they found new ways to kill" (p. 188). They might have thrown rocks or spears [Wrangham, 2009]. Although cooking plays no role in my explanatory proposal, the concept that hominins used throwing weapons is central to its structure.

Why did thoracic form change from conical to cylindrical? Narrowing of the base of the cone-shaped rib cage might have been due to shrinkage of the digestive tract to allow for a larger brain [Aiello and Wheeler, 1995] or resulted from cooking food [Wrangham, 2009]. The chimpanzee gut is larger than that of humans [Milton, 1999], but most of the digestive tract is in the abdomen, not the thorax, and these hypotheses say little about the upper thorax.

In its upper part, the thorax expanded in *Homo*, increasing the breadth of the shoulders. The broader shoulders along with reduction of the bulging expanse of the lower rib cage increased the range of motion of the upper limb in *Homo*. The modern human shoulder is the most mobile of all primates [Oxnard, 1969]. Why did mobility increase at the shoulder joint in *H. erectus*, a creature who no longer used its arms for locomotion? Perhaps natural selection acted to increase mobility in the shoulder of *Homo* due to a "greater dependence on manipulatory abilities" involving larger brain size and more sophisticated tool technology; or, alternatively, it might have been due to the cessation of life in trees [Larson, 2007].

Kirschmann [1999] clearly and explicitly accounted for the change in shape of the thorax and the increased breadth of the shoulders in *Homo* by attributing it to natural selection for throwing ability. He wrote that the demands of throwing explain the evolution of chest and shoulder anatomy during hominin evolution. Because upper body rotation increases the acceleration distance during a throw, natural selection produced the broad shoulders that are male sexual characteristics in modern humans. The barrel shape of the human thorax contributes to broadening the shoulders that are needed as levers for the rotation of the upper body around its long axis with a low moment of inertia [Kirschmann, 1999].

Why did the axis of the glenoid fossa change from an upward to a lateral orientation? All authorities agree that the upward orientation of the glenoid fossa in arboreal primates and in *Australopithecus* is due to the importance of vertical climbing, below-branch locomotion, and arm-hanging, all of which employ the arm in a fully abducted position, raised above the head, parallel with the long axis of the body. In arborealists the shoulder joint faces upward to aid reaching above the head [Oxnard, 1963; Stern and Susman, 1983; Fleagle, 1999; Larson, 2007]. This orientation was perpetuated until the transition to *Homo*.

In *Homo*, this changed. The advantage no longer accrued to those who fully abducted their arms vertically, but rather to those who abducted their arms *half-way*, to 90° instead of 180°. The shoulder joint became oriented *laterally*. This has routinely [1963-2007] been attributed to "the typical lowered position of the upper limb" in *Homo* (references above). However, if the behavior acted upon by natural selection exerted its effect when the arm was in a lowered position, it seems the glenoid fossa would have been reoriented *downward*; that is, caudally, not laterally. The lateral orientation indicates that the reproductively advantageous behavior was facilitated when the humerus was abducted at a right angle to the long axis of the body, parallel with the plane of the shoulders. This is readily explained by the reproductive advantages associated with improved throwing and club-swinging, as described below.

Why did *Homo* acquire a waist permitting independent rotation of thorax and pelvis? In the transition to *Homo*, the thorax narrowed near its junction with the pelvis. The lower rib cage also became more widely separated from the pelvis, due to lengthening of the lumbar spine and rearrangement of muscles attached to the ribs and pelvis. This produced a distinctive waist, which is absent in apes. The result was that the thorax and pelvis gained an unprecedented ability to rotate independently around the body's long axis. Schmid [2004] reports that in *Australopithecus* the rigid, ape-like midsection with its lack of a waist would be inefficient for torsion, which would have made running in a modern human manner impossible. A slight (about 4°) independent rotation of the pelvis occurs during human bipedal locomotion (Chapter 14). It will be shown below that in modern humans during the act of throwing, independent rotation of the thorax and pelvis occurs through an enormously greater range than this, and at a much faster rate.

Why did the chewing apparatus diminish in size?
Early hominins had ape-like faces with projecting upper and lower jaws. During the australopithecine era, tooth enamel thickened as molars and jaws increased in size. In the robust australopithecines, the molars became still larger, suggesting an adaptation to eating food that required even more powerful chewing. Thickened enamel may have been a way to counteract tooth fracture after a dietary shift to harder foods [Ward, et al., 1999b; Teaford and Ungar, 2000]. The australopithecines might have had difficulty masticating tough, pliant foods, such as meat and leaves, but could process hard, coarse and brittle plant parts [Teaford and Ungar, 2000; Foley, 2001; Schoeninger, et al., 2001; Ungar and Sponheimer, 2011]. This might be related somehow to the expansion of open, dry habitats after 3.0 Mya [Kimbel, et al., 2006].

A reduction in tooth material appeared in the transition to *H. erectus* and continued in that lineage, implying a reduced requirement for chewing. All experts agree that this had something to do with adaptation to a new diet. According to Tobias [1983], the crown areas of mandibular teeth diminished markedly from *A. africanus* to *H. erectus*. In Wrangham's view [2009], the benefits of cooked food could account for this, although as noted above, the effects of cooking may have begun later. Others believe that early *Homo* gained access to a new, easier-to-chew, nutritious dietary staple, namely animal tissues.

Meat was becoming abundant as browsing and grazing animals proliferated on the expanding grasslands. It is easily assimilated and requires relatively little mastication or digestion [Foley, 2001]. Attributing diminution of the chewing apparatus to an increased intake of the edible parts of animals is consistent with the idea that meat was the quality food required by a significantly enlarged body in *Homo* [Bunn, 2001; O'Connell, et al., 2002; Leonard, et al., 2007].

AN ALTERNATIVE EXPLANATION OF THE TRANSITION TO *HOMO* BASED ON THE BIPEDAL USE OF WEAPONS

Introduction. I will now examine the changes associated with the transition to *Homo erectus* using an explanatory method that seeks to identify the reproductive advantages which accompanied each evolutionary modification. It will be shown that most of the evidence can be explained by the proposition that hominins underwent a prolonged evolutionary adaptation to throwing and clubbing. Natural selection acted to improve prowess in the bipedal use of weapons because those most proficient in that behavior obtained more food and mating opportunities, thereby gaining reproductive benefits.

***Australopithecus* was arboreal. *Homo erectus* abandoned the trees.** *Ar. ramidus* occupied relatively high altitude, closed-canopy woodlands of Ethiopia. *A. anamensis*, a younger hominin, inhabited riverine woodlands and gallery forests of the Turkana Basin. *A. afarensis*, in turn, lived in forests, woodlands, and more open country across East and possibly Central Africa [Ward, et al., 1999b]. About midway through the australopithecine period the climate began to change in Africa. Cooler and drier conditions promoted the expansion of open habitats at the expense of woodlands and forests, providing new opportunities for species that could adapt to more arid environments. I suggest that one branch of the *Australopithecus* lineage began to apply its adaptation to hand-held weapons in a new arena–the grasslands–where it proved to yield even greater reproductive dividends.

Australopithecine species and their hominin ancestors expressed features related to bipedalism on the ground, but maintained others associated with arborealism for approximately four million years. Natural selection acts to maintain behaviors and the physical features that facilitate them as long as they support reproduction of the species. Because the early hominin body plan is a recognizable adaptation to arboreal habitats and hominin fossils from this era are associated with woodland environments, one can assume that the reproductive advantages of continuing an arboreal adaptation were similar to those of their predecessors. In short, it met their biological requirements for survival and reproduction and no better alternative had emerged. Why, then, did a branch of the hominin lineage set off in a new direction, dedicated to life on the ground, forsaking their ancient arboreal heritage?

The new lifestyle must have provided *more* reproductive benefits than the old. Indeed, *Homo* has gone forth and multiplied all over the earth while its arboreal ancestors became extinct and today's woodland-dwelling apes are endangered. So then, what was the secret of *Homo's* success?

There is widespread agreement that *Homo* gained access to a nutritious food source that was lacking among the trees. The new diet featured *animal tissue*, abundant in the herds of herbivores on the expanding savannas. Early *Homo* began to eat meat in unprecedented quantities (Chapter 8). An opportunity had opened for an aggressive predator adapted to the use of weapons and a willingness to employ them against unsuspecting ungulates such as horse, zebra, cattle, antelopes, giraffes and other animals who could be hunted or whose carcasses could be obtained by driving predators from their prey. This concentrated, nutritious food enhanced their ability to proliferate, due to the link between nutrition and reproduction (Chapter 3). How did they acquire animal tissue? They made use of their long heritage of natural selection for improved bipedal use of weapons (Chapter 8).

Why body size increased in the transition from *Australopithecus* to *Homo*. Males who were taller and more massive gained an advantage in fighting among themselves for access to food and females, promoting natural selection for these traits. This traditional argument for gender size dimorphism in hominins can be applied here, incorporated into the explanation that males were fighting with hand-held weapons from a bipedal stance. It is asserted (Chapter 3) that throughout hominin history males did most of the fighting. Combat with predators and prey as well as with each other would have favored larger body size. (Boxing or wrestling, using the forelimbs [Carrier, 2004], seems an unpromising strategy in fighting with carnivores over meat or in hunting to acquire it). The interpretation proposed here is that they fought with hand-held weapons such as rocks, clubs and spears.

Among modern humans, the most proficient throwers and club-swingers are characterized by large size and musculoskeletal robusticity (Chapter 12). They are taller and heavier than males of comparable age in the general population. Their bulk, strength and sturdiness enable them to apply greater force to the hand-held implement, accelerating it to greater velocity. Early hominins who could apply greater force to their weapons would have been more successful fighters, leading to an increase in their chances of eating and mating. Consequently, they passed on more copies of their genes.

The increase in female size may in part be due to selection for larger males. It could be the kind of trait selected in one gender but affecting both, because of the commonality of the genes which participate in its expression. This is a likely explanation of why modern females are able to throw and swing clubs more effectively than either gender in any other species, although lagging behind modern males in this skill as a general rule (Chapter 13).

However, the fact that in *Homo erectus* females became disproportionately larger suggests that the advantages of increased body size differed in the two genders and the greater increase in female size requires an additional explanation. That extra benefit may have resided in the reproductive process itself. In many living species larger females have increased success in gestating, birthing, lactating and rearing larger and healthier offspring. They have larger birth canals and more nutritional resources that can be transferred to their children through the placenta or after birth in milk. Typically they have more surviving offspring. The genetic benefit of larger size in females seems at least partially linked to successful production and nurturing of babies.

Meat was the nutritional source that enabled a larger body mass and fostered reproductive success. As emphasized above, there is a consensus of opinion that the high-quality source of nutrition *Homo* exploited for growing and maintaining a more massive body was meat. Evidence indicates that enhanced harvesting of animal tissues coincided with the emergence of *Homo,* but the question of how they obtained such food has long remained a puzzle. The answer (groups of males used hand-held weapons from a bipedal stance) is discussed in the next chapter. Both genders and all members of early *Homo* populations, due to their increased size, needed more food. *H. erectus* mothers would have incurred extra costs from gathering sufficient food for their own daily needs, augmenting it during pregnancy and lactation, and providing it to their immature children. Did early *Homo* males who acquired meat share some of it with females? Nobody knows. There is no evidence.

The thorax changed from conical to cylindrical. Early hominins had an ape-like funnel-shaped thorax, narrow at the top. Expansion of the upper rib cage in early *H. erectus* increased the separation of the shoulder joints. Subsequently, the clavicle also lengthened, further augmenting shoulder breadth (Chapter 6, Figure 3). This, along with reduction in diameter of the lower thorax, increased the range of motion of the humerus. Why did mobility increase at the shoulder joint after *Homo* became fully terrestrial? Increased shoulder mobility has been attributed to manipulatory abilities promoted by a larger brain and more sophisticated tool technology, but the tool techniques, their dependence on shoulder mobility, or the possible reproductive benefits have not been described.

The proposal elaborated in this book provides an evolutionary explanation. Increased mobility of the shoulder was the result of natural selection for improved use of weapons employed by throwing and club-swinging. Mobility of the shoulder is a crucial element in this behavior. As noted above, Kirschmann [1999] first enunciated this concept. The bipedal use of weapons offered numerous ways of enhancing access to mates and food, defense against predators and belligerent conspecifics (Chapter 3). In the process that produced the modern throwing and clubbing motions, increased mobility at the shoulder joint was one of many anatomical features acted on by natural selection to improve this behavior.

The axis of the glenoid fossa was redirected laterally. In *Homo*, the orientation of the shoulder joint changed from the upward direction observed in arboreal animals (including pre-*Homo* hominins) to a lateral orientation. The idea that it was a result of the "typical lowered arm position" in *Homo* is not compelling. This hypothesis apparently is based on the notion that the lateral orientation was naturally selected because it was used by *Homo* for a typical arm position, presumably meaning the fully adducted one used when sitting, standing or walking, but full adduction implies that a downward-facing joint would be appropriate. In any event, no genetic benefits are cited.

The lateral orientation indicates that natural selection acted to improve a behavior in which the humerus was oriented *laterally*, abducted at right angles to the vertical axis of the upper body and aligned with the plane of the shoulders. This is readily explained by the orientation of the humerus in the throwing motion, when the pulse of energy is transmitted through the shoulder joint to the upper arm (Chapter 13; Figure 12). During the acceleration phase of the throw, starting when the hand-held missile begins its forward motion until it is released, the humerus is held in a position of 90° abduction, in line with the plane of the shoulders. It rotates medially about its long axis at an extremely high rate, but its orientation in the glenoid fossa does not change. The alignment of the humerus with the shoulders and with the forearm and fingers at missile release assures the greatest possible biomechanical lever-arm action for transfer of the maximum amount of energy to the hand-held weapon. Whether a missile is thrown (or a club is swung) "overarm", "sidearm" or somewhere in between, the upper body is always tilted to maintain the humerus in a lateral orientation in line with the shoulders.

The reproductive advantage of the lateral orientation of the glenoid fossa lies in its role in the act of throwing missiles or swinging clubs at adversaries during hominin evolution. Those throwers and clubbers whose shoulder joints were more mobile, and whose glenoid fossae were oriented more laterally, were able to transmit more energy from the torso to the arm and hand. Thereby they reaped benefits in the competition to transmit more copies of their genes to succeeding generations.

The waist was modified to permit independent rotation of the thorax and pelvis. The same general explanation can be applied to changes in the junction of the thorax and pelvis during the transition to *Homo erectus*. Mobility at the waist in *Australopithecus* was limited by a short lumbar spine, firm binding of the lower thorax to the pelvis, and by the arrangement of muscular attachments in this region, a complex believed to aid arboreal travel. In early *Homo*, lengthening of the lumbar spine separated the lower rib cage from the pelvis and the involved muscle attachments were modified. The result was a distinctive waist region and an unprecedented ability to rotate the pelvis and thorax independently, which acted to improve throwing and club-swinging ability.

Analysis of these full-body motions in modern humans has revealed how a pulse of energy that begins in the lower extremities is ultimately transmitted to the hand-held weapon. It begins with leg thrusts into the ground, is subsequently converted to pelvic rotation, that is followed by rotation of the thorax. This rapid, coordinated sequence progressively augments the amount of kinetic energy and transfers it from the legs to the pelvis, then to the thorax, then to the upper limbs, and lastly to the weapon (Chapter 13). Independent rotation of the pelvis and thorax is a crucial link in this chain.

Why did the lower limbs become longer and more powerful? One plausible explanation is that when *Homo erectus* abandoned the trees for life on the ground, large, powerful upper limbs lost their utility for arboreal pursuits whereas longer, more muscular and robust lower limbs became advantageous for upright gait. This explanation, however, requires closer scrutiny.

Longer, more massive lower limbs might have been selected to increase the velocity of bipedal locomotion because longer stride length would increase gait velocity "at only a slight increase in energy cost" [Jungers, 1982; Cartmill and Smith, 2009]. Others believe that "elongation of the lower limb does not reduce energy costs, carries substantial disadvantages, and would not have occurred without substantial positive selection" [Lovejoy, 2005, p. 119].

There has been considerable analysis and debate concerning the energetic effects of longer lower limbs on bipedal locomotion. Some conclude it saved energy because increased stride length means fewer strides for a given distance; others suppose it would increase energy consumption because longer limbs have a greater moment of inertia [Steudel-Numbers, et al., 2007]. Biomechanical considerations and mathematical analyses have suggested that longer limbs promote an increase in the efficiency of bipedal walking (commonly attributed to stride length) [Preuschoft and Witte, 1991; Preuschoft, 2004; Pontzer, et al., 2009] whereas Kramer and Eck [2000] proposed that the short hindlimbs of *A. afarensis* could have been more efficient for walking due to the reduced energy required to swing them. Modern human elite long-distance runners are short and slender, but the long lower limbs of *Homo erectus* were heavy, muscular and robust (Chapter 12).

Steudel [1996] reviewed published evidence on human bipedal locomotion compared to that of quadrupedal mammals without finding a significant correlation between cost of locomotion and limb length in either group. Although increased lower limb length is widely discussed as an indicator of locomotor efficiency in early hominins, the evidence failed to demonstrate any relationship at all, except that both are correlated with body mass. "It may be that increases in locomotor efficiency produced by longer limbs are balanced by decreases in efficiency due to increased moment of inertia" [Steudel, 1996, p. 352]. The change in relative hindlimb length from *A. afarensis* to modern *Homo* may have had little effect on locomotor energetics, although the increase in body size would have increased the cost [Steudel, 1996].

Subsequent experiments on a treadmill revealed that subjects with longer limbs relative to body mass had locomotor costs slightly less than 10% of the variation in cost [Steudel-Numbers and Tilkens, 2004]. Among subjects with relatively long or short lower limb-lengths, those with long limbs relative to their body mass tended to have lower locomotor costs than those with relatively short lower limbs. However, a relationship between lower-limb length and stride length was not confirmed. This failed to support the belief that longer limbs reduce energy requirements by increasing stride length [Steudel-Numbers, et al., 2007]. Evidently, the advantage of longer lower limbs has yet to be identified (but see below).

Why did the upper limbs become less massive? Compared to living great apes, we have the heaviest lower limbs. All limb segments, except for the foot, are more massive [Isler, et al., 2006]. In contrast, we have less massive upper limbs. In *Pan*, for example, the upper arm averages 21.3% of total limb weight; in *Homo* it is 10.9%. In *Pan* the forearm averages 16.4%; in *Homo* it is 6.5%. In *Pan* the hand averages 5.7%; In *Homo* it is 2.5% [Isler, et al., 2006]. Thus, compared to chimpanzees, modern humans have less massive upper limbs with relatively light forearms and very small hands. Why did the upper limbs became less massive? Part of the explanation presumably involves the loss of their participation in arboreal pursuits and their relatively insignificant role in bipedal walking. Nevertheless, human upper limbs share many similarities with those of apes. Aiello and Dean [1990, p. 344] observe that the common features of the upper limb in apes and humans are inherited from an arboreal ancestor, but in humans these features have been "maintained in a slightly altered form by selection for upper-limb mobility associated with manipulation and tool use." What kind of manipulation? What kind of tools? What were the reproductive advantages?

A behavior that accounts for evolutionary change in both upper and lower limbs. In light of doubts that bipedal locomotion is the only behavior that drove natural selection for lower limb structure in the transition to *Homo erectus*, and the absence of a clear identification of selective effects acting on the upper limb, an opportunity is provided to test further the utility of the bipedal use of hand-held weapons proposal. Can it help account for these features of the new hominin bauplan that appeared in *Homo erectus*?

The answer is affirmative. The longer, more muscular lower limbs and the less massive upper limbs were at least partially the result of natural selection acting to improve throwing and club-swinging prowess, due to the reproductive benefits this behavior provided. The evidence on which this assertion is based is derived importantly from analysis of the biomechanics of the human throwing motion, discussed in more detail elsewhere (Chapter 13) [Young, 2009], but partially summarized here in relation to the present context.

The highly evolved modern throwing motion begins with a windup, in which the body faces sideways, then shifts the center of mass over the rear leg as the front leg is raised and the body rotates rearward, away from the target. In *H. erectus,* as Kirschmann [1999] noted, when the lower limbs became longer and heavier and the upper body became more gracile, the body's center of gravity was lowered. This redistribution of mass facilitated initiation of the throwing motion because the heavier lower limbs served as a counterweight for a full-body windup.

The windup is followed by two successive leg thrusts into the ground, first by the rear leg which accelerates the body towards the target, then by the raised leg when it strikes the ground. Foot strike terminates the stride and converts the energy of forward motion into rotation of the pelvis. Increasing stride length seems not to be correlated with locomotor costs (see above), but it is positively correlated with throwing velocity. Advanced throwers step forward as much as 80% of their standing height. Ground reaction forces from rear-foot push-off and forward-foot back-thrust strongly correlate with missile speed. About half of the velocity of the overhand throw or club swing is developed in the legs, hip and trunk. Longer, more muscular legs produce more powerful leg thrusts generating greater linear velocity, resulting in faster pelvis and thorax rotation. This sequence augments the kinetic energy transmitted vertically to the upper extremity where it yields increased missile or club velocity (Chapter 13, Figure 11). Modern male elite throwers and clubbers are taller and more muscular than the average male in their age group (Chapter 12).

The same explanation can be applied to the reduction in mass and musculature of the upper limb. As the energy of the throw or club-swing is generated in the powerful muscles of the lower body, then transmitted and supplemented through sequentially rotating hips and torso, the pulse of kinetic energy is conveyed successively through faster-moving body segments of smaller mass. When energy is passed from the shoulder to the arm, it is transmitted to an upper limb that has been reduced in mass and muscularity by natural selection so that it can *move faster*, transmitting greater kinetic energy to the weapon gripped in the hand. Both the upper and lower limbs are modified in *Homo erectus* in ways that increase the efficiency of throwing and clubbing behavior.

The reduction in length of the forearm compared to the upper arm is reflected in the brachial index, which was significantly reduced in *Homo erectus* (Chapter 6). The same explanation that accounts for the relative changes in size of the lower and upper limbs can be applied to the diminution of forearm length relative to upper arm length. They are the result of millions of years of adaptation to throwing and striking that was naturally selected to yield a unique human motion in which progressively faster-moving, less massive segments apply maximum energy to a hand-held weapon. From the upper arm, kinetic energy is transmitted through the elbow joint to the forearm, and then through the wrist joint to the hand. This accounts for the diminution of the forearm in *Homo erectus* and predicts the diminution of the hand, a forecast that is fully realized [Young, 2003] (Chapter 9). A reduction in hand size took place from *Ardipithecus* to *Australopithecus* to *Homo*. (Only the thumb was enlarged). As the size of the hand diminished in hominins, two new grips emerged: one for throwing spheroids the other for gripping a club.

The chewing apparatus diminished in size. From *A. anamensis* onward, the australopithecine jaw and dentition were marked by a massiveness which indicates an adaptation to a diet that required powerful and prolonged chewing. In *A./H. habilis,* a reduction of molar size and enamel thickness becomes evident, an attenuation that continued during the *H. erectus* era. Apparently, in the transition to *Homo* there was a reduced need for intensive mastication due to a change in diet. Many scholars believe that this occurred when *Homo* gained access to a high-quality, easier-to-chew dietary staple–specifically, meat and other animal tissues. Later, processing of food by cooking may well have accentuated this trend [Wrangham, 2009]. The reduction in size of molars and jaws is consistent with evidence that more frequent ingestion of animal tissues contributed to the nutritional base that made it possible to support a significantly larger body in *Homo* [Bunn, 2001; O'Connell, et al., 2002; Leonard, et al., 2007]. According to this commonly held conclusion, greater consumption of meat to supplement a diet that formerly relied on ripe fruit and low-quality plant foods led to selection of a smaller masticatory apparatus. How did hominins obtain meat? The answer (they used weapons) is elaborated in the next chapter.

Summary. In the transition from *Australopithecus* to *Homo*, hominins abandoned the trees and lost their ancestral arboreal adaptations because they obtained greater reproductive benefits from a dedicated terrestrial life. They were lured from the woodland fringes out onto the grasslands in search of food and there they found herds of browsing and grazing animals that were proliferating on the expanding savannas. Hominins brought to the confrontation with this potential food source several million years of natural selection for improved bipedal use of hand-held weapons combined with a long history of male coalitions that protected home territories from marauding outgroups. Now this heritage could be put to use in procuring a nutritious, energy-rich food. Hominins who remained bound to the trees became extinct.

Abandonment of the trees removed the stabilizing effect of natural selection that had long maintained arboreal adaptations. It also unfettered selection for throwing and club-swinging prowess which accompanied hominin origins and had continued unabated since then. (Selection for bipedal locomotion was part of the mix: Chapter 14). Four million years of natural selection for bipedal use of hand-held weapons had yielded a creature equal to the task of acquiring animal tissues in sufficient quantity to provide the nutritional basis for survival and reproduction. According to this scenario, it was access to meat that ended the ancestral dependence on a woodlands habitat and led to a commitment to life on the ground.

In the transition from *Australopithecus* to *Homo*, the greatest transformation in the hominin lineage, natural selection for improved throwing and club-swinging prowess played a major role. Most of the anatomical changes preserved in the fossil record can be explained in this manner. The larger body, more cylindrical thorax, broader shoulders, enhanced mobility of the shoulder joint, lateral reorientation of the glenoid fossa, longer lumbar region, appearance of a waist, independent rotation of the thorax and pelvis, longer, more massive and muscular legs, a less massive arm, a forearm diminished in length, smaller hands and a masticatory apparatus reduced in size can all be accounted for by this explanatory principle.

CHAPTER 8

Weapons and the Acquisition of Meat

"The combined evidence suggests that behavioral changes associated with lithic technology and enhanced carnivory may have been coincident with the emergence of the Homo clade from Australopithecus afarensis in eastern Africa" [de Heinzelin, et al., 1999, p. 625].

"Homo erectus in the Middle Pleistocene was fully capable of organising, coordinating and successfully executing the hunting of big game animals in a group using long distance weapons" [Thieme, 2005, p. 129].

Introduction. Earlier chapters support the proposition that ancient hominins had access to weapons, used them, and were becoming adapted to their use by natural selection. This process had continued for at least four million years when it began to accelerate about two million years ago, associated with the abandonment of life in the trees. Within a relatively short period, the hominin lineage underwent a profound physical transformation with multiple new features which enhanced throwing and club-swinging prowess. A new life strategy began, based on bipedal use of weapons and dedication to living on the ground.

The metamorphosis included a significant increase in body size, which must have required some sort of nutritional augmentation to grow and maintain. Available evidence and current opinion concur that the new dietary staple was *meat* (animal tissue) (Chapter 7). Compared to most plant foods, the high nutritional value of meat indicates that increasing its content in the diet would have yielded reproductive benefits (Chapter 3). Other possible effects attributed to increased meat consumption are brain enlargement, dental reduction, gut modification, augmented cooperation, food-sharing, changes in growth and development, locomotion, habitat preferences, activity patterns, population size and structure, social behavior, predator avoidance, technology and cognition. Meat seems to be a pivotal part of the evolutionary transition to *Homo*. How did hominins *acquire* meat? This chapter addresses that question.

Evolution of meat-eating. Wild chimpanzees hunt small mammals without using weapons. This supplies less than 5% of their calories [Stanford, 2001]. In contrast, the median percentage of calories derived from meat in modern human foraging societies is about 35% [Lee, 1968; Eaton and Konner, 1985; Leonard and Robertson, 1994; Stanford, 1999]. This suggests that beginning with a mainly vegetarian diet hominins progressively consumed increasing amounts of animal food [Zihlman, 1981; G. Isaac, 1984; Klein, 1987, 1999; Shipman and Walker, 1989; Leonard and Robertson, 1994; Walker and Shipman, 1997]. According to Bunn and Stanford [2001, p. 356], "...there is compelling evidence that meat-eating had a major, influential role in making us human."

Stone tool cut marks on fossilized bones show that *A. garhi* ate meat and marrow [de Heinzelin, et al., 1999] and *A./H. habilis* was processing animal parts to a degree previously unknown among primates [Potts, 1988; Klein, 1999]. Still more meat was eaten by *H. erectus* [G. Isaac, 1977]. Middle Paleolithic hominins in Europe consumed prime-aged large animals [Churchill, 1993; Roebroeks, 2001] and this trend continued after the emergence of modern humans [Klein, 1987, 1999].

Evolution of meat acquisition. It was long believed that hominins acquired meat by hunting [Darwin, 1871; Dart, 1949a, 1953; Washburn and Lancaster, 1968; Laughlin, 1968; Foley, 2001]. Hunting implies killing animals then eating them. However, scavenging was another option [Bartholomew and Birdsell, 1953] in which the killing was carried out by another carnivore after which hominins participated in the eating. *Confrontational* scavenging (also called power or aggressive scavenging) refers to displacing a predator from prey it has killed and is still eating. *Passive* scavenging denotes eating parts of carcasses abandoned by predators, or from animals that died of disease, accidents or other natural causes.

The scavenging hypothesis was tested by analysis of faunal body parts. By this method, Potts [1983] concluded that hominins at Olduvai Gorge (~1.9-1.8 Mya) probably engaged in both hunting and scavenging, whereas Binford [1981, 1985, 1987] maintained that hominins there and elsewhere were passive scavengers of bone marrow. Subsequent analyses of cut marks and tooth marks on bone surfaces [Bunn, et al., 1980; Bunn, 1981; Potts and Shipman, 1981; Shipman, 1986] led to renewed emphasis on hunting and confrontational scavenging as the main routes to meat acquisition. Cut marks on four fossil ungulate bones found at a site near Dikika, Ethiopia, dated at 3.4 Mya (~0.8 My *before* hominins began to *make* stone tools), suggest that *A. afarensis* was already harvesting meat from goat- to cow-sized animals [McPherron, et al., 2010], possibly obtained by some kind of scavenging [Braun, 2010]. No stone tools were found, and the nearest suitable raw materials were 6 km away [McPherron, et al., 2010]. The cut marks could have been made by tools taken from the site by hominins who had carried them in.

Scavenging from abandoned carcasses is generally thought to be unproductive because there is little left that is edible [Bunn, 2001; O'Connell, et al., 2002, Stanford, 2003]. Scavenging may have been an option after hominins became able to take meat away from carnivores [DeVore and Washburn, 1963], but it can be a useful strategy for obtaining meat only if it is confrontational [Domínguez-Rodrigo, 2002]. Clearly, both hunting and confrontational scavenging are dangerous pursuits.

Evidence of hunting and confrontational scavenging. There is substantial evidence for carnivory in *Australopithecus*. The earliest evidence has been associated with *A. afarensis*, as shown by marks on ungulate bones found at Dikika [McPherron, et al., 2010]; however, the identification of these marks as evidence of butchery has been questioned by

Domínguez-Rodrigo, et al. [2010]. Sites at Gona and Bouri in Ethiopia have yielded verified cut-marked bones (2.6 to 2.1 Mya). It is at Gona that the earliest known flaked stone tools have been found (2.6 Mya). These might be the work of *A. garhi* [Asfaw, et al., 1999; de Heinzelin, et al., 1999]. According to Domínguez-Rodrigo and coworkers [2005], most of the cut marks at Gona are due to hominin butchery of a nature that indicates early access to large ungulate carcasses, which implies hunting and/or confrontational scavenging. The butchered bones show that hominins exploited a wide range of animals, including equids and medium-to-large sized bovids, which were disarticulated and defleshed.

Outside of Africa, at the Dmanisi site in the Caucusus (1.8 Mya), it seems that the "...hominins were involved in meat acquisition, and that they had early access to carcasses, which suggests hunting or power scavenging" [Lordkipanidze, et al., 2007, p. 305]. Cut marks were found on large bone specimens. Carnivores were still a factor at Dmanisi (five genera were documented) but evidence suggests that hominins had early access to carcasses. The hominins were small (australopithecine-size), and had a primitive tool kit (the Oldowan). They also brought manuports to the site [Lordkipanidze, et al., 2007, Supplementary material, S5]. The Dmanisi site, in an open, temperate habitat outside Africa, strengthens the view that carnivory was involved in the expansion of early *Homo* throughout Africa and into Asia [Pontzer, et al., 2010]

In Africa, during the transition to *Homo erectus* (1.9-1.7 Mya), cut marks were made on 200 bovid, suid, equid and giraffid bones at the Olduvai Bed I FLK Zinj site in Tanzania [Bunn and Kroll, 1986]. There was an abundance of meatier limb elements among these bones, indicating that hominins had priority of access to carcasses. The prevalence of adult animals probably rules out death from natural causes or passive scavenging. It seems that hunting or confrontational scavenging were the strategies Olduvai hominins used to gain access to meat [Bunn & Kroll, 1986]. (They might have been confrontational scavengers of large animals and hunters of smaller ones [Bunn, 2001]). At Koobi Fora in Kenya (1.6 Mya), where most faunal specimens were large mammals (115-910 kg), cut marks were found on more than 60 bones [Bunn, et al., 1980; Bunn, 1981;

Bunn, 1994]. The presence of defleshing marks on small bovid limb bones may indicate hunting, because small bones can be consumed entirely by large carnivores in a matter of minutes [Bunn, 1997].

Domínguez-Rodrigo, et al. [2002] reexamined cut marks from site FLK Zinj (Olduvai) and a site at Koobi Fora in Kenya. They agreed that hominins were processing carcasses with substantial amounts of meat, evidently obtained by hunting or confrontational scavenging. Further analysis [Domínguez-Rodrigo and Barba, 2006] led to the conclusion that at FLK Zinj hunting was a reasonable explanation [Domínguez-Rodrigo, et al., 2002]. Gaudzinski [2004] proposed that cut marks on horse and cervid bones at 'Ubeidiya, Israel (1.4 Mya) show that these animals were being hunted.

At the BK site at the top of Olduvai Bed II (1.2 Mya), abundant cut marks on fossil bones have been reported which indicate that hominins were actively engaged there in acquiring animals of both small and large sizes, using strategies other than passive scavenging of carnivore kills [Domínguez-Rodrigo, et al., 2009]. About 2,900 faunal remains were unearthed, with bovids, equids and suids being particularly abundant. Exploitation of large game (>500 kg) such as *Pelorovis*, first reported by M. Leakey [1971], was confirmed. A minimum of 24 individuals of this species were present at the site [Domínguez-Rodrigo, et al., 2009]. (*Pelorovis* resembled a buffalo with long, curved horns). At the Olduvai FLK site (1.8 Mya) and at Koobi Fora where cut-marked hippopotamus and giraffe bones were found [Bunn, 1994], hominins evidently incorporated large animals in their diet [Domínguez-Rodrigo, et al., 2009].

Butchery-marked bones from three sites at Koobi Fora (~1.5 Mya) indicate that hominins carried out defleshing, disarticulation, tongue extirpation, evisceration, filleting, skinning, and periosteum-stripping prior to marrow extraction of both meaty and lower quality skeletal elements from small rats and monkeys to large hippos, giraffes and elephants with no apparent preference for prey size, skeletal region, limb class or portion. They exploited animal resources in a variety of habitats including those located near shallow water, woodlands and grasslands [Pobiner, et al., 2008]. They were not scavenging felid kills. Only four specimens bore tooth marks. However, there were 292 cut-marked and 27 percussion-marked faunal specimens out of a total of 6,039 examined [Pobiner, et al., 2008]. As reported by Bunn [1994], there were no stone tools at the sites and the nearest stone raw materials are 5 km from one site (Ileret) and 15 km from another (Koobi Fora Ridge). Hominins must have brought in stones to use for butchering, then carried them away when they left. The hominin responsible was presumably *H. erectus* [Bunn, 1994; Pobiner, et al., 2008].

Hunting versus Confrontational Scavenging. Confrontational scavenging would have been an option in habitats where there was a significant carnivore presence, as was the case at East African sites as early as 4.2 Mya [Ward, et al., 1999; White, et al., 2006], and afterwards at the South African cave sites [Brain, 1981; Pickering, et al., 2004] at Olduvai Gorge [Shipman, 1986; Potts, 1988], at Dmanisi (1.8 Mya) [Lordkipanidze, et al., 2007, S5], but perhaps not at Koobi Fora by 1.6 Mya [Bunn, et al., 1980, Bunn, 1981]. After 1.5 Mya, carnivore interaction was greatly reduced at hominin sites [Potts, 1989; Klein, 1999; Roebroeks, 2001]. By 1.2 Mya at Olduvai Gorge it seems to have been limited to hyenas scavenging epiphyses of bones abandoned by hominins [Domínguez-Rodrigo, et al., 2009].

At sites earlier than 1.5 Mya, it is uncertain whether hunting or confrontational scavenging was the route to meat acquisition. O'Connell and coworkers [2002] find the data consistent with confrontational scavenging; others support both options [Bunn and Ezzo, 1993; Domínguez-Rodrigo, 2003]. In either case, repeated participation in a dangerous activity was involved [Bunn and Kroll, 1986; Bunn, 2001]. Hominins capable of confrontational scavenging should also have been good at hunting [O'Connell, et al., 2002], but the former may have been more hazardous because kill sites are dangerous places when carnivores are present and hungry [Rose and Marshal, 1996; Domínguez-Rodrigo, 2002; Domínguez-Rodrigo and Barba, 2006]. Conflicts over carcass possession may prove fatal [Walker and Shipman, 1997].

When Did Hunting/Confrontational Scavenging begin? Earlier writers hypothesized that hunting began during the australopithecine era [Broom, 1951; Dart and Craig, 1959; Washburn and Lancaster, 1968; Washburn and Moore, 1974]. *Australopithicus* may have been prey at some sites in South Africa [Brain, 1981], but elsewhere in Africa, as noted above, the australopithecines were harvesting meat from a variety of animals, some of which were much larger than themselves. Cut-marked bones from sites at Gona and Bouri, Ethiopia (2.6-2.1 Mya) may have been due to *A. garhi*. According to Domínguez-Rodrigo, et al. [2005], most of the cut marks at Gona are due to hominin butchery of a type which implies hunting or confrontational scavenging. Butchered bones at Bouri confirm the defleshing of large animal carcasses by *Australopithecus* [Bunn and Stanford, 2001].

Hunting or confrontational access to animal carcasses must have had a long prior history if by 3.4 Mya (at Dikika) hominins were butchering large animals and by 2.6 Mya were making their own butchering tools by knapping flakes from cores. Many experts agree that with the advent of *Homo erectus* there was a major increase in meat-eating [Bunn & Kroll, 1986; Shipman and Walker, 1989; Walker & Shipman, 1997; Leonard & Robertson, 1997; de Heinzelin, et al., 1999; O'Connell, et al., 2002; Bunn, 1997, 2001; Foley, 2001]. In *Homo* "hunting became efficient enough to ensure regular access to high-protein animal foods" [Walker and Shipman, 1997, p. 297]. How did they hunt these animals or drive carnivores from their kills?

Hominins used weapons to obtain meat. Chimpanzees grab and bite when they hunt small mammals, but it seems unlikely that hominins used this method to gain access to large animal carcasses. The lack of evidence of ancient hand-made weapons has led some authors to use neutral terms (e.g., early hominins "acquired", "obtained", "caught", "procured" or "captured" prey). Others acknowledge that the strategies they used are unknown. Still others have noted that weapons available in the hominin environment may be unrecognizable because they were unmodified natural objects or were made of materials that were not preserved (Chapter 4). The use of weapons for dispatching prey or fending off carnivores would seem to factor into any hominin carcass-acquisition strategy [Kirschmann, 1999; Van Valkenburgh, 2001; O'Connell, et al.,

2002; Domínguez-Rodrigo & Pickering, 2003; Young, 2003, 2010].

This view has a venerable history. Dart [1949a, 1953] and Broom [1951] proposed that *Australopithecus* hunted with stones and bones. According to Dart [1953], australopithecines used stones for missiles and clubs formed by the long bones of antelopes and the maxillae of carnivores. He proposed that these weapons were dispatched from an upright throwing or clubbing motion. Dart [1949a] determined that most of more than 50 baboon skulls from South African sites had depressed fractures caused by bludgeons of wood, stone or bone, although some might have been caused by stones in the matrix in which the skulls were buried [Brain, 1981]. Howell [1965] speculated that *A. robustus* and *A. africanus* may have used rocks and sticks for hunting. Over the millions of years during which hominin hunting evolved, Laughlin [1968] conjectured, game animals were probably killed with clubs, stones, knives and spears. Washburn and Moore [1974] thought hominins were able to kill small or young animals with hand-held stones, or by throwing rocks might cripple an animal to facilitate capture. Kirschmann [1999] proposed that by 2.6 Mya hominins made sharp-edged tools to cut up carcasses obtained by use of throwing stones to drive away the carnivores who had killed them. This was a probable development, he wrote, since they had been throwing stones at predators for millions of years by then.

Homo erectus likely used weapons [Boehm, 1999]. Throwing stones or using rocks, bones or wood as clubs, throwing or thrusting wooden spears have been postulated by several writers [Oakley, 1964; Churchill, 1993; Schick and Toth, 1993; Schmitt, et al., 2003]. At Olduvai, depressed fractures on three skulls of *Parmularius*, a large antelope, indicated to M. Leakey [1971] that the animals were killed by means of a blow delivered at close quarters. The manuports found there could have functioned in this capacity or as defensive weapons [Bunn and Ezzo, 1993] (Chapter 4). Phytolith analyses of handaxes from Peninj in Tanzania (1.5 Mya) provides evidence of their use as woodworking tools, perhaps for producing spears [Domínguez-Rodrigo, et al., 2001]. Throwing stones and swinging wooden clubs as weapons could account for the accumulation of baboon bones and of unmodified missile stones found at Olorgesailie (0.7-0.9 Mya) [G. Isaac, 1977]. Howell [1965] suggested that *H. erectus* killed elephants mired in mud by using wooden spears and stones at Ambrona and Torralba in Spain (0.4-0.5 Mya). At Zhoukoudian, China (0.5 Mya) Weidenreich [1939, 1943] saw evidence of antemortem depressed fractures, trauma and scars in *H. erectus* skulls that had been caused by clubs, knives or axes, although some might have resulted from falling stones. A circular hole in a rhinoceros scapula found at the 0.5 Mya *Homo heidelbergensis* site at Boxgrove, England, may have been made by a spear [Pitts and Roberts, 1998; Stringer, 2012]. Cut marks and tooth marks on large animals show that the hominins had primary access to them [Stringer, 2012]. Stone artifacts from South Africa dated at ~0.5 Mya have features suggesting they were fabricated for use as points on thrusting spears [Wilkins, et al., 2012].

Three complete throwing spears dated at 0.4 Mya were discovered by Thieme [1997] at Schöningen in northern Germany. Since then more than six well-preserved spears with lengths between 1.8 and 2.5 m and maximum diameters between 29 and 50 mm have been excavated [Thieme, 2005]. The spears are contemporaneous with *H. heidelbergensis* or late *H. erectus*. Associated with the spears were stone tools and wooden implements, including a throwing stick. Thousands of faunal specimens, mainly horses, were represented by the cut-marked and intentionally fractured bones of at least ten individuals. The spears were fashioned so that the maximum thickness and weight was near the front and with long tails that taper towards the end like modern javelins. They were clearly designed as projectile weapons rather than thrusting spears or lances and show an advanced knowledge of aerodynamic principles [Thieme, 1997, 2005].

Male hunting and team aggression. Early hominins were small in size, slow afoot, and their use of weapons was in a primitive stage. It would have been useful in defense against predators and fights with each other, but not in combat with large animals, particularly carnivores. Later, as natural selection led to greater prowess in throwing and club-swinging, this skill would eventually be applied to hunting and confrontational scavenging. Based on ethnographic studies [Watanabe, 1968;

Coon, 1971; Stanford, 1999; Hawkes, 2001] and other considerations (Chapters 3 and 5), it can be assumed that these were essentially male pursuits. Another assumption with a high probability is that males often hunted in groups, as is observed in chimpanzees when they hunt monkeys [Stanford, 2001]. Groups of chimpanzee males also carry out aggression against conspecific outgroups. In modern human hunting societies, success in the hunt is more likely when several men hunt together [Coon, 1971]. Cooperative hunting by early hominin males has often been proposed [Washburn and Lancaster, 1968, G. Isaac, 1977; Bunn and Kroll, 1986; Rose and Marshall, 1996; Roebroeks, 2001; Rose, 2001; Alvard, 2001].

The relationship between male group aggression in chimpanzee and hominin societies was first examined by Wrangham and colleagues, who identified *coalition aggression* (also called team aggression): the proclivity of young adult males to band together to fight a common enemy. This confers a major advantage against a single individual or any outnumbered group of outsiders [Manson and Wrangham, 1991; Wrangham, 1999; Wrangham and Peterson, 1996]. These authors extrapolated behavior observed in chimpanzees, connected it with behavior observed in modern males, and provided a modern Darwinian explanation based on reproductive benefits: dominant groups achieve increased access to reproductive females, food, or other material resources [Manson and Wrangham, 1991; Wrangham, 1999]. In hominins, team aggression would have increased access to animal carcasses through hunting and confrontational scavenging. This strategy may be a retention from the common ancestor of hominins and chimpanzees.

Potts and Hayden [2008] elaborated this line of thought and also underlined the importance of reproductive benefits. "Team aggression and killing members of an outgroup was a relatively low-risk way for the males who evolved the behavior to increase their access to territory and resources, and those who exhibited this behavior were more likely to pass on their genes to succeeding generations than those who did not" [Potts and Hayden, 2008, p. 366].

Group aggression against a prey animal, or a predator who had killed one, greatly increased the chances of obtaining a highly prized source of food: meat. It is proposed that groups of male hominins, armed with rocks, clubs and spears began to add increasing amounts of meat to their diet during the australopithecine era. Their first prey were presumably small animals. Eventually, hunting weaker members (young, sick, injured, crippled, pregnant, lone or peripheral) of larger species would have been attempted. Flight distance, the instinctive recognition of a predator which reflexly initiates flight when a threshold distance is violated, may have played a role. Darwin [1839] found that naive animals lacking prior experience with humans approach with curiosity or do nothing. Thus, a new predator may wreak havoc [Darwin, 1839; Diamond, 1982; Martin, 2005]. Hominin skill with weapons may have been primitive, yet sufficient to kill small ungulates who did not flee [Brues, 1959; Washburn and Lancaster, 1968]. When predation occurs rarely in a large population, it has a negligible effect on the gene pool of the prey species [Dawkins, 1996] as would have been the case when hominins approached large herds of naive animals. As hominin hunting and confrontational scavenging prowess increased with the advent of *Homo*, competing carnivores were driven away or avoided contact by 1.5 Mya. If prey animals became wary, hominins could migrate to another area where similar prey remained unaware of the danger of vertically oriented predators bearing weapons.

The rise of hominins in the predator guild. As hominin abilities to use hand-held weapons from a bipedal stance increased, the size of the prey animal or carnivore competitor that could be overcome would rise commensurately. This began to be evident before the transition to *Homo erectus* and must have accelerated in that taller, heavier and stronger new hominin. Prey size, over hundreds of thousands of years, gradually grew larger, as hominins worked their way up the ladder of the carnivore guild they would ultimately dominate. When throwing spears became part of their arsenal of weapons, the distance from which the prey could be killed or wounded increased. Wounded animals could be tracked until they succumbed or fell and could be dispatched. The continued improvement of throwing and clubbing prowess would have ultimately led to the capacity to kill animals of any size then in existence. Early *Homo* (1.9-1.5 Mya) was already gaining access

CHAPTER 8: WEAPONS AND THE ACQUISITION OF MEAT

to animals of impressively large size, such as giraffe, hippopotamus and elephant. By the Middle Paleolithic, animals such as mammoths, hundreds of times larger than the throwing-and-clubbing hominins were being brought down. Hominin hunters in northern Europe and Asia during the Pleistocene ice ages subsisted almost entirely on meat. How could they overcome such large prey animals and ward off scavengers approaching the carcass? How did they "acquire", "obtain", "catch", "procure" or "capture" the animals they wanted to eat? The answer proposed here is that groups of males used their multi-million-year heritage of natural selection for bipedal use of weapons. In this manner they gained dominion over all the edible creatures on earth.

CHAPTER 9

The Human Hand

"Man could not have attained his present dominant position in the world without the use of his hands which are so admirably adapted to act in obedience to his will" [Darwin, 1871, p. 141].

"...adaptations for manual dexterity appear well before the earliest known evidence of modified stone tools. These adaptations suggest that early hominins used or made tools...before the earliest-known stone tool record" [Panger, et al., 2002, p. 239].

Introduction. If natural selection for bipedal use of hand-held weapons played a major role in human origins and evolution, then our hands should bear the crucial evidence. Natural selection would have acted to improve the hand's grip of those weapons. The prediction that from the earliest hominins until the emergence of *Homo* there will prove to be an ineluctable remodeling of the hands to enhance the prehension of spheroidal missiles and clubs with cylindroidal handles is a crucial test which could falsify the throwing-and-clubbing explanation or provide it with a powerful stamp of confirmation.

It was once thought that the human hand was the preeminent example of benevolent design in nature—the most remarkable, most perfect combination of bones and muscles [Bell, 1837]. Darwin [1871], like Bell, marveled at the structure and capacities of the human hand, but imagined an evolutionary explanation which involved the bipedal use of hand-held weapons: "To throw a stone with as true an aim as can a Fuegian in defending himself, or in killing birds, requires the most consummate perfection in the correlated action of the muscles of the hand, arm and shoulder, not to mention a fine sense of touch. In throwing a stone or spear, and in many other actions, a man must stand firmly on his feet; and this again demands the perfect coadaptation of numerous muscles" [Darwin, 1871, p. 138]. "But the hands and arms could hardly have become perfect enough to have manufactured weapons, or to have hurled stones and spears with a true aim, as long as they were habitually used for locomotion and for supporting the whole weight of the body, or as long as they were especially well adapted...for climbing trees" [1871, p. 141].

These ideas hold true today. The evolution of the human hand can be explained in remarkable detail by the simple proposal that it is the result of natural selection for improved gripping of spheres and cylinders which for millions of years were used as weapons from a bipedal stance, a strategy that promoted greater reproductive success for those who were most adept at it. Although our signature behavior is not much use in the modern world, we are still the master throwers and strikers of the animal kingdom. Furthermore, our hands, naturally selected for a firm grip on a club and a fingertip-pad grasp on a spheroidal missile, have proved to be capable of all of the manipulatory behaviors displayed in our modern culture. The hands we inherited from our ancient ancestors, still bearing the imprint of the missiles and clubs they once used in the competition to survive and reproduce, are among the greatest gifts of our hominin legacy.

That our hand structure is the result of genetic inheritance is uncontested. It arises from a developmental process which differs from that in all other known species, living or extinct. The hands of our nearest relatives, the great apes, are adapted for an arboreal lifestyle. Ours must be adapted for something else—an innovative behavior that enhanced survival and reproduction.

What was the hand like in the hominin ancestor?

The first hominins were arboreal creatures (Chapter 2); their hands were "ape-like" [Tocheri, et al., 2008] and our ancestors retained arboreal traits in their hands for several million years. In the absence of fossil hands of our last common ancestor with chimpanzees, the chimpanzee's hand has often been used as a model for the ancestral hand [Young, 2003]. Recently discovered fossils, however, have raised the possibility that after the hominin divergence chimpanzee hands may have undergone selection which accentuated their arboreal adaptation, shrinking and weakening their thumbs, lengthening their palms, elongating and increasing the curvature of their fingers, and losing their fleshy fingertip pads. Features of the *Orrorin* distal thumb phalanx suggested to Gommery and Senut [2006] that this bone in chimpanzees was a poor model for that of the hominin ancestor. Short metacarpals may be more of an ancestral condition than the long metacarpals in *Pan* [Drapeau, et al., 2005; Drapeau and Ward, 2007]. Lovejoy et al. [2009b, d] think the *Ardipithecus* hand indicates that fingers of modern apes became elongated and the thumb atrophied. Human hand proportions may be more like those of Miocene apes than modern apes [Alméija, et al., 2010]. These recent conjectures emphasize the caution required when there is insufficient fossil evidence. The hand of the hominin ancestor was ape-like, but which ape's hand it was like is unknown. With this caveat, I shall use the hand of the chimpanzee.

The chimpanzee hand. The major features of the chimpanzee hand are depicted in Figures 4 and 5. Compared to humans, the range of wrist flexion is large in the chimpanzee, but extension is constrained by bony ridges, ligaments and the long flexors of the fingers [Napier, 1960; Tuttle, 1967; Richmond et al., 2001]. (This stiffens the wrist for knuckle-walking). Both the palm and the fingers are elongated. The third and fourth metacarpals are especially robust [Lewis, 1977; Susman, 1979]. Proximal and middle phalanges are curved toward the palm [Susman and Creel, 1979; Susman, 1994], the distal phalanges lack apical tufts [Napier, 1960; Susman, 1988b, 1991], and both thenar and hypothenar muscles at the base of the hand are poorly developed [Napier, 1960]. Bones of the thumb are slender and diminutive relative to the fingers; intrinsic thumb muscles are small and weak [Susman, 1994; Marzke, et al., 1992,

CHAPTER 9: THE HUMAN HAND

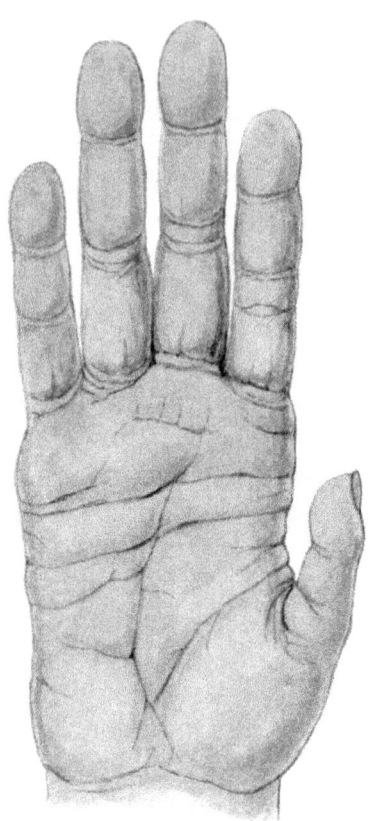
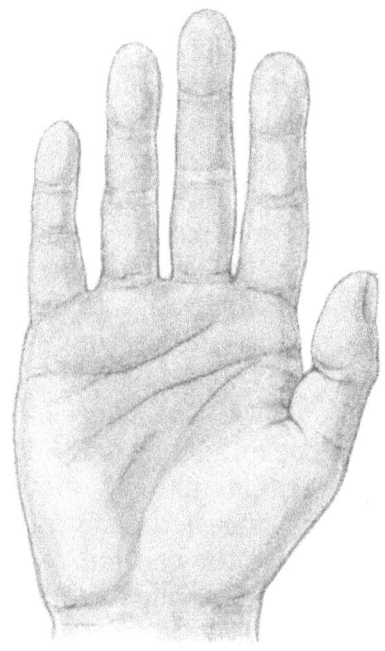

Figure 4. The chimpanzee hand (left) serves as a model of the ancestral ape hand for comparison with the modern human hand (right). The chimpanzee hand is adapted for arborealism. The human hand has a shorter palm, shorter fingers, and an enlarged, mobile thumb which can be rotated into opposition to the palmar surface of the other fingers. The thenar eminence is larger, due to increased size of intrinsic thumb muscles. There is a shift in finger robusticity from fingers 4 and 3 in the chimpanzee to fingers 3 and 2 in the human hand. The distal creases in the palm take a more oblique course in humans. These and other features of the evolution of the human hand are explicable as an adaptation to gripping missiles and clubs (from Young, 2010).

1999]. The long, curved fingers in conjunction with a small, weak thumb are characteristics of arboreal apes who are adapted to climbing vertical supports and suspension below tree limbs.

Two grips are commonly used by chimpanzees. Small objects are held by the thumb and the side of the index finger [Napier, 1960; Marzke and Wullstein, 1996]. Large objects are secured by hook-like flexion of the fingers, a grip which is also used for hanging from horizontal supports. The hook grip reflects the hinge action at the metacarpo-phalangeal joints which fold horizontally across the palm [Lewis, 1977]. With vertical supports, a diagonal hook grip is employed in which the thumb does not play an active role [Susman, 1979; Marzke, et al., 1992]. This grip is also used when chimpanzees swing sticks [Marzke et al., 1992; Marzke and Wullstein, 1996]. Cylindrical objects of large diameter cannot be controlled by the small thumb when they are swung towards a target [Marzke, et al., 1992]. Because the thumb is short, weak, relatively immobile, and its distal pad cannot be applied to those of the fingers, it is unable to generate a firm pinch or squeeze [Marzke, 1992a, 1997; Marzke and Wullstein, 1996].

The human hand. Extension of the modern human wrist exceeds the range of that in chimpanzees, aided by reduced dorsal ridges on the radius [Richmond, et al., 2001]. Humans can also extend their fingers and wrist at the same time; in great apes, this is impossible [Jones, 1946]. The thumb is longer, the palm and fingers shorter, and the fingers are straight (Figures 4, 5). Large apical tufts are present on the distal phalanges. These support broad, palmar, fibrofatty pads that distribute pressure during forceful grasping and accommodate the fingertips to uneven surfaces [Napier, 1965; Susman, 1979; Shrewsbury and Johnson, 1983; Marzke and Shackley, 1986]. These digital pads have a fatty, mobile proximal portion and a more static distal part which ensure friction and accommodation of the object between the pulp of the thumb and those of the fingers during precision grasping [Alméija, et al., 2010].

Strength and robusticity in the hand has shifted from the third and fourth digits towards the thumb, second and third fingers. The fifth metacarpal is also more robust [Marzke, et al., 1992]. The thumb metacarpal articulates with the carpals in

CHAPTER 9: THE HUMAN HAND

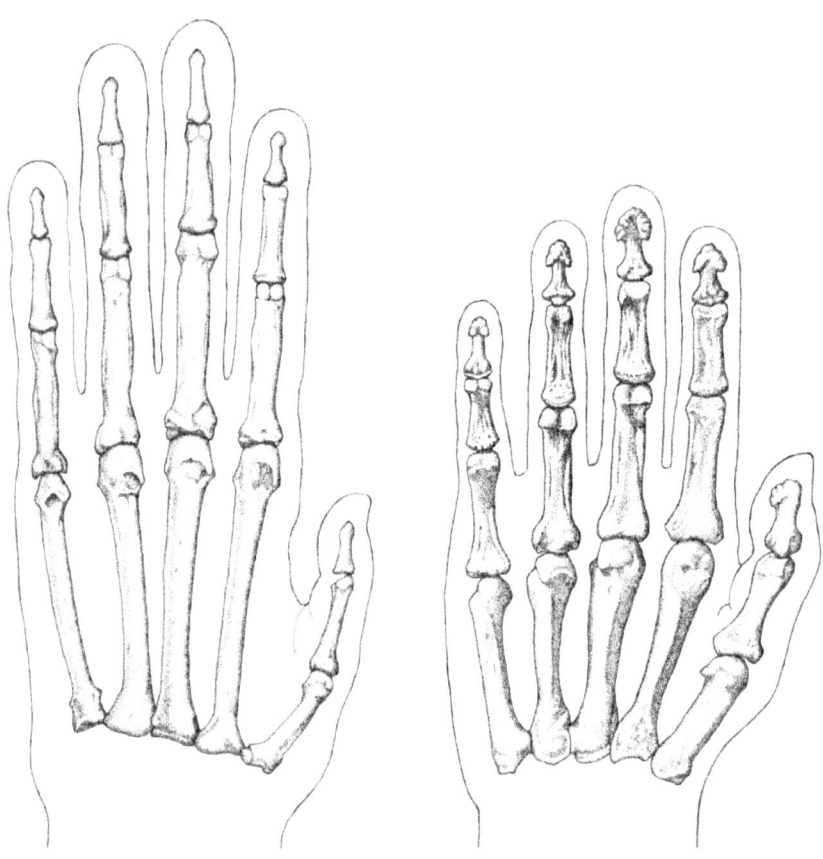

Figure 5. Chimpanzee (left) and modern human hand bones, palmar view. In the chimpanzee the metacarpals and phalanges are elongated, but the thumb is small and its bones are slender. The finger tips lack broad apical tufts. In the human, the thumb is longer and more robust. The metacarpals and phalanges are shorter, and the distal phalanges have large apical tufts. The proximal and distal phalanges have lost their curvature (not shown in this view). Strength and robusticity in the hand has shifted from the third and fourth digits towards the thumb, second and third fingers (from Young, 2003).

a saddle joint which together with remodeling in the metacarpal-phalangeal joint allows the distal pad to be opposed to those of the other fingers [Napier, 1955, 1993]. Intrinsic thumb muscles are larger [Marzke, 1992b, 1997; Marzke, et al., 1999] and three new muscles add control to thumb motion. The flexor pollicis longus, a fully or partially independent muscle [Shrewsbury, et al., 2003], is the most powerful thumb muscle in humans, but is absent in apes. Humans have a distinctive insertion for the flexor pollicis longus on the distal phalanx of the thumb [Alméija, et al., 2010]. The muscle flexes the distal phalanx and maintains orientation of its pad toward the fingers against pressure [Susman, 1994; Hamrick, et al., 1998]. Remodeling of the radial carpal/metacarpal region resists stress from the increased thumb muscle forces [Tocheri, 2007].

The palm also has several new features. Because the fourth and fifth metacarpals are progressively shorter, there is an obliquity to the hand when it is flexed, producing palmar creases that run obliquely, from the lower ulnar to the upper radial side. The thenar and hypothenar eminences are enlarged by fat pads which overlie the muscles. Contraction of the palmaris brevis muscle, small or lacking in chimpanzees, stiffens the hypothenar pad, deepens the cupping of the palm and protects the ulnar neurovascular bundle [Shrewsbury, et al., 1972]. Other features increase the ability of the center of the palm to withstand stress along the second and third fingers [Marzke and Marzke, 1987]. The metacarpals and bases of the proximal phalanges of these fingers are robust. A palmar fat pad in the third metacarpal region protects the deep branch of the ulnar nerve. Stability of the third metacarpal base is enhanced by a styloid process on its dorsal radial aspect. When the finger is extended, the styloid process locks the carpal and metacarpal bones together, preventing hyperextension, which is further restrained by a ligament between the pisiform bone and third metacarpal [Marzke and Marzke, 1987]. The fingers rotate towards the central axis when they are flexed (Figure 8). Rotation is more pronounced for the two outermost fingers [Susman, 1979; Marzke, 1983, 1997].

Precision and power grips. The derived structure of the hand yields two unique grips, identified by Napier [1956], who asserted that they provide the basis for all prehension [Napier, 1960, 1961, 1965]. "In spite of the multiplicity of activities of the hand, involving countless objects of various shapes and sizes, there are only two prehensile *actions*: these are called the *precision grip* and the *power grip*" [1965, p. 550]. The precision grip is a fingertip-pad grip, used for precise movement and the grip of a sphere. The terminal pad of the thumb forms one jaw of a clamp; the other is formed by one or more of the other fingertip pads (Figure 6). The power grip is used for application of force. A cylinder lying obliquely against the palm of the hand is squeezed firmly in a clamp formed by the flexed fingers and the palm, with the thumb wrapped over the fingers for reinforcement (Figure 7).

The evolutionary explanation of the human precision and power grips to be introduced below refers specifically to Napier's concept of a fingertip-pad grip of a sphere and the grip of a cylinder squeezed against the palm by the flexed fingers buttressed by the thumb. It is proposed that the precision grip was evolved due to selection for gripping and throwing spheroidal rocks used as missiles and the power grip arose to enhance gripping implements with cylindroidal handles that were used as clubs. The two grips were employed in the innovative hominin behavior of bipedal use of weapons.

Linking precision and power grips with throwing and club-swinging requires a strict definition of terms to avoid confusion, because these appellations have acquired a variety of meanings. In the present context, the names will be used to denote *distinctive human* grips and will be specified as *hominin* precision and power grips. Due to the structure of their hands, apes cannot form these grips, although these grip names have sometimes been applied to ape grips. Napier [1960] emphasized that execution of the two grips differs profoundly in humans and apes, primarily because of the disproportion in length of the thumb and fingers in apes, and the ineffectiveness of the ape thumb in prehension. The chimpanzee four-finger flexion hook grip, (sometimes called a power grip) is a thumbless grip that does not involve the palm [Marzke and Wullstein, 1996]. So-called chimpanzee precision grips are not thumb-pad-to-fingertip-pad grips [Christel, et al., 1998]. Humans can form these grips, which are retentions from an ape ancestry.

CHAPTER 9: THE HUMAN HAND

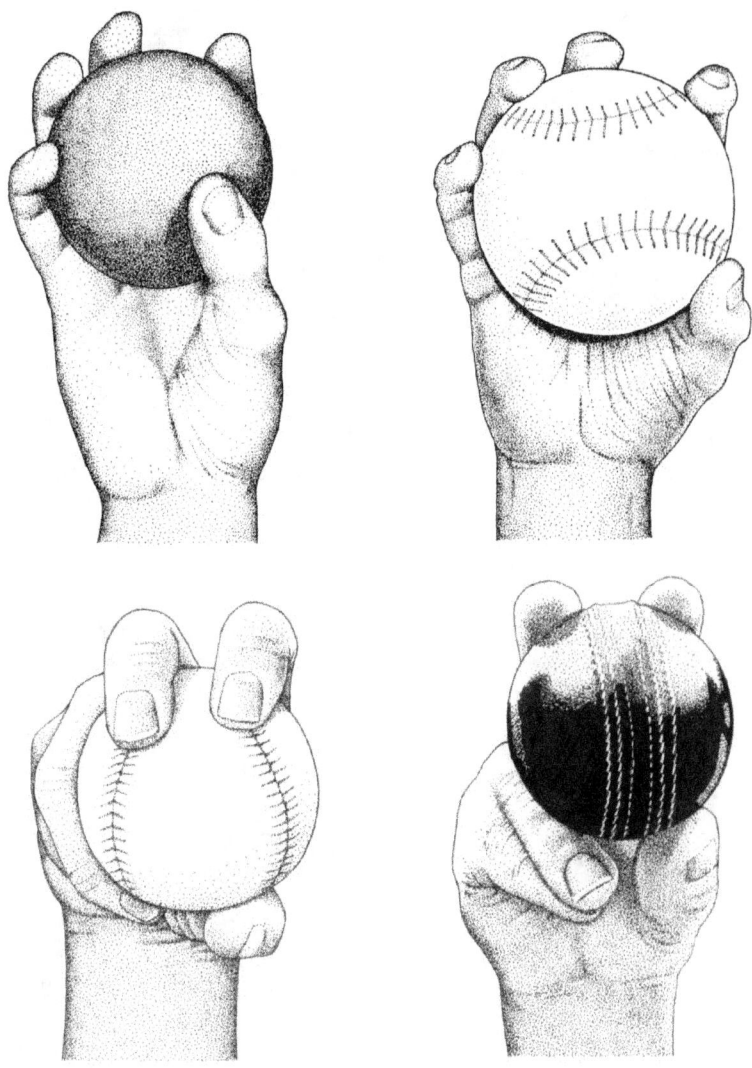

Figure 6. The human precision grip, first identified by Napier [1965], who illustrated it with the grip of a sphere (top left), is depicted above by hands gripping missiles: a softball, top right, a baseball and a cricket ball below. This unique human grip was named by Napier because its fingertip-pad grasp is commonly used for precision movements. It can also be called a throwing grip due to its evolutionary origins (from Young, 2003).

CHAPTER 9: THE HUMAN HAND

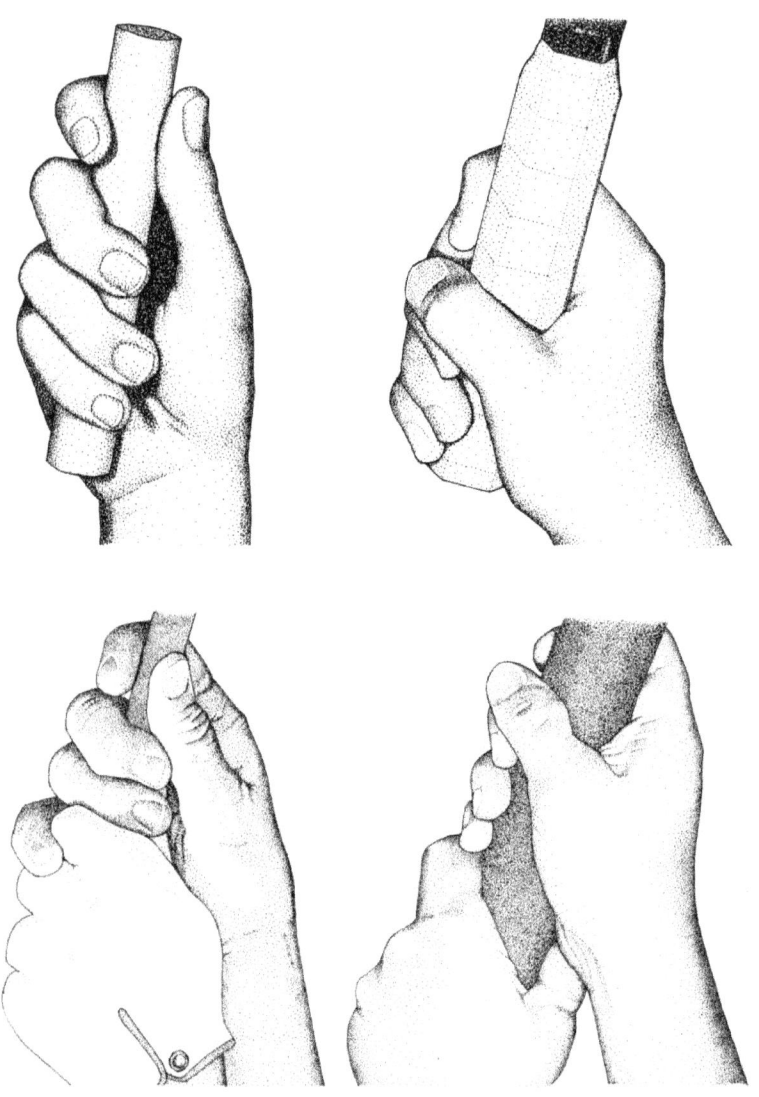

Figure 7. The human power grip, identified by Napier [1965] who illustrated it with the grip of a cylinder (top left), is depicted by hands gripping clubhandles: a tennis racquet, golf club and cricket bat. Napier named this the power grip because the firm grasp against the palm is commonly used for the application of force. It can also be called a clubbing grip because of its evolutionary origins (from Young, 2003).

CHAPTER 9: THE HUMAN HAND

The *hominin precision grip* is a fingertip-pad grip, employed for precise movement. The terminal pad of the thumb forms one jaw of a clamp; the other is formed by the remaining fingertip pads. Large spheres held in this way involve all of the fingers. The grip can encompass spheres of decreasing diameter by eliminating the fifth (or fifth and fourth) rays from the grip and flexing them completely to provide lateral stability [Napier, 1960]. Napier [1956, 1965, 1993], Marzke [1983] and Young [2003, 2010] illustrate the hominin precision grip with a hand grasping a ball or sphere. The grip is aided by the distensible fingertip pads and the lengthened, more fully opposable thumb in which opposition is enhanced by flexion of the terminal phalanx. Also contributing to the grip are fingers which rotate as they flex (supination on the ulnar side, pronation on the radial side) exactly as needed for a fingertip-pad grip on a sphere (Figure 8).

The *hominin power grip* is used for application of force [Napier, 1960]. A cylinder lying obliquely against the palm is squeezed firmly in a clamp formed by the palmar surface of the flexed fingers and the palm, with the thumb clasped over the fingers for added grip strength. By reducing finger flexion, this grip can be adjusted for cylinders of increasing diameters. (If the thumb is positioned parallel with the cylinder, some control of the direction of force is gained, but the grip loses strength). Napier [1956, 1965, 1993] and Young [2010] illustrated the grip with a hand grasping a cylindrical rod, as in Figure 7. It has also been depicted by the grip used with the handle of a hammer [Marzke, 1992b] and, as in Figure 7, various sports implements [Young, 2003]. Because of the obliquity of the channel in the flexed hand, a cylinder squeezed against the palm assumes a diagonal position. The fourth and fifth fingers rotate medially as they flex, lodging the tool against the ulnar base of the palm while the opposed thumb buttresses the tool on its radial side. The flexor pollicis longus contracts strongly [Hamrick, et al., 1998; Marzke, 2005]. Thenar and hypothenar regions cushion pressure from the cylinder and the palmaris brevis stiffens the hypothenar fat pad [Napier, 1965; Marzke, et al., 1992].

There are two positions of the carpometacarpal joint of the thumb when it is stabilized by congruent articular surfaces and tension of ligaments. These coincide with the two grips

CHAPTER 9: THE HUMAN HAND

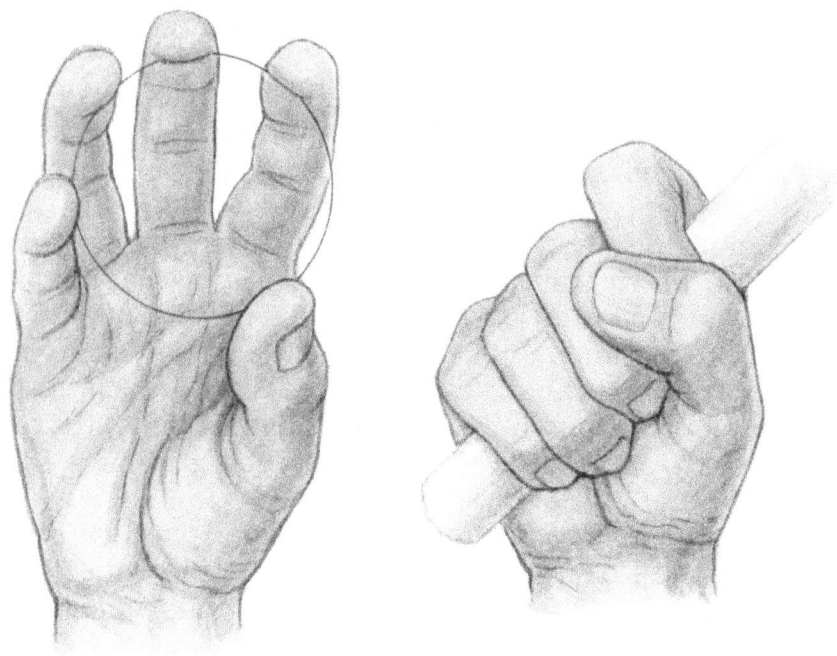

Figure 8. The precision (throwing) grip (left) and the power (clubbing) grip (right) [from Young, 2010]. The throwing grip is depicted with the fingertips grasping the outline of a transparent sphere so that rotation of the partially flexed fingers can be seen. The fourth and fifth fingers rotate towards the center of the hand, the second finger rotates in the opposite direction towards the center, and the third finger does not rotate. The thumb moves into opposition to form a fingertip-pad clamp on the sphere. Fingertip control is necessary for precise release of a missile to achieve throwing accuracy. In the clubbing grip, a cylinder is clasped firmly against the palm by the flexed fingers buttressed by the muscular thumb. The structure of the hand causes the cylinder to be oriented obliquely. This yields the maximum leverage when the wrist deviates in the ulnar direction as the cylinder (clubhandle) swings towards the target.

[Napier, 1955]. In the hominin power grip, when the thumb exerts pressure towards the palm, the lateral ligament of the carpometacarpal joint is taut, and the joint is in one close-packed position. In the hominin precision grip, as the thumb abducts and moves into opposition, the anterior and posterior oblique ligaments are tightened, and the joint is in its second close-packed position [Lewis, 1977].

Movement of the hand during human throwing and club-swinging. Overarm throwing involves use of the fingertip-pad precision grip. The hand and wrist make an essential contribution to the velocity and accuracy of the throw. During acceleration of the missile, the palmar surface of the hand faces the target. As the forearm accelerates forward, the hand at the wrist joint moves backwards into full extension. Wrist flexor muscle activity begins at this time, combining with a flexion torque from forearm deceleration [Hirashima, et al., 2007]. At missile release, flexion velocity reaches its peak, the hand is aligned with the forearm [Atwater, 1979; Debicki, et al., 2004; Gray et al., 2006], the thumb abducts and the fingers extend, opening the grip and causing the missile to roll distally along the palmar surface of the fingers as force continues to be applied until final release occurs from the fingertips [Hore, et al., 1996a, b; 1999b] (Chapter 13, Figures 11, 12). The major factor determining accuracy is the timing of finger opening; hitting the baseball strike zone consistently requires a timing precision of 1-2 ms [Hore, et al., 2002; Hore and Watts, 2005].

Fast throws are associated with large back forces from the missile on the fingers that must be controlled if accuracy is to be achieved and injury prevented [Hore, et al., 1999b]. Back forces begin with the start of the acceleration phase and increase progressively throughout it. During release, the back force on the third finger's distal phalanx increases during extension and peaks just before release as kinetic energy transferred to the missile is channeled primarily through the extended index and middle fingers. This acts to hyperextend these fingers, particularly the third finger, which being the longest is the last to lose contact with the missile. During the process of release, the fingers do not extend fully, because the central nervous system (CNS) anticipates the size of the back forces and generates appropriate finger flexor torque to oppose them [Hore, et al., 1999b, 2001] (Chapter 15). The styloid process on the base of the third metacarpal and other structural features in the central palm, described above, also increase the resistance of the center of the hand to injury [Marzke and Marzke, 1987]. Additional injury-prevention factors protect the arm from damage due to the intense forces associated with power throwing (Chapter 13).

In clubbing (also called striking) the clubhandle is held in a firm power grip. When a club is swung, the wrist is first radially deviated; the ulnar side of the hand faces the target. As the elbow extends, aligning arm and club, the wrist accelerates in the ulnar direction (Chapter 13, Figures 13, 14) the fingers exert muscular force to maintain the grip, tightening just before impact [Stretch, et al., 1995; Hamrick, et al., 1998; Marzke, et al., 1998] so that the amount of energy applied to the target is maximized, elastic recoil is minimized, and ejection of the club from the hand by the force of impact is prevented. For added force and grip strength, both hands may be used.

Adaptation of the human hand for throwing and club-swinging. The chimpanzee hand, adapted for knuckle-walking on the ground and arboreal behavior in trees, is capable of a rudimentary form of throwing and clubbing but is not adapted for it. In contrast, the human hand facilitates the grasp of spherical and cylindrical weapons and the transfer to them of kinetic energy [Young, 2003, 2010]. Critical to this function are the long, opposable thumb, the apical fingertip pads, and short, straight fingers that rotate towards the central axis when flexed. The robusticity and muscularity of the thumb is especially important for preventing dislodgement of a club due to the back force at impact with the target [Ohman, et al., 1995]. Simultaneous extension of the fingers and wrist, anatomically impossible in apes, occurs during the acceleration phase of human throwing. Several features which prevent hyperextension of the third finger and protect the center of the palm against stress play an important role in resisting the back forces generated during throwing.

The significance of these many changes is that the human hand gained two new grips [Napier, 1956, 1965]: a fingertip-pad grasp of spheroidally shaped objects and a firm clamp of a cylindrical implement held obliquely against the palm. Napier named these grips according to their modern uses. On the basis of their evolutionary origins, however, the precision grip can be referred to as a *throwing grip*, and the power grip a *clubbing grip* (Figures 6-8).

These two specialized grips support the conclusion that hominins used two types of hand-held weapons—spheroidal missiles (rocks) that required a precisely timed release and clubs with cylindroidal handles that were not released, even after collision with the target. These are the two classic categories identified by weapon specialists [Burton, 1884; Keeley, 1996]. Another weapon that depends upon a strong grip is the thrusting spear. Ethnographic studies indicate that thrusting spears are used by modern humans for hunting [Churchill, 1993] and experimental work reveals that this requires a firm hominin power grip [Schmitt, et al., 2003]. As in clubbing, this grip resists displacement of the weapon from the hand (or hands) by the reaction force of striking the target and permits application of additional force after contact is made. A measure of the duration of the adaptive process that led to these changes can be obtained from examination of the record of hominin hand evolution.

Paleontological evidence. The fossil evidence in this and other chapters is presented chronologically to avoid the difficult and controversial issue of the phylogenetic relations among ancient hominins, such as which species are on the main line leading to *Homo* and which may be extinct side branches without descendants.

Orrorin tugenensis. The earliest known hominin hand bones are assigned to this taxon from Kenya, dated at ~6.0 Mya. One fossil is a long, curved, proximal phalanx [Senut, et al., 2001]. The terminal phalanx of a thumb resembles that of the Miocene ape *Oreopithecus* (~8 Mya) in several features, such as its channel for a flexor pollicis longus muscle, its two-part fossa for fingertip pads, and robust apical tuberosity. This suggests that these features could be inherited from pre-hominin apes [Gommery and Senut, 2006]. Markings on a humerus fragment and the long, curved proximal phalanx connote an arboreal adaptation [Senut, et al., 2001; Gommery and Senut, 2006].

Ardipithecus kadabba. Also dated near the presumed origins of the hominin lineage are hand bones found in Ethiopia (5.8-5.2 Mya) [Haile-Selassie, 2001], assigned to *Ardipithecus kadabba* [Haile-Selassie, et al., 2004], which include the distal portions of an intermediate and a proximal phalanx. The curvature of the bones is said to resemble those of a younger hominin species, *Australopithicus afarensis* (described below), although the intermediate phalanx may be larger and longer than the *A. afarensis* homologue [Haile-Selassie, 2001].

Ardipithecus ramidus. The first hand fossils of this species, found at Gona, Ethiopia, were dated at 4.5-4.3 Mya. Three phalanges were similar to those from *A. afarensis*, but longer [Semaw, et al., 2005]. Then, in 2009, a trove of additional fossils was reported from the Aramis region (4.4 Mya). *Ar. ramidus* had an ape-like hand with long, curved fingers and a diminutive thumb. Although small, the thumb is more robust than that of an ape [Lovejoy, et al., 2009a, e] with a flexor pollicis longus insertion site on the terminal phalanx [Lovejoy, et al., 2009e], a bone which seems more ape-like than that of *Orrorin* [Alméija, et al., 2010]. The thumb metacarpal is shorter than that in *A. afarensis* [Lovejoy, 2009f, Fig. S16]; those of digits II-V are shorter than in extant apes [Lovejoy, et al., 2009b, e]. The fifth metacarpal had a greater range of flexion-extension than in extant apes and modern humans, suggesting a degree of dorsiflexion like that seen in some monkeys who walk on their palms [White, et al., 2009; Lovejoy, et al., 2009b, e].

The phalanges of the *Ar. ramidus* hand (except for the thumb) are very long. When body size is considered, they are longer than in *Homo* or *Gorilla,* overlapping the low end of the range in chimpanzees [Lovejoy, et al., 2009c, Fig. S7] and longer than in *A. afarensis* [Haile-Selassie, 2001; Semaw, et al., 2005; Lovejoy, et al., 2009b]. There is no mention or depiction of apical tufts on the distal phalanges [Lovejoy, et al., 2009d, e, f]. The proximal and intermediate hand phalanges are curved inward toward the palm [Lovejoy, et al., 2009e, Fig. 1, p. 70e2; Lovejoy, et al., 2009e*, Fig. S7]. The manual left proximal phalanx found at Gona, Ethiopia, also appears to be curved [Fig. 3 in Semaw, et al., 2005]. The proximal phalanges are slightly more curved than those in an adult male chimpanzee depicted by Susman [1979, Fig. 5]. Curvature of hand phalanges is believed to be an adaptation for dissipating stress produced by suspensory arboreal behavior [Jouffroy and Lessertisseur, 1960; Oxnard, 1963; Susman and Creel, 1979; Susman, et al., 1984; Stern and Susman,1983; Hunt, 1996; Fleagle, 1999; Richmond, 2007]. This indicates the *Ardipithecus* hand was adapted for suspensory behavior, although it has been said to lack any adaptations to below-limb suspension [Lovejoy, et al., 2009a, b, d, e; White, et al., 2009].

In *Ar. ramidus*, the phalanges of rays III and IV are not only longer than ray II, they are also thicker in the mediolateral dimension [Lovejoy, et al., 2009e*, Fig. S7]. (Note: Diameters were measured on the photos, then reduced to match the cm line). The greater phalangeal robusticity of rays III and IV compared to II, and of metacarpal IV compared to II, is also shown in a digital rendering [Fig. 1 in Lovejoy, et al., 2009e]. The greater robusticity on the ulnar side of the hand may be related to selection for climbing vertical supports, in which the ulnar side of the hand provides friction against the downward force of the individual's weight [Susman, 1979a; Marzke and Wullstein, 1996], although this adaptation was also denied by the *Ardipithecus* team [Lovejoy, et al., 2009a, b, d, e; White, et al., 2009]. The presence of a diminutive thumb, elongated, curved proximal and intermediate phalanges, apparent absence of apical tufts on the distal phalanges, and greater robusticity in the third and fourth digits, all features of the chimpanzee hand, suggest that *Ardipithecus* handgrips would be similar to those of chimpanzees.

Australopithecus anamensis. Hominin hand bones of this species found at Kanapoi and Allia Bay in Kenya (4.2-3.9 Mya) include parts of a capitate and a phalanx [M. G. Leakey, et al., 1998; Ward, et al., 1999b, 2001]. The capitate is similar in length to that of a chimpanzee, lacks a facet for the third metacarpal styloid process, and retains the primitive condition of a facet for the second metacarpal facing laterally, unlike *A. afarensis* and later hominins [Ward, et al., 1999b]. Rotation at this joint would not have been possible. A curved phalanx is another primitive character [Ward, et al., 1999b, 2001]. The partial manual phalanx of *A. anamensis* is as large as the largest of the *A. afarensis* specimens and is similar in shape, curvature and development of the flexor sheath ridges [Ward, et al., 1999b]. The *A. anamensis* intermediate hand phalanx from the nearby Asa Issie site also resembles those of *A. afarensis*, although longer relative to its breadth [White, et al., 2006].

Australopithecus afarensis. This species is known from Hadar in Ethiopia (3.7-3.0 Mya) and Laetoli in Tanzania (3.7-3.5 Mya) [Kimbel and Delezene, 2009; Haile-Selassie, 2010]. In specimens of the hand from Hadar [White, et al., 1993; Johanson, et al., 1994] an improved capacity to form the hominin power and precision grips is unmistakable. The carpometacarpal joint of the thumb allows the range of movement needed for both grips [Marzke, 1983] but may have been less than it is in modern humans [Tocheri, et al., 2003; Tocheri, 2007]. Relative to *Pan* and *Gorilla*, the *A. afarensis* metacarpals II-V tend to be short (relative to metacarpal I and estimated body size) as they are in the modern human hand. The fingers are shorter than those of chimpanzees and the thumb is longer [Marzke, 1983, 1992b; Stern and Susman, 1983], although not yet fully in proportion to modern humans [Marzke, et al., 1992; Alba, et al., 2003]. The thumb is less robust than in later hominins and may have lacked some muscles [Stern and Susman, 1983; Susman, 1994]. In distinction to *A. anamensis*, the second metacarpal could be pronated during flexion as in modern humans, based on the shape of the capitate-MC2 joint [Marzke, 1983, 1997] and the trapezium and capitate morphology [Tocheri, et al., 2003; Tocheri, 2007]. This would have facilitated the hominin precision grip.

In additional hand specimens from Hadar (~3 Mya), the metacarpals are short, nearly straight as in humans, the thumb is well developed, and the second metacarpal pronates during flexion [Drapeau, et al., 2005; Drapeau and Ward, 2007]. Restricted supination with flexion of the fifth metacarpal in *A. afarensis* may have precluded an effective hominin power grip [Marzke, 1983, 1992b, 2005; Marzke and Shackley, 1986], although Ricklan [1987] has questioned this conclusion, because the fifth finger exerts only 16% of the total grip force of rays II-V in modern humans [Hazelton et al., 1975]. The third metacarpal lacks a styloid process [Bush, et al., 1982; Drapeau, et al., 2005], but there are signs of ligamentous changes that stabilize the palm [Marzke, 1983, 1992a; Marzke and Marzke, 1987].

Proximal phalanges are curved in a palmar direction [Bush et al., 1982; Stern and Susman, 1983; Marzke, 1983; Alemseged, et al., 2006]. Apical tufts on the distal phalanges are expanded [Bush, et al., 1982], although less so than in humans [Stern and Susman, 1983; Marzke, et al., 1992].

Hand remains of an australopithecine from South Turkwel, Kenya (dated at ~ 3.5 Mya) which resemble those of *A. afarensis* indicate that some form of hominin precision grip would have been possible, given an inferred rotatory capacity of the second metacarpal. The hamulus and other hand specimens, suggest that the functional capabilities of the hand may have been comparable to *A. afarensis* [Ward, et al., 1999a].

Australopithecus africanus. In earlier australopithecines there are possible (but disputed) remnants of wrist-locking mechanisms. By 3.2 Mya, in *A. africanus,* these were gone [Richmond and Strait, 2000; Richmond, et al., 2001]. In a specimen from Sterkfontein, dated between 4 Mya [Partridge, et al., 2003] and 2 Mya [Walker, et al., 2006], the first metacarpal and proximal phalanx indicate a robust, opposable thumb. The metacarpal is long as in humans but slender as in apes. Compared to the other metacarpals, the length proportions are like those of *A. afarensis* [Green and Gordon, 2008]. The metacarpals of an australopithecine of uncertain species from Sterkfontein (3.3 Mya) are similar in length to those of modern humans. The first metacarpal and proximal phalanx of the thumb indicate that relative to the rest of the hand, it had a proportion and disposition similar to that of modern humans. The proximal phalanges of the thumb and forefinger, however, are curved, indicating some arboreality [Clarke, 1999, 2002].

Ricklan [1987, 1990] studied 16 *A. africanus* hand specimens from Sterkfontein and found all lacked evidence of arboreal adaptation [Ricklan, 1990]. The insertion site of the flexor pollicis longus on the base of a thumb distal phalanx indicates that this muscle was as well developed as in modern humans. A styloid process on the third metacarpal is present, as is the capacity to rotate the fifth metacarpal [Ricklan, 1987]. Ability of the thumb to oppose the other digits was similar to (or slightly less) than in modern *Homo.* An expanded tuft on the distal thumb phalanx signifies a broad fingertip pad and increased grip capacity [Ricklan, 1990]. *A. africanus* had a firm hominin power grip and a strong muscular capacity for radial and ulnar deviation of the wrist, as occurs in striking or hammering [Ricklan, 1987]. The hominin precision grip was also well developed, although not yet completely modern [Ricklan, 1990].

Australopithecus sediba. Two partial skeletons of this hominin, dated at 2.0 Mya, are from Malapa, South Africa [Berger, et al., 2010a; Pickering, et al., 2011]. A nearly complete hand of an adult female is smaller than that of any other hominin [Kivell, et al., 2011b]. The thumb is relatively long. Compared to the fingers it is longer than those in other hominins—even greater than the upper range in modern humans. The shaft is also narrower and taller than in other hominins [Kivell, et al., 2011a]. The extreme length of the thumb is due mainly to the first metacarpal; the proximal phalanx is short and the distal phalanx is very short [Kivell, et al., 2011b]. A broad apical tuft is present on the distal phalanx, but is less flattened than in other hominins. It has a palmar pad with a mobile proximal pulp and indications of a flexor pollicis longus muscle. The base of the thumb metacarpal has the narrowest relative breadth of any fossil hominin and is outside the range of modern humans [Kivell, et al., 2011b]. Its head is flanked by depressions for large sesamoid bones, suggesting that some intrinsic thumb muscles were strong; others appear to have been weak [Kivell, et al., 2011a]. The proximal and intermediate phalanges of rays 2-5 in *A. sediba* are proportionally shorter than those of *A. afarensis* and *A./H. habilis* (OH 7). Their curvature is also reduced, but is greater than in modern *Homo* [Kivell, et al., 2011b]. Both sets of phalanges have bilateral flexor sheath ridges typical of australopithecines. Combined with evidence from the pelvis and ulna, this indicates that *A. sediba* retained arboreal adaptations [Kivell, et al., 2011a, b].

A./H. habilis. This species inhabited East Africa from about 1.9 to 1.4 Mya [L. Leakey, et al.,1964; Spoor, et al., 2007]. The hand has a sturdy thumb, an advanced saddle joint and marked apical tufts on the distal phalanges of all fingers. Proximal and middle phalanges curve toward the palm and there is a marked insertion for the flexor digitorum superficialis [Napier, 1962; L. Leakey, et al., 1964; Tuttle, 1967; Susman and Creel, 1979]. Curvature and cortical thickness of two proximal phalanges indicate an adept climber [Susman, 1998].

Australopithecus robustus. Susman [1988a, b, 1991, 1993, 1994] and Brain, et al. [1988] described hand bones from Swartkrans, South Africa, attributed to *A. robustus* (~1.8-1.2 Mya). The distal phalanx of the robust, muscular thumb has an enlarged apical tuft and an expanded proximal joint surface. The metacarpal head is also expanded and the carpometacarpal joint supports opposition of the thumb. The fingers are straighter, with large apical tufts. These features are consistent with refined precision grasping [Susman, 1988a, b; Brain, et al., 1988], although the evidence may be insufficient to show that the precision and power grips were present in fully modern human form [Marzke, et al., 1992; Shrewsbury, et al., 2003]. Verified hand bones are unavailable from *Homo erectus*. Those of *H. neanderthalensis* are similar to those of modern humans except for greater robusticity [Trinkaus, 1983].

As described above and in earlier publications [Young, 2003, 2010], the evolution of the hominin hand can be explained by a process of natural selection that acted to improve bipedal use of hand-held weapons, resulting in the remodeling of an ape-like hand into one specialized for gripping missiles and clubs. These are the so-called precision and power grips that characterize the modern human hand. Before summarizing the central features of that proposal, however, other hypotheses for hominin hand evolution will be examined.

The stone tool-making explanation. Tuttle [1967] first suggested that evolutionary changes in the hominin hand were due to selection to facilitate the fabrication of stone tools, citing pulp-to-pulp opposition of thumb and fingers, broadening of terminal phalanges and increased finger mobility. This idea survived for many years, and still may have some adherents. It was supported by Hill [1982], Napier [1993], and Lovejoy, et al. [2009b], who hypothesize that reduction in length of the phalanges resulted from selection for tool-making. The apical tufts on human fingers are often attributed to stone-tool manufacture [Green and Gordon, 2008]. Williams, et al. [2010] remarked that knapping was an important factor in the evolution of the human wrist and Kivell, et al. [2011a] opined that the modern hand arose from committed stone tool production.

Marzke [1992a] stated that most of the distinctive aspects of the modern human thumb can be explained by "the use and manufacture of stone tools" and most of the distinguishing features of modern human hands are consistent with "habitual stone-tool-use and stone-tool-making" [Marzke, 1992b; see also Marzke and Marzke, 2000].

Marzke's own research, however, exposed problems with this concept. The power grip does not expedite stone tool-making [Marzke, 1997], so attention was directed to the precision grip. Fingertip pads, a longer thumb and shorter fingers contribute to gripping stones during stone tool manufacture [Marzke and Shackley, 1986; Marzke, 1992a,b; 1997; Marzke, et al., 1998]. The flexor pollicis longus muscle helps maintain a secure grip of the hammerstone during percussion [Hamrick, et al., 1998; Marzke, et al., 1998] (Figure 9), and a fingertip pad "cradle" grip is useful for holding and turning the core [Marzke and Shackley, 1986]. These "forceful precision grips" [Marzke and Wullstein, 1996; Marzke and Marzke, 2000] might have become more forceful from their repetitive use in stone knapping [Marzke, et al., 1998, 1999], although *A. afarensis* [Drapeau, et al., 2005] and *A. africanus* [Ricklan, 1987] had already evolved firm precision pinch grips. Features of the third carpometacarpal joint that protect against hyperextension might be related to making stone tools [Marzke and Marzke, 1987] but in the grip used with hammerstones, which are typically flattened ovoids [Whittaker, 1994], the third finger is shifted to the side and the impact is absorbed more by the index finger than the third finger [Marzke, 1992a, b; Marzke, et al., 1998] (Figure 9).

Analysis of forces and pressures acting on the hand during Oldowan stone tool production do not support the idea that the long and robust hominin thumb experiences relatively high loads during Oldowan stone tool-making [Williams, et al., 2012]. These authors propose that the robust thumb in early *Homo* was selected for "tool use". (I propose that the tools were spheroids and cylindroids used as weapons from a bipedal stance). Thus, neither the classic hominin power grip or precision grip is used for stone tool-making. The core and the hammerstone are grasped in two different modification of the precision grip. Both are examples of the countless potential

CHAPTER 9: THE HUMAN HAND

Figure 9. A hammerstone gripped in the manner used for percussion flaking to produce sharp-edged stone tools [after Marzke, et al., 1998]. The two unique and basic human grips identified by Napier [1956] are not used in this mode of tool-making. It is the second finger that absorbs the most force at impact while the middle finger is displaced to the side of the hammerstone. The styloid process on the base of the third finger thus cannot be attributed to the stress of stone-knapping, but is readily explained by the selective benefits of throwing stone spheroids. This is one of several reasons why the behavior of stone tool-making cannot explain the evolution of the hominin hand.

variations of a grip that was evolved to facilitate accurate and forceful throwing.

Furthermore, fabricated stone tools appear too late in the chronological record to account for the distinctive features of the human hand. Stone tool-making began about 2.6 Mya [de Heinzelin, et al., 1999; Semaw, 2000, Semaw, et al., 2003]. Prior to this, a long period of hand evolution had occurred, yielding unmistakable signs of the hominin precision and power grips in *A. afarensis* and *A. africanus*. The earliest stone tool-makers *already* possessed hands with derived features that closely approached their current form, a conclusion reached by many authorities [Susman, 1988a, b, 1991, 1993; Panger, et al., 2002; Alba, et al., 2003; Drapeau, et al., 2005; Moyà-Solà, et al., 2008; Tocheri, et al., 2008; Green and Gordon, 2008; Alméija, et al., 2010; Kivell, et al., 2011a]. By then the hand had also been adapted for precise, forceful throwing [B. Isaac, 1987]. The grips used for stone tool-*making* were not naturally selected for that behavior; rather, they resulted from preceding *tool-using* of unmodified stones, bones and pieces of wood [Marzke, 1997; Hamrick, et al., 1998] (Chapter 4)—specifically, the use of stones for throwing.

Why did hominins begin to make stone tools? The answer is that they needed sharp-edged flakes to butcher animals so large they could no longer be pulled apart—animals they were acquiring by use of clubs, throwing-stones and possibly thrusting spears, dispatched from a bipedal stance. (The first hominin-made flakes may have been produced by throwing one rock against another, a method that would be archaeologically invisible). Another advantage of making sharp flakes by hand-held percussion was that this technique was more portable and controllable. Flakes are held for cutting, butchering and scraping primarily with a "pad-to-side" grip (the fingertip-pad of the thumb pressing the flake against the side of the second digit), a primitive grip in the chimpanzee repertoire.

The evolved hand was exapted for stone tool-making. Traits that arise from an adaptation for one use may subsequently be employed for another. This idea was developed by Gould [1982], who noted that structures now indispensable may have evolved earlier for other reasons and been co-opted for a new function. (This is sometimes called "preadaptation", a term that can be misleading because natural selection does not produce adaptations to future situations). Gould [1982] suggested that these co-opted uses be called "exaptations." Accordingly, handgrips that were initially adapted for gripping clubs and missiles were subsequently exapted for stone toolmaking [Young, 2003, 2010]. The human hand contains twenty-seven bones each of which articulates with at least one other bone in a movable joint. It is capable of forming thousands of different grips, given the numerous degrees of freedom among its skeletal elements, but this does not mean that these grips require thousands of different evolutionary explanations.

Rather, it is argued here, they are all made possible by a modern hand shaped by natural selection for gripping spheres and cylinders that were used as weapons. The modern hand has been exapted for typing on a keyboard, playing the piano, holding a pencil, fingering violin strings, holding an artist's brush and milking a cow. The profusion of modern grips does not require an equal number of explanations. One will do.

The human hand might have been "fine tuned" by stone tool-making because its modern structure was completed after stone tool-making began [Tocheri, et al., 2008]. This speculation, however, lacks an argument based on reproductive benefits. Chipping stones did not by itself yield any; no author has yet attempted to make the case that it did. Although throwing is innate (Chapter 13), stone-knapping is not. It has to be learned and requires a long period of practice and experience [Whittaker, 1994].

In summary, stone tool-making cannot explain hominin hand evolution because neither the hominin power grip or the hominin precision grip is used for stone tool-making; fabricated stone tools appear too late in the chronological record to account for the distinctive features of the human hand; chipping stones *per se* does not yield reproductive benefits; it is not an inherited behavior; it has to be carefully taught. Rather, the evidence is consistent with the conclusion that the hand, adapted for throwing, was exapted for stone tool-making.

The throwing motion was also exapted for stone tool-making. Savage-Rumbaugh was the first to note that the movement of the forearm, wrist and hand in stone tool-making was a form of throwing motion when she observed Toth knapping flakes: "he was throwing the rock in his right hand against the edge of the rock in his left hand, letting the force of the controlled throw knock off the flake. The 'hammer rock' never really left his right hand, but it was nonetheless thrown, as a missile, against the 'core' or the rock held in place in his left hand....[It was then] I grasped the profound similarity between the activities of throwing and stone knapping" [Savage-Rumbaugh and Lewin, 1994, p. 3]. Other research supports this linkage. During stone-knapping, a proximal-to-distal sequence involving the wrist plays an important role. In this manner knapping is similar to complex upper limb activities involved in striking or throwing [Wolfe, et al., 2006; Williams, et al., 2010].

Throwing and clubbing: Marzke's perspective. The concept that evolutionary changes in the human hand would have enhanced throwing and club-swinging is represented in Marzke's publications. In her view, adjustments to bipedal posture in the earliest hominins may have promoted the use of the trunk in accelerating the hand during tool use [Marzke, 1992b, 1998]. The power grip would have been effective for wielding branches, long bones, and antlers as clubs and hammers [Marzke, et al., 1992; Marzke, 1992b]. The grip provides an advantage in securing the grasp of a cylinder used to increase the leverage of the forearm [Marzke and Wullstein, 1996]. Marzke called the precision grip a "baseball" grip [1992a] and conjectured that it might have been used by *A. afarensis* to throw stones [1983, 1992b, 1997, 2005]. Trunk and hindlimb movements enhance the force and acceleration of tools held in the hand or thrown by it, suggesting that hominin energy linkage systems should be examined in search of patterns reflective of tool use in throwing and clubbing [Marzke, 2005]. These statements are all compatible with the bipedal use of weapons proposal. An examination of the energy linkage systems in throwing and club-swinging appears in Young [2009, 2010] and in Chapter 13.

Additional explanations for hominin hand evolution. *Manipulation and manual dexterity.* The human hand testifies to an evolutionary history of selection for "manual dexterity" [Green and Gordon, 2008] such as tool-related "manipulative behaviors" [Tocheri, et al., 2008]. Early hominins had to enlarge their thumbs and shorten their fingers to improve dexterity for tool-using or tool-making [Lovejoy, et al., 2009c, d, e]. Ape hand proportions and the thumb distal phalanx were co-opted by early bipedal hominins for refined manipulative purposes, later to be co-opted again for stone tool-making in *Homo* [Alméija, et al., 2010]. Reduced use of the hand in trees caused hominin fingers to shorten, the thumb to lengthen, and a flexor pollicis longus muscle to appear, probably for manipulative behavior. Later, a more mobile, robust thumb resulted from forceful manipulation [Kivell, et al., 2011a].

To offset loss of gripping capacity by the foot. A more powerful handgrip may have been selected to compensate for reduced arboreal performance in the foot [Crompton, et al., 2008].

Selection for climbing and balancing. The thumb of *Orrorin* reflects "a deeper adaptation concerning the precision grip essential for climbing and balancing, different from that of apes" [Gommery and Senut, 2006, p. 372].

Relaxed selection; new selection. The relatively short fingers in *A. afarensis* show that these hominins had largely abandoned arboreal behaviors, relaxing selection on the hand for arborealism while increasing it on the foot for bipedality. The fingers were then selected for better tool-using. The trend toward modern proportions could have been due to "manipulation selection pressures" such as harvesting, food processing and grooming [Alba, et al., 2003; Alméija, et al., 2010].

Food-gathering and processing. Alba et al. [2003] and Moyà-Solà et al. [2008] have suggested that the hominin hand was the result of adaptation to a variety of complex manipulative activities essentially related to food gathering and processing. Marzke [2005, p. 252] expressed a similar view when she wrote that "stresses associated with extractive, forceful retrieval and processing of tough vegetation and fauna may have been a factor in the evolution of features in ancestral hominin hands that were preadapted to the requirement of forceful precision grips in tool making".

Foot-hand pleiotropy. Selection for a shorter foot could have resulted in shortening of the hand because of genetic processes that link these two developmentally. It could be an example of pleiotropy, in which a single gene influences multiple phenotypic traits [Alba, et al., 2003; Rolian, et al., 2010]. Selection of the foot for bipedality might have led to changes in the fingers that facilitated the emergence of stone tool-making. If so, this would be "the first empirically validated case of a preadaptive morphological feature in human evolution" [Rolian, et al., 2010, p. 1566].

Summary and conclusion. Compared to the hand of modern apes and early hominins, the fingers and palm in the modern human hand are shorter, the fingers are straighter, the thumb is larger, more robust, mobile and muscular, fingertip pads are expanded, the fingers rotate during flexion, and the fingertip pad of the thumb can be applied to those of the other digits. These changes yield two distinctive human grips: A fingertip-pad grip of spheres (the hominin precision grip), which is a *throwing grip*, and a firm clamp of cylinders against the palm by flexed fingers and thumb (the hominin power grip), which is a *clubbing grip*.

The onset of remodeling of the hominin hand can be detected near the estimated origin of the hominin lineage. Through all subsequent species of hominins there is a continuing trend toward development of the two distinctive handgrips, which existed in recognizable form before the onset of stone tool-making. The evidence is consistent with an ancient origin and a prolonged course of natural selection driven by the reproductive advantages derived from throwing spheroidal rocks and swinging clubs with cylindroidal handles as weapons in conflict situations.

The fossil evidence shows that the hominin hand in our earliest ancestors was ape-like. The curvature, length, and ulnar robusticity of the *Ardipithecus* phalanges, and the diminutive size of the thumb indicate that at this early stage of hominin evolution, the hand was adapted for vertical climbing, below-branch suspension, above-branch palmigrady and other actions associated with arboreal pursuits. The hand continued to retain signs of an arboreal adaptation during several million years, but concurrently experienced a remodeling that affected all five fingers. The evidence is explained by the proposition that our ancestors remained arboreal animals until the transition to *Homo*, even as they exploited the possibilities of bipedal locomotion on the ground and their hands were simultaneously selected for gripping spheroids with their fingertip pads and cylinders in a powerful grasp against the palm. The assertion that our ancestors were becoming adapted to the bipedal behavior of throwing and clubbing because of the reproductive benefits that ensued explains all of these changes.

CHAPTER 10

Human Handedness

"It seems that humans routinely use a single 'preferred' hand to play the leading role in tool use, and that nearly nine times out of ten the choice is of the right hand" [Steele and Uomini, 2005, p. 234].

"The exact mechanisms that selected for increased preferential use of one hand or another remain unclear" [Hopkins, 2006, p. 554].

Introduction. If the bipedal use of weapons provided advantages to early hominins who used this behavior in conflicts over scarce resources, natural selection would have acted to incorporate into the hominin genome inherited variations that led to its improvement. Handedness, which means one of the two hands is preferably used, may be an example of this effect. About 90% of modern humans are right-handed and the remaining 10% are left-handed [Coren, 1993; McGrew and Marchant, 1997]. There are some geographical differences in this ratio; these may be cultural or genetic in origin [Raymond and Pontier 2004, Faurie, et al., 2005]. Ethological studies also reveal variability [Marchant, et al., 1995, Marchant and McGrew, 1998]. However, all human societies, whatever the method of investigation, show right-hand predominance. There are no populations with anything close to a majority of left-handers.

Is human handedness inherited or learned? There

is a genetic component, but it is complex and no simple genetic model seems possible [Annett, 1973; McManus, 1991; Corballis, 1991; Coren, 1993; Raymond and Pontier, 2004]. Right- or left-handedness runs in families, it develops in all children without teaching, and in all ethnic and geographically dispersed populations where it has been studied. Prehistoric populations also show a dominance of right-handed individuals. Nevertheless, a minority of left-handers persists. Natural selection must be maintaining this lateral bias; otherwise, selection for right-handedness would have eliminated it [Llaurens, et al., 2009].

Handedness emerges from asymmetrical neural organization of human motor systems [Sainberg, 2002, 2005; Bagesteiro and Sainburg 2002]. It appears early in childhood, with a tendency to use a preferred hand beginning to appear by about two years of age. After four years, right-handedness predominates [Halverson, 1940; Gesell and Ames, 1947; Annett, 1970; Rat-Fischer, et al., 2012]. During development of over-hand throwing, the preferred arm always outperforms the other one [Teixeira and Gasparetto, 2002].

Is handedness a unique human characteristic?
"Historically, population-level handedness has been considered a hallmark of human evolution" [Hopkins, 2006, p. 538]. This is the received view—that handedness is a unique trait which developed in the hominin lineage [Wile, 1934; Annett, 1970; Corballis, 1991; Coren, 1993]. According to McGrew and Marchant [1997, p. 226] "the biggest, simplest conclusion is that there is yet no compelling evidence that nonhuman primates are lateralized at the population level" for any species, task or setting. Nonhuman primate hand function therefore seemed unable to provide a model for the evolution of the pronounced handedness of modern humans.

Some recent studies have reported evidence that certain anthropoid ape populations lack a preferred hand, lending support to the view that humans alone exhibit this property. Corp and Byrne [2004] found no population-level hand preference in wild chimpanzees. Fletcher and Weghorst [2005], who studied 28 captive chimpanzees, found little evidence of hand specialization. The subjects mainly used either hand for tool use (yoghurt fishing). They observed either left-handedness or no preference. Humle and Matsuzawa [2009] studied four tool-using behaviors in wild chimpanzees. Nut-cracking, the most cognitively challenging, yielded the greatest incidence of handedness, but no directional preference. Chapelain and Hogervorts [2009] used the "tube task" to study handedness in 29 captive bonobos. (In this task the ape must use one hand to hold a tube so that a tasty treat can be retrieved by the fingers of the other, preferred hand). The frequency of unilaterality was 11 right-handed, 15 left-handed and 3 no preference. The authors conjectured that hand preferences in non-human primates are about equally right and left handed. Harrison and Nystrom [2010] examined handedness in 22 captive gorillas by noting their preferences in a range of behaviors in their daily routine. Ten showed no hand preferences and most of the others were handed for only a single behavior. For tool use, two showed a left-hand bias, two a right-hand bias and six showed no preference.

Other reports indicate that there *is* population-level handedness in ape populations, but the bias toward one side or the other (more often to the right) is much less than in human populations and many individuals show no preference. No one has been more assiduous than Hopkins in researching this topic of population-level handedness in apes. Together with his colleagues, he has amassed evidence demonstrating that the received view of human uniqueness in possessing this property needs to be revised.

In 2004 Hopkins and coworkers reviewed their earlier studies in captive chimpanzees which revealed population-level right-handedness for throwing, gestural communication, and coordinated bimanual actions. To this they added a new study of 467 captive chimpanzees in three separate colonies. A bias toward right-handedness was found that was unrelated to whether the subjects were raised by humans or not [Hopkins, et al., 2004]. They used the "tube task" which revealed that in this large sample, the animals displayed an approximate 2:1 ratio of right-handed to left-handed individuals and many animals were ambidextrous. Lonsdorf and Hopkins [2005] found a 2:1 ratio of dominant to non-dominant handed individuals within captive and wild communities of chimpanzees, depending on the task:

termite-fishing elicited left-handedness; nut-cracking was mainly right-handed. Hopkins and coauthors [2005] reported that right-handed throwing was more pronounced than other measures of handedness in captive chimpanzees. Such throwing (which involves only the arm) was observed in about half of the apes in two captive colonies. Among the most active throwers (n=89) 56 threw right-handed, 23 left-handed, and 16 were ambiguous. Throws were underhanded (25), mixed posture (27) and overhand (38). Most of the overhand throws were from an erect stance.

In 2006 Hopkins summarized all the published data on handedness in great apes then available. Comparative analysis indicated that chimpanzees (n = 1,044) and bonobos (97) show population-level right-handedness, whereas gorillas (n = 280) and orangutans (n = 103) do not. In the genus *Pan*, throwing and hand gestures were the two most lateralized behaviors, and both the strength and direction of hand preferences ran in families [Hopkins, 2006]. Handedness was more pronounced for some measures than for others, suggesting that hand preferences in apes are task specific [Hopkins, 2006]. "At a minimum, the evidence of population level handedness in *Pan* suggests that hemispheric specialization for motor skill can be dated back to 6 or 7 million years ago when the *Pan* and *Homo* genuses split" [Hopkins, 2006, p. 554].

Later, in 2011, Hopkins and his coworkers examined investigations of 536 chimpanzees, 76 gorillas, 118 bonobos and 47 orangutans by means of the "tube task". Table 2 of this publication shows that among 324 adult chimpanzees, 50% used their right hand to retrieve the treat, 29% used their left hand, and 20% had no preferred hand; thus 80% were handed, and the ratio of right-to-left preference was 1.7 to 1 (compared to ~9 to 1 in humans). Smaller groups of gorillas and bonobos also favored use of the right hand for retrieving the treat, but orangutans did the reverse. The authors conclude that apes show population-level handedness, but the magnitude of expression is lower than in humans, and handedness varies in direction between species. They suggest that "the antecedents to human right-handedness developed in African apes, after they split from the common ancestor with orangutans" (p. 609).

Forrester and colleagues [2011, 2012, 2013] studied three

small groups (12 gorillas, 9 chimpanzees, and 10 human children) using the same video sample approach in each case. They found a slight right-hand bias for hand movements from a position of rest directed towards inanimate objects and a lack of bias of similar movements to animate targets (self or social partners). Although the sample sizes were small, the investigators agreed they supported Hopkins' conclusion that human handedness is not a new or unique human characteristic.

Nevertheless, there is something about human handedness that *is* unique. No other primates show frequencies even close to the right bias typical of modern humans [Frayer, et al., 2011]. What is also unique is that not only are we more strongly handed, we are essentially *all* handed. There are exceedingly few ambidextrous individuals. Essentially everyone has a preferred hand, the vast majority of us being right-handed (~90%) and the remainder are left-handed (~10%). Something unprecedented must have happened in the hominin lineage [Chapelain and Hogervorts, 2009] which activated natural selection to exaggerate markedly the weak right-handed bias that was present in our ape ancestors—something that also preserved a modicum of left-handedness. How can this be explained? What was the reproductive advantage? No attempt has been made so far to account for the tendency toward right-handed bias in apes, but some writers have attempted to explain this phenomenon in hominins.

Previous explanations of handedness. Earlier writers sought to link handedness with asymmetries in anatomy, blood pressure, or vascularization [Wile, 1934]. Some noted that since the heart lies to the left of center it would be an advantage when fighting to hold a weapon in the right hand so the left hand could protect the heart (the right hand was the "sword hand"; the left was the "shield hand"). Calvin's [1983] hypothesis also invokes heart position and weapons: when mothers hunted they held their infants against the left breast, where the infant was calmed by hearing the heartbeat. This meant mothers had to throw right-handed, which would lead to handedness.

More recent authors have "tried their hand" at human handedness: Lonsdorf and Hopkins [2005] proposed that there might have been a genetic mutation that enhanced preferential use of the right hand. Another idea [MacNeilage, et al., 2009] is that human right-handedness arose through modification of feeding behavior inherited from ancestral primates. (Note the distinction between a *behavior*, which involves actions such as talking, walking, or throwing, and *handedness*, which is a tendency to use one of the two hands preferentially for certain behaviors). Hopkins and coworkers [2011] think tool use, bipedalism or spoken language may have played an important role in shaping human handedness. An emphasis on tool use was reiterated by Forrester, et al. [2013]. What kind of tools and how they were used was not specified.

Because the left hemisphere of the brain controls both language and right-handedness in most modern humans, other explanations link these two asymmetries [Corballis, 1991; Coren, 1993]. This remains a "hotly debated topic" [Forrester, et al., 2013]. Which came first is disputed. Several researchers have proposed that handedness initiated left-hemisphere specialization [Humle and Matsuzawa, 2009]. The first to make this argument was Frost [1980], who noted that most authors place the onset of tool behavior much earlier than the origin of language. Lateralization of speech in the left hemisphere might be due to coupling of linguistic mechanisms to motor mechanisms already lateralized by tool behavior. Forrester, et al. [2013] supported this view: tool-use is a cognitive skill that could have supported the evolution of human grammar capabilities; if not tool use, it could have been "object use" [2012] or "object manipulation skills" [2011] that were precursors to the emergence of human language.

Calvin [1983] suggested that the tool behavior might have been *throwing*. He proposed that sequential motor processing circuits were first situated in the left hemisphere of right-handed throwers and language subsequently made use of it. Kirschmann [1999] thought the evolved thrower's abilities to process sequential data may play a role in the development of language. Hopkins, et al. [2005] reiterated that right-handed throwing may have been a preadaptation for the emergence of left hemisphere specialization for speech. Asymmetries in the homologues to human language areas in the left hemisphere of chimpanzees were found by Hopkins and coauthors [2007] to be associated with handedness for tool use. They suggested "that the neural substrates of tool use may have served as a preadaptation for the evolution of language and speech in modern humans" (p. 971). Later, Hopkins, et al. [2012] revisited this concept when they reported brain studies in chimpanzees which indicated a link between right-handed throwing and its left hemisphere localization in the same region that in humans underlies the capacity for language.

In contrast, Falk [1980] asserted that language came first and left-hemisphere asymmetry of speech control spread to include the motor area controlling the right hand. McGrew and Marchant [1997] also speculated that human handedness was facilitated by prior lateralization for language. Braccini and colleagues [2010] similarly concluded that right-handedness in humans might have been driven by the evolution of language. However, today most authorities agree with Frost [1980] that tool use and right-handedness long preceded spoken language. Modern-style, rapidly spoken speech may have arisen less than 100,000 years ago [Klein and Edgar, 2002].

Why did hominins become handed in a symmetrical world? As yet there is no answer to this query. "Population-level right handedness is a human universal, whose evolutionary origins are the source of considerable empirical and theoretical debate" [Humle and Matsuzawa, 2009, p. 40]. "Despite extensive research, the origins and functions of right-handedness in humans remain an unsolved issue" [Chapelain and Hogervorst, 2009, p. 15]. The existence of any departure from symmetry suggests that notable evolutionary advantages must have been gained to counteract the apparent handicap. Where are the reproductive benefits? No existing theoretical framework seems able to explain it [McGrew and Marchant, 1997]. The exact mechanisms that selected for handedness remain unclear [Hopkins, 2006].

If handedness is part of our hominin genetic inheritance, it must tell us something about our ancestors' behavior—but what behavior would that be? It almost certainly is due to a behavior involving the *hand*. Evidence presented in Chapter 9 suggests that the hominin hand was undergoing selection for gripping spheroidal and cylindrical tools used as weapons. A parsimonious explanation based on that assertion might also apply to handedness, an idea that will be explored below.

The role of task complexity. If natural selection acted to produce behavioral handedness in the hominin lineage, it may have been a "complicated" behavior that was involved. Captive monkeys show increased individual hand laterality when the difficulty of manipulative tasks is increased [McGrew and Marchant, 1997]. During primate evolution, performing ever more complex and elaborate tasks made it increasingly necessary for the control signals from the brain to pass as directly as possible to the more skilled hand [MacNeilage, et al., 2009]. In contrast to simple reaching, more complex manual actions, particularly those requiring coordinated actions of both hands, often elicit strong hand preferences [Hopkins, 2006]. Recent studies reveal that hand preference is also more often observed in complex, tool-using actions in bonobos [Chapelain and Hogervorst, 2009] and gorillas [Harrison and Nystrom, 2010]. Experimentally-induced bipedalism in chimpanzees causes them to become more lateralized, but not in any particular direction [Braccini, et al., 2010]. The concept that the brain should be lateralized for complex processes that are important for survival means that behavioral laterality should appear in such tasks [Chapelain and Hogervorst, 2009]. Simple tasks such as reaching for food have no such asymmetrical effect [Fagot and Vauclair, 1991].

The role of skill. One possible advantage of lateralization in performing manual tasks is that one hand can specialize in perfecting a particularly difficult skill, making it more efficient than using either hand indiscriminately. Washburn and Lancaster [1968] were the first to make this argument. Handedness, they proposed, is a part of the biology of skill: "To be ambidextrous might seem to be ideal, but in fact the highest level of skill is attained by concentrating both biological ability and practice primarily on one hand. The evolution of handedness reflects the importance of skill, rather than mere use" (p. 299). This view has received support [Toth, 1985; Corballis, 1991; Corp and Byrne, 2004].

Lateralization might also improve brain and behavioral efficiency by saving neural space, minimizing replication of functions, allowing simultaneous activity of different processes, and avoiding hemisphere competition, but would not require aligning the direction of the asymmetry in a majority of individuals [Vallortigara and Rogers, 2005].

A satisfactory evolutionary explanation seems to require the identification of a complex, bipedal, skillful, one-handed behavior that favored the right hand, while accounting for the persistence of a 90/10 ratio in right/left handedness, and demonstrating how this network of unique human characteristics provided reproductive advantages to our ancestors. Additional evidence on this topic will now be summarized.

Handedness is an ancient hominin trait. In modern humans there is a relationship between right-handedness and enlarged left occipital and right frontal cerebral lobes. This asymmetry pattern occurs in robust australopithecines and *H. erectus,* possibly related to handedness in throwing and hammering [Begun and Walker, 1993; Walker, 1993]. However, the link of brain asymmetry with handedness is weak [Steele, 2000]. Based on analysis of baboon and *Australopithecus* skull fractures, Dart [1949a] concluded that hominins gripped clubs in the right hand. Other interpretations are possible [Brain, 1981].

Every tool-making site examined by Schick and Toth [1993] from the Lower Paleolithic shows the flaking pattern that characterizes right-handed stone knappers. Stone artifacts from Ambrona, Spain (0.5-0.4 Mya) reveal a mode of stone core rotation also consistent with right-handedness [Toth, 1985]. In general, both microwear and asymmetry of stone implements in the Paleolithic indicate predominant right-handedness [Corballis, 1991; Stringer and Gamble, 1993; Stringer and McKie, 1996]. In hominin specimens from Sima de los Huesos (Atapuerca, Spain) dated at >500,000 years ago and others from their likely descendants, the Neanderthals (~130,000-30,000 ya), scratches on incisors and canines resulting from cutting meat held by the teeth show a pattern of right-handedness [Frayer, et al., 2011].

Neanderthal humeral asymmetry in bone robusticity, especially in males, supports this contention [Ben-Itzhak et al., 1988; Aiello and Dean, 1990]. Bilateral asymmetry in the humerus has also been documented in Upper Paleolithic *H. sapiens* [Trinkaus, et al., 1994; Churchill, et al., 1996; Trinkaus and Churchill, 1999]. It is similar to what is observed in the dominant and nondominant arms of modern throwers and clubbers [Trinkaus, et al., 1994]. Elite tennis players display greater robusticity in the humerus of the dominant arm [Jones et al., 1977; Trinkaus, et al., 1994; Haapasalo, et al., 1996] and also in the forearm [Montoye, et al., 1980; Etherington, et al., 1996; Kontulainen, et al., 1999]. There is a comparable humeral diaphyseal asymmetry in male baseball pitchers [King, et al., 1969]. Others have suggested that humeral bilateral asymmetry in Neanderthal and Upper Paleolithic *H. sapiens* may be due to use of throwing and thrusting spears [Aiello and Dean, 1990; Churchill, et al., 1996; Churchill and Schmitt, 2000]. These observations can all be explained by assuming that Middle and Upper Paleolithic hominins were predominantly right-handed throwers and clubbers, supporting the proposition that handedness may have had its origins in the bipedal use of weapons [Young, 2010].

Throwing, Striking and Handedness. Throwing and club-swinging in the human manner is also a unique feature of our species. Stone-throwing in macaques is mainly underarm, performed from a tripedal posture, often accompanied by vertical leaps, and most throws are backwards, without signs of handedness [Leca, et al., 2008]. Chimpanzees are the most advanced throwers and strikers among the great apes. They use a variety of one and two-handed throwing motions and throw quadrupedally, tripedally and bipedally. With either arm they hurl underarm, sidearm, overarm and back-handed [Goodall, 1964, 1984, 1986; Kortlandt, 1972; Brewer, 1978; deWaal, 1989; Marzke and Wullstein, 1996]. When they throw an object or strike with a stick, only the arm acts to accelerate the hand, and if the act is performed bipedally, due to vertical instability they may fall to the ground [van Lawick, 1991]. The stick is often released prior to contact [van Lawick-Goodall, 1968; Goodall, 1986], perhaps due to weakness of the grip [Marzke and Wullstein, 1996]. Throws are feeble and accuracy is lacking [van Lawick, 1991; Goodall, 1986].

From a rudimentary beginning no better and possibly

worse than this, additional muscle groups and body segments must have been recruited by natural selection in the hominin lineage to standardize the act, improve the handgrips, increase the force exerted on the hand-held weapon, the kinetic energy transmitted to it and the accuracy of its trajectory. In our species, throwing and striking are inherited behaviors involving the entire body [Chapter 13; Young, 2009, 2010].

The throwing and striking motion that evolved in hominins is asymmetrical. As noted above, Calvin [1983] linked throwing with handedness and cerebral asymmetry. Kirschmann [1999] connected the rotation of the upper body around its long axis with one-armed throwing in the origin of handedness. Evidence of right-handedness in stone tool fabrication in the Lower Paleolithic led him to conclude that by 2.5 Mya australopithecines were capable of rotating the upper body to increase the acceleration distance in throwing.

In the throwing motion left and right sides of the body move differently, as each segment contributes to the unified, integrated act (Chapter 13). Rotation of the hips and torso around the vertical axis is inherently sided—the body spins to the left or to the right. If the right side of the body rotates forward, it delivers energy to the right arm and hand. If the body rotates leftward, the left upper limb is energized.

These observations support a proposal that handedness may have resulted from the asymmetry of the throwing and striking motions. The antiquity of handedness is consistent with the concept that the onset of a specialization for throwing and striking began very early in the hominin lineage [Young, 2003, 2009, 2010]. It also is consonant with evidence that handedness is most evident in throwing and club-swinging skills. The right-handedness of hammering (88.2%) is exceeded only by swinging a racket (89.3%) and throwing a ball (89.57%) [Annett, 1970; Calvin, 1983; Corballis, 1991]

This explanation also meets the criteria of complexity and evolutionary significance noted above, as well as the crucial factor of reproductive benefits. When primates stand on their hind legs to perform a one-handed act, without using the other hand to grasp a support for balance, it greatly increases the difficulty and the complexity of the task. If, in addition, the structure of their hands is not adapted to the act performed, this multiplies the challenge. One-handed throwing or club-swinging by our early ancestors would seem to be the most complex and difficult behavior involving the hand that became a staple part of their repertoire, and thus would produce the strongest effect on laterality of hand use. Something unprecedented during hominin evolution linked handedness with a novel behavior that had strong effects on reproductive success. The evolution of throwing and striking provides a possible explanation.

The Persistence of Left-Handedness. Why has a 10% minority of left-handers survived? What benefit does it yield? The advantage seems to depend upon its being relatively rare and must compensate for health problems that may occur with greater frequency in left-handers [Wile, 1934; Coren and Halpern, 1991; Coren, 1993]. Benefits may be skewed towards males, since 3-4% more women than men are right-handed [Coren, 1993; Raymond, et al., 1996].

The persistence of left-handed individuals has been clarified by analysis of sports in which they are represented by more than 10% of the participants. The higher proportion of left-handers is not due to an inherent, neurological advantage. Rather, in certain sports the right-hander is relatively unaccustomed to facing a left-handed adversary and must reverse his or her usual strategies [Wood and Aggleton, 1989]. This advantage occurs in sports which involve face-to-face competition. A high proportion of left-handed individuals occurs among baseball pitchers, boxers, and fencers, for example [Coren, 1993]. Left-handed athletes may have an advantage because their throws, punches, or thrusts come from directions that differ from those of the more prevalent right-handers, giving them the benefit of surprise.

Raymond and coworkers [1996] confirmed and extended this concept. They found that only in sports involving interactive competition was an increase of left-handers observed, suggesting a frequency-dependent advantage. As the frequency of left-handed individuals increased, the element of surprise would decrease during face-to-face combat. This would maintain the frequency at a low level. Grouios, et al. [2000] also supported this proposal. They noted that the excess of left-handers among sporting competitors applies only to interactive sports such as basketball, boxing, fencing, football, handball, judo, karate, table tennis, tennis, and volleyball, and not to sports like cycling, discus throwing, diving, gymnastics, rifle shooting, rowing, running, skiing, swimming, and weight lifting, where there is no direct interaction between competitors. Moreover, the closer the physical interaction of the opponents, such as in boxing, fencing, judo, or karate, the greater the prevalence of left-handers [Grouios, et al., 2000].

The left-handed advantage in face-to-face interactive sports can be explained tactically, and is a marker of a strong selective advantage during fights. This could be the source of natural selection involved in the persistence of handedness polymorphism in humans [Llaurens, et al., 2009]. This idea, "the fighting hypothesis" [Faurie and Raymond, 2005], is perfectly compatible with the proposal that early hominins underwent natural selection for bipedal fighting prowess with hand-held weapons. Here lies the reproductive advantage. In the evolutionary context, those who win fights over scarce resources gain reproductive benefits [Raymond, et al., 1996, Grouios, et al., 2000; Raymond and Pontier, 2004; Llaurens, et al., 2009].

CHAPTER 11

Diminution of the Canine Teeth

"Canine tooth size reduction is one of the few defining features of the hominin clade...and is recognized as a signal of important behavioural and adaptive changes". "[S]ubstantial reduction in male canine crown size and loss of significant dimorphism probably occurred near the origin of hominins" [Ward, et al., 2010, pp. 3334, 3336].

Introduction. As the epigraph states, diminution of canine tooth size is a trend in hominin evolution which began close to the onset of our lineage. Its subsequent progression during millions of years indicates it was associated with a behavior that brought reproductive benefits for an extended duration of time. Because size reduction commenced so early and followed a course that is seemingly unique among the anthropoids, it has become a marker of hominin status [Richmond, et al., 2001; Haile-Selassie, et al., 2004; Brunet et al., 2005; White, et al., 2009a; Ward, et al., 2010].

As the canines began to shrink, there were also changes in their shape. Simultaneously, the premolars and molars enlarged (Chapter 6). It now seems possible to account for the evidence with an explanation based upon use of hand-held weapons coupled with changes in diet. The basic idea (to be developed below) is that hominin canine teeth became reduced in size because their long, dagger-like form, exaggerated in males, lost the ability to promote reproductive success by use in threat and fighting. They were rendered obsolete for this function due to the development of a new mode of combat involving hand-held weapons. Natural selection then favored smaller canines with a new shape that improved the processing of food.

Canine teeth in anthropoids. In anthropoids, the length of canines (crown height) and their gender size dimorphism is strongly associated with fighting between males [Greenfield, 1992; Plavcan and van Schaik, 1992, 1997a, b; Plavcan, 1993, 2000; White, et al., 2009a]. They are significantly larger in males of species that engage in combat to gain reproductive advantages. The enlarged canines function as weapons both in fights among males and conflicts with predators when their length and sharp tips (particularly in the upper jaw) facilitate puncturing and slashing [Greenfield and Washburn, 1991; Greenfield, 1992; Plavcan and van Schaik, 1992; Plavcan, 1993]. The smaller adult female canines and the deciduous canines of both sexes exhibit incisor-like traits, indicating selection for dental functions other than fighting [Greenfield, 1992].

In chimpanzee males, the canine teeth are large, conical, sharply pointed, and project well beyond the incisors and premolars, as shown in Figure 10. There is a wide diastema (gap) in the mandibular and maxillary arches to accommodate them during occlusion. A facet of wear develops on the crown of the maxillary canine from contact with the third premolar in the opposing jaw. In modern humans, the situation is quite different. The canines are relatively small, incisiform (incisor-like), the tip is blunt, prominent gender dimorphism is lacking, no diastemata are present, and the canines wear down from the tip to the level of the premolars [Le Gros Clark, 1964; Aiello and Dean, 1990].

Paleontological evidence of canine evolution in hominins. Specimens of the earliest putative hominin, *Sahelanthropus tchadensis* from Chad, are dated near the estimated time of hominin origins (7 Mya). The crown of a mandibular canine tooth is short, compared to its counterpart in modern chimpanzees. (This suggests that the canines might have lengthened in chimpanzees after their divergence from

CHAPTER 11: DIMINUTION OF THE CANINE TEETH

Figure 10. Male anthropoids, such as the chimpanzee shown here, use their long, pointed canine teeth as weapons for threat and fighting [from Young, 2010]. Diminution of human canines, which began early in hominin evolution and continued during four million years, is likely due to the origin and subsequent selection of a more effective way of fighting—the bipedal use of hand-held weapons. The shrinking canines became adapted to another function: food processing.

hominins). The pattern of wear is consistent with the absence of a functional canine/third premolar (C/P_3) honing complex, by which apes sharpen the distal edge of the upper canine [Brunet, et al., 2002, 2005].

Remains of *Orrorin tugenensis* from Kenya (6.0-5.8 Mya) include an upper canine tooth said to be "short" but "large for a hominid" with a pointed apex recalling those of female chimpanzees [Senut, et al., 2001].

CHAPTER 11: DIMINUTION OF THE CANINE TEETH

Canine teeth of *Ardipithecus kadabba* from Ethiopia (5.7 Mya) are shorter and more incisiform than in apes [Haile-Selassie 2001] although one upper canine lies at the margin of the female chimpanzee distribution [Haile-Selassie, et al., 2004]. A small facet occurs on a left lower P_3, indicating a C/P_3 honing complex. There is no evidence of this feature in subsequent hominins.

Ar. ramidus (4.4 Mya) also from Ethiopia, had canines that are more incisor-like and protrude less than in chimpanzees [White, et al., 1994; Semaw, et al., 2005]. They are similar to those from female apes. Gender size dimorphism is not apparent [White, et al., 2009a]. *Ar. ramidus* canines have thin tooth enamel and are large relative to the premolars and molars [Ward, et al., 1999b]. The lower canines retain a more apelike morphology than do the uppers, and are unlike those in other anthropoids insofar as the height of the maxillary canine crown is less than that of the mandibular [Suwa, et al., 2009b].

Fossils of *Australopithecus anamensis* (4.2-3.9 Mya) have been found in Kenya, [Ward, et al., 1999, 2001] and Ethiopia [White, et al., 2006]. All australopithecines, beginning with *A. anamensis*, have a thicker enamel layer on their teeth than earlier hominins, with that in robust species being especially thick [Ward, et al., 1999b]. *A. anamensis* upper canines are the same size or slightly smaller than those in *Ar. ramidus* [Ward, et al., 1999b, 2001] and gender dimorphism is reduced [Ward, et al., 2001; White, et al., 2009a]. Diastemata are still present, but there are changes in the maxillary canine-mandibular P_3 relationship. Both teeth are shorter mesiodistally, canine tooth crowns are more symmetrical in profile, third premolars are broader, canine roots are smaller, gender dimorphism is further reduced and molar crowns are higher [Ward, et al., 2010]. Two maxillary canines from Ethiopia are at or above the upper range of size in *A. afarensis*; the remainder are within that range. When the canines are compared to the postcanine dentition, *A. anamensis* has relatively larger canines than *A. afarensis* [Ward et al., 1999b, 2001].

A. afarensis is known from Ethiopia and Tanzania, 3.8-3.2 Mya [White, et al., 1993; Haile-Selassie, 2010]. In this hominin the canine teeth are further reduced in size [M. G.

Leakey, et al., 1998; Ward, et al., 1999; Kimbel, et al., 2006]. The crown tips were blunted as they came into occlusion [Greenfield, 1992]. There are some diastemata between the lateral incisors and canines, but they are less common than in earlier hominins [Johanson and White, 1979; White, et al., 1981; Greenfield, 1992]. The maxillary and mandibular canine teeth project slightly beyond the tooth row; their roots are large and long [Johanson and White, 1979]. Gender size dimorphism is reduced, but exceeds that in *H. sapiens* [McHenry, 1991b]. The shape of the canine crown is more incisiform in *A. afarensis* than in *A. anamensis* [Kimbel, et al., 2006] and there is less wear on the anterior teeth [Ward, et al., 2010]. *A. garhi*, a 2.5 Mya hominin from Ethiopia, had canines similar to those of *A. Africanus* [Asfaw, et al., 1999].

The canines of *A. africanus* (3.2-1.8 Mya) from South Africa are incisiform, do not project markedly beyond the level of the premolars and molars, and there is no evidence of sexual dimorphism [Le Gros Clark, 1950]. Only one of twelve specimens shows a mandibular diastema [White, et al., 1981]. In *A. sediba*, also from South Africa (2.0 Mya), canines are small, but protrude well beyond the incisors and premolars, as seen in a virtual reconstruction of the maxilla [Berger, et al., 2010b, Fig. S1]. Despite having enlarged molars, the upper and lower canines of *A. robustus* (*Paranthropus*) (1.8-1.2 Mya) are smaller than those of earlier australopithecines [Aiello and Dean, 1990].

Hominins from Dmanisi, Republic of Georgia (~1.8 Mya), thought to be a primitive form of *Homo*, have larger upper canine crowns than other members of that genus [Gabunia, et al., 2002; Vekua, et al., 2002], but are less ape-like than in some species of *Australopithecus* [Rightmire, et al., 2006]. In Javanese *H. erectus*, some upper canines retain primitive features such as projecting beyond the level of the premolars and a separation from the adjacent incisor, but these are lacking in *H. erectus* canines from China, which are modern in size and form [Le Gros Clark, 1964].

The fossil record shows that the earliest known hominins had canine teeth smaller than those of chimpanzees and a continuing diminution proceeded during several million years. In addition to a reduction in size (most notably in crown height) there were also changes in shape, much of which took place after significant size reduction had occurred. Reduction in male canine teeth had the effect of reducing gender differences and was accompanied by a loss of canine honing during contact with the adjacent premolar. The changes in crown size and shape indicate that two different factors affected natural selection of the canines [Ward, et al., 2010].

Changes in other parts of the hominin dentition. In *Australopithecus* there was a progressive incorporation of the premolars into the grinding mechanism of chewing, a trend that continued into the early Pleistocene (~1.8 Mya). By that time it was part of a highly derived masticatory complex associated with enlargement in the basal dimension of premolars and molars ("megadontia") [Teaford and Ungar, 2000; Kimbel, et al., 2006; Suwa, et al., 2009b]. Anterior teeth of *A. anamensis,* but not *A. afarensis*, are heavily worn. The transition between these two species was associated with adaptations for increasing masticatory loads on the postcanine dentition and with changes in the canine/P_3 honing complex: the third premolar was progressively molarized as the canine crown became more incisiform [Kimbel, et al., 2006; Ward, et al., 2010]. Large cheek teeth with thickened enamel imply a greater emphasis on the ability to process tougher or more abrasive foods [Teaford and Ungar, 2000; Ward, et al., 2001, 2010; Kimbel, et al., 2006; Suwa, et al., 2009b]. It was probably the result of natural selection that facilitated exploitation of an altered diet [Ward, et al., 2010 Ungar and Sponheimer 2011] which endured until the transition to *Homo,* when adaptation to yet another new diet apparently took place (Chapter 7).

How many explanations are required to account for these complex changes in the hominin dentition: the reduction of canine height, the transformation of canine shape from dagger-like to incisor-like, the loss of gender size dimorphism, modifications of wear patterns of the front teeth and the appearance of massive cheek teeth? What were the reproductive advantages?

Explanations of diminished canine size. One hypothesis is that long, pointed canine teeth were abandoned because of the evolutionary onset of male pacifism. If males

CHAPTER 11: DIMINUTION OF THE CANINE TEETH

stopped fighting with each other for access to females, so the argument goes, then canine size would have become diminished. This concept was proposed by Holloway [1967] and Zihlman [1981] who attributed the loss of protruding canines to reduced competition among males and increased collaboration with other group members. Lovejoy [1981, 1993, 2009a, b] has persistently supported this explanation, elaborating it to include a dramatic change in cooperation among males, and between males and females, involving male parental investment and monogamous pair-bonding. Others have recently supported the view that large canine teeth lost their role in threat and fighting because males gave up these pursuits [White, et al., 2009a; Suwa, et al., 2009b]. These explanations do not address the question of why males who stopped fighting with each other and with predators would have left more descendants than those who were inclined to engage in aggressive combat. This certainly is not a Darwinian view insofar as he imagined that they would be using stones, clubs, or other weapons, for fighting with their enemies (see below).

A second explanation for canine shrinkage attributes it to fighting with forelimbs. According to Carrier [2004], the trend towards reduction in canine size suggests that biting and "opponent manipulation" were not important functions of the head and neck during fighting in early hominins. His idea is that biting was replaced by "swings, strikes, and pushes from the forelimb, and...craniad directed forelimb punches or pushes" [pp 144-145]. This hypothesis reaffirms McHenry's [1991a] statement that early bipedal hominids used their forelimbs in place of their canine teeth.

The third explanation, currently the majority opinion, is that hominin canine teeth diminished in size because their role in threat and fighting was replaced by hand-held weapons—a view that can be traced back to Darwin [1871]. Because large canine teeth are used by male anthropoids for threat and fighting, Darwin concluded that they had been replaced by hand-held weapons in humans. (He had no sympathy for the idea that hominin males simply stopped fighting).

Early hominins, Darwin wrote, were "probably furnished with great canine teeth; but as they gradually acquired the habit of using stones, clubs, or other weapons, for fighting with their enemies.the teeth would have become reduced in size" [Darwin, 1871, p. 144]. Most writers have supported the link between canine size reduction and the use of hand-held weapons [Sollas, 1924; Dart, 1925; Keith, 1949; Washburn, 1950, 1967; Bartholomew and Birdsell, 1953; Washburn and Howell, 1960; Devore, 1964; Brace, 1967; Laughlin, 1968; Pilbeam, 1970; Washburn and Moore, 1974; Wolpoff, 1976; Symons, 1979; Hill, 1982; Parker, 1987; McHenry, 1991b; Greenfield, 1992; Kirschmann, 1999; Boehm, 1999; Ward, et al., 2001; Young, 2009, 2010]. This conclusion, consistent with the theme of this book, would imply that the emergence of weapons behavior was closely associated with the origin of the hominin lineage.

Plavcan and van Schaik [1997b] agreed that weapons could help in fights, but observed that large canines would still be useful for close-in fighting (see next paragraph); secondly, they noted that canine reduction is not associated with the appearance of stone tools—the "no hand-made weapons = no weapons" argument (Chapter 4); and third, even if use of weapons can explain canine size reduction, it does not resolve the paradox of high gender dimorphism in body size in *A. afarensis* (this issue is resolved in Chapter 5).

Explanations of the change in canine shape. Beginning as a pointed dagger, the hominin maxillary canine became remodeled into a tooth shaped more like the adjacent incisors. Pilbeam [1970] was the first to address this topic. He asked why not retain large canines *and* use weapons? His answer was that when selection for large canines decreased because of a shift to throwing and striking for combat, it may have promoted smaller, more incisiform canines that improved ingestion of food. As he put it, "Canine reduction may also be associated with the improved slicing and cutting ability of the hominid front teeth" [p. 91]. This concept was later applied to meat by Szalay [1975]. "Another function of natural selection for reduction of canines in both sexes may have been to increase the effectiveness of the food-grinding mechanism for processing large quantities of tough, gritty foods" [Zihlman, 1981, p. 102].

Greenfield [1992] proposed that natural selection of large, dagger-like canines for use as weapons in anthropoid males *overrides* selection for incisor-like traits of the canines which appear in females and in the deciduous teeth of both genders. The reduction in canine size in hominins due to replacement by hand-held weapons explains how canine reduction would have been "permitted". If the cost of developing and maintaining large canines was not offset by the advantages of weapon use, the benefit of efficiency and economy would have favored their reduction [Greenfield, 1992]. This is a form of the "expensive tissue" argument: There is a cost to producing large teeth that can be put to better use if their function becomes outmoded. The "better use" in Greenfield's conception is similar to the one imagined by Pilbeam. Large, projecting canines preclude their application in pulling, grasping, and gripping like incisors. When there was a decline in the evolutionary value of canines as weapons, they evolved toward a more incisor-like form in subsequent hominins, because of the improvement in processing food before it was swallowed [Greenfield, 1992]. Ward and colleagues [2010] reached the same conclusion.

Another factor affecting hominin tooth evolution seems to have been a dietary shift incorporating increased amounts of foodstuffs that required powerful mastication. This trend can be detected in *A. anamensis* [Ward, et al., 1999b]. It led to the natural selection of a smaller, incisor-like canine tooth in *A. afarensis* and changes in the premolars and molars that (along with reduced wear on anterior teeth) signal increased crushing and grinding ability of the posterior teeth. The ultimate expression of this adaptation is seen in the robust australopithecine, who exhibit craniofacial and dental features that apparently reflect adaptation to a diet requiring heavy mastication involving great vertical occlusal force [Johanson and White, 1979]. The nature of the food source has yet to be identified in either branch of *Australopithecus* [Teaford and Ungar, 2000; Ward, et al., 2001, 2010; Kimbel, et al., 2006; Suwa, et al., 2009b; Ungar and Sponheimer 2011]. However, it is noteworthy that this changed in the transition to *Homo*, when reduction of the massive chewing apparatus and an increase in meat-eating seem to be linked (Chapter 7).

Summary and conclusion. Darwin [1871] asserted that small canines must be an evolved feature in our lineage and created an explanation for it (well before there was any fossil

CHAPTER 11: DIMINUTION OF THE CANINE TEETH

evidence on this topic). He suggested that dagger-like canines were replaced as the hominin weapon of choice by rocks, clubs and spears held in the hand and employed from a bipedal stance. From then until now the vast majority of scientists who have addressed this issue have agreed with Darwin. When the weapons function was transferred from the face to the hand, the result profoundly affected the evolution of many parts of the body, including the canine teeth.

As fossil evidence accumulated, it became clear that reduction in size was only part of the story. Also requiring explanation was a significant change in the shape of the canines. To address this question, Greenfield [1992] expanded Darwin's hypothesis to encompass both size and shape. If big canines are weapons in anthropoid species, he reasoned, then the use of hand-held weapons by hominins would stop the selective process that maintained them. This would allow natural selection to eliminate the expense of growing and maintaining these cumbersome elements and provide an opportunity for selection of inherited variations that would enhance another behavior that contributed to reproductive success—namely, food-processing, an idea introduced by Pilbeam [1970]. (Food and reproduction are fundamentally linked: Chapter 3). Males with more incisor-like canines could pull, grasp, tear and separate morsels into pieces suitable for crushing and grinding by the cheek teeth to yield fragments that could be swallowed. The result was a tooth with a smaller, more incisor-like shape, aligned with the incisors beside it in the jaw.

The change in weaponry exerted a *negative* effect on natural selection for large canines in males because they no longer provided reproductive benefits resulting from their large size. Together with the simultaneous *positive* effect of natural selection for an incisor-like shape, and the advantages of reducing the cost of growing and maintaining them, they diminished in size, reducing the gender size dimorphism; the gaps in the tooth row that had made room for them in the opposing jaw gradually disappeared; the third premolar no longer rubbed against the opposite canine, thereby losing its honing properties. These are all indirect effects of the new and improved weapons system.

Other changes appear to be explicable by natural selection acting to improve access to new food sources. An increase in the thickness of tooth enamel began in *A. anamensis.*

Then, in *A. afarensis,* a major enlargement of the cheek teeth occurred. The premolars became molarized, contributing to the megadontia of the back of the jaw. This escalation of a massive crushing apparatus included both jaws and the associated muscles of mastication, apparently due to an adaptation to chewing harder, grittier, more fibrous foods. Perhaps changing demands of new environments encountered by hominins moving closer to the woodland fringe was a factor. When they abandoned the trees, during the transition from *Australopithecus* to *Homo,* the massive crushing and chewing apparatus began to shrink as hominins with weapons increasingly acquired a more easily masticated, enriched nutritional source—namely, animal tissues (Chapter 7).

In this chapter the role of the bipedal use of weapons in hominin evolution has been narrowly focused on the teeth, particularly the canine teeth. In the following chapter, the explanatory scheme is expanded to encompass the entire postcranial skeleton.

CHAPTER 12

Robusticity of Bone and Muscle

"General massiveness apparently is a primitive hominid character" [Weidenreich, 1943, p. 175].

"All of the postcranial elements [of A. afarensis] indicate high levels of skeletal robustness with regard to muscular and tendinous insertions" [Johanson and White, 1979, p. 324].

"The presence of increased cortical [bone] tissue in the early hominines, while widely recognized, has elicited very few explanatory models" [Kennedy, 1985].

Introduction. Increased robusticity of the skeleton refers to strengthening by addition of compact bone tissue to the outer (cortical) layer that forms the main component of most bones. Robusticity of the muscular system denotes increased muscular mass, typified by modern body-builders. Bone robusticity in fossil hominins can be determined by measurements of the bones themselves, but statements about muscular robusticity are based on evaluation of rugosities on the surface of bones where muscles were attached.

Some authors have based their assessment of skeletal robusticity on visual inspection of bones. Others have used quantitative methods, such as determinations of the percent of cortical area in cross sections at specific sites on bone shafts [e.g., Ruff, 2009; Berger, et al., 2010b]. The general conclusion that can be drawn from these reports is that bones of most if not all extinct hominin species were very robust compared to ours. Such generalized, thickened, cortical bone is found in very few animals [Kennedy, 1985]. Apparently the musculature was also massive in fossil hominins, although the evidence for this is more limited.

Why were our early hominin ancestors so robust? The evidence will first be presented, followed by the explanations that have been proposed to explain it. The discussion will be focused on the postcranial part of the body; that is, everything from the neck to the feet.

Paleontological evidence. The earliest hominin postcranial specimens are from *Orrorin tugenensis*, dated at 6 Mya, represented by three fossil femora discovered in Kenya by Senut, Pickford and colleagues [Senut, et al., 2001]. In one (BAR 1002'00) the femoral neck is complete. The cortex of the neck is thickest inferiorly and thinnest superiorly (a condition linked to bipedalism). However, superiorly, anteriorly and posteriorly the cortex is relatively thicker than it is in modern humans, although thinner than in African apes. Another of the specimens (BAR 1003'00) is larger and more robust [Pickford, et al., 2002]. Richmond and Jungers [2008] concluded that the width of the shaft of BAR 1002'00 is similar to early hominin femora in general, all of which have mediolaterally broad shafts. The robusticity exhibited by this ancient *Orrorin* femur appeared near the time of origin of the hominin lineage and persisted for more than 4 million years.

The degree of postcranial bone robusticity has not been reported for *Ardipithecus ramidus*. The bony remains are poorly fossilized and all large limb bones are variably crushed. Some of the smaller, better preserved bones of the hand and foot are described as robust [White, et al., 2009; Lovejoy, et al., 2009j].

Australopithecus had robust bones. A femur shaft of *A. anamensis* (from Kenya; 4.2-3.9 Mya) is notable for its thick cortex [White, et al., 2006]. The humerus also had an extremely thick cortex with a cortical area of about 86%–more robust than in tree-climbing African apes [Ward, et al., 2001].

A. afarensis (~3.9 to 3.0 Mya) maintained a robust postcranial skeleton during the nearly one million years of its existence. The proximal ends of a radius and ulna from Sterkfontein, South Africa (3.3 Mya) are robustly structured, as are the first metacarpal and proximal phalanx of the thumb [Clarke, 1999, 2002]. In a partial *A. afarensis* skeleton from Hadar, Ethiopia, dated at 3 Mya, the ulna is large and robust. Muscle markings on the humerus are "quite dramatic" as are those on other bones. Even the hand is robust. The metacarpal shafts are thicker for their length than those of chimpanzees [Drapeau, et al., 2005] and some of the foot bones are described as having markedly thick cortical bone [Latimer, et al., 1982]

Australopithecus africanus was also very robust, similar to *A. afarensis* [McHenry, 1986].

The femora of *Paranthropus* (*A. robustus* ~1.8-1.0 Mya) [Klein, 1999] have large transverse shaft diameters, as do all non-*Homo* femora and other postcranial elements in hominins from the Pliocene and Pleistocene eras [McHenry 1991a, b].

Robusticity continued during the transition to Homo. Three hominin species associated with the transition from *Australopithecus* to *Homo* about 2 Mya (Chapters 6, 7) show a high degree of postcranial robusticity. In *A./H. habilis,* this is well documented in the femur and humerus; the phalanges of the hands are also exceptionally robust, with thick mid-shaft diameters with thick walls and features indicating powerful musculature [Susman and Creel, 1979; Susman and Stern, 1982; Trinkaus, 1984]. The relative cortical thickness (% CA) of the OH 62 femur (83.2%) is similar to other early *Homo* femora near mid-diaphysis (about 80-85%) and is elevated relative to modern humans (71.5%) [Ruff, 2009]. The humerus in OH 62 also has relatively thick cortices (% CA, 79.7%), similar to the KNM-ER 1808 mid-distal humerus (81.7%) and again elevated relative to modern humans (59.8%). "Both the humerus and femur of OH 62 have relatively thick cortices...although they are not unusually thick when compared to other early *Homo* adult specimens" [Ruff, 2009, p. 96].

Australopithecus sediba from the Malapa site in South Africa (2.0 Mya) [Berger, et al., 2010a; Woodhead, et al., 2011] had a robust skeleton, including the humerus, femur and phalanges of the hand, which had strong attachment sites for the flexor digitorum superficialis muscle [Berger, et al., 2010b,

Table S2]. Based on the percentage of cortical area (% CA) in cross-sections at or near midshaft in specimen MH1 (a juvenile), the humerus had a % CA of 75, which exceeded that of *A. africanus* (63-74%) and was somewhat less than in *A./H. habilis* (80%) and *H. erectus* (82%). (The latter two estimates are from Ruff [2009]). The % CA of the femur was 84% for the *A. sediba* juvenile; 80% for *A. africanus,* and (from Ruff [2009]) 83% for *A./H. habilis*. By this measure, the *A. sediba* femoral shaft was near the top of the *H. erectus* range.

Metatarsals found at the Dmanisi hominin site (1.8 Mya) have elevated robusticity indices based on body mass estimates. A complete adult femur shaft is markedly robust, with a narrow medullary canal. A less-robust tibia has mid-shaft proportions like those of early *Homo* [Lordkipanidze, et al., 2007].

Early African *Homo erectus* femora have very thick shafts [Walker and Leakey, 1993]. Robusticity also characterized the femur of *H. erectus* in China, exceeding that in any other living primate [Weidenreich, 1947]. The same is true of the humerus. *Homo heidelbergensis* ~500 kya had massively robust bones [Roberts, et al., 1994; Stringer, et al., 1998; Wood, 2010; Stringer, 2012]. In fact, all premodern people were characterized by extraordinary muscularity and robust bones, including hominins from Atapuerca, Spain, dated at about 0.3 Mya [Klein, 1999] and *H. neanderthalensis*, who was very robust throughout the postcranial skeleton, including the hand [Trinkaus, 1983].

Children were robust. Robusticity also characterized the skeletons of hominin children. The femoral morphology of three *Homo* juveniles (one *H. erectus* specimen and two Neanderthals) was studied by Ruff and coworkers [1994]. The midshaft femoral cortical bone thickness in these children (ages about 4-12 years) all fell in the upper part of the range for modern humans of a similar age. Increased diaphyseal robusticity relative to modern humans was apparently a characteristic of immature as well as adult *H. erectus* and *H. neanderthalensis* [Ruff, et al., 1994]. Even *infants* were robust. Skeletal robusticity was documented in a Neanderthal child found in Ukraine whose estimated age was 5-7 months [Vlcek, 1973]. The bones display pronounced robusticity compared to a modern child of the same age, according to Vlcek, who asserted that the robusticity of ribs and long bones in the infant may be regarded as "phylogenetically ingrained"—that is, genetic, due to evolutionary selection.

Robusticity eventually declined in *Homo*. There has been a decline in shaft robusticity in the hominin femur and humerus during the last 2 million years, beginning in *H. erectus* and continuing through the Neanderthals and early modern *H. sapiens* but most of the decrease occurred recently, in our species, after 0.03 Mya. [Ruff, et al., 1993; Trinkaus, 1997].

Why were hominins so robust? Kennedy [1985] suggested that the increase in bone thickness in early hominins might be related to diet. Early hominins shared a long history of primate subsistence on fruit, foliage and vegetable foods, all excellent sources of calcium. When they went from a mainly plant-based diet to one including periodic ingestion of meat, a low-calcium food, this might have led to variable blood levels of calcium, conceivably resulting in a hormonal reaction which reduced bone resorption [Kennedy, 1985].

Kirschmann [1999; 4.1] wrote that the massiveness of hominin bones is due to natural selection that enhanced the ability to withstand forceful blows from stone missiles. "The robusticity of the long bones of *Homo erectus* is...a reaction to intraspecies confrontations involving the use of stones."

Another view is that hominin robusticity was not a result of natural selection, but rather was due to the physiological effects of vigorous physical activity producing "mechanical loading" [Ruff, et al., 1993]. Its eventual decline in recent *Homo sapiens* might have been linked to an increase in brain size and improved technology. Two different explanations might be required for the upper and lower limbs.

The decrease in humeral robusticity is difficult to correlate with the archeological record [Trinkaus, 1997], but might be due to the invention by modern humans of improved throwing projectiles, such as spears with bone and stone points [Trinkaus, 1986]. The legs should reflect body mass and locomotor activity and may not correspond closely with technological or cognitive processes [Ruff, et al., 1993]. Robusticity of the legs in *H. neanderthalensis* might be the result of prolonged locomotion over irregular terrain, perhaps in the pursuit of mobile prey [Trinkaus, 1986, 1987]. "In fact, the level of diaphyseal hypertrophy of their femora and tibiae suggests that they spent a sufficient portion of their waking hours moving across the landscape to require an exaggerated level of endurance, far more than did early modern humans" [Trinkaus, 1986, p. 205]. Wood and Lonergan [2008] write that the robust long bones and large lower limb joints of *H. heidelbergensis* were well suited to long-distance bipedal walking. Trinkaus and Hilton [1996] remarked that in the robust foot bones of Neanderthals, the middle three toes have the widest proximal phalangeal shafts, suggesting higher mediolateral loading on these phalanges due to greater locomotor activity, reiterating the idea that this might have been due to habitual locomotion over irregular terrain. They also suggested that it could result from inefficient carrying techniques or bracing of the body during stationary activities such as hunting with thrusting spears [Trinkaus and Hilton, 1996].

Bramble and Lieberman [2004] addressed the origins of long-distance running ability observed in some modern humans. They speculated that such behavior might have begun in early *Homo* but were unable to identify the reproductive advantages that might account for it. There seemed to be no advantage, since walking is easier, safer and less costly. Endurance running might have been used for hunting or scavenging, yet it seems the energy cost would have outweighed the benefits [Bramble and Lieberman, 2004].

Later, these authors suggested that long-distance running could have had net benefits if used for persistence hunting during midday heat to drive animals into hyperthermia and exhaustion so they could be killed with a spear or club [Lieberman, et al., 2009]. Men, women, and children who were good endurance runners would have been effective persistence hunters with little risk. This type of hunting is demanding not just because of its energy cost. It is also metabolically and physically more stressful than other methods, and requires large quantities of water for temperature regulation, dietary sources of salt, and high levels of glycogen and triglycerides. Furthermore, it is suited primarily to hot, open habitats such as short grass savannas, semi-desert or scrubland and it depends upon a cooperative social system and food-sharing [Lieberman, et al., 2009; Lieberman, 2012b].

This scenario has its critics: African environments occupied by hominins during the transition to *Homo* were savanna-woodlands, a difficult habitat in which to practice persistence hunting. Ethnographically it is a rare hunting method, attempted only in hot, open habitats [Pickering and Bunn, 2007]. It would be extremely costly energetically [Steudel-Numbers and Wall-Scheffler, 2009]. Even if used by walking rather than running, it requires advanced tracking skills with an added risk of dehydration and heat exhaustion, and it demands locomotive efficiency, sweating rates and areas of hairless skin similar to modern humans [Ruxton and Wilkinson, 2011]. Because it would be inappropriate in cold climates, it would have been of little use to Neanderthals [Raichlin, et al., 2011]. Their robusticity might be due to habitual locomotion over irregular terrain, but probably not for persistence hunting.

Modern endurance runners have reduced musculoskeletal robusticity. The body structure of athletes who specialize in endurance running conflicts with the idea that this activity can account for the robusticity of the hominin lineage. Elite distance runners are slight of build. In 2005, the top-rated male long-distance runners in the world (5,000 m, 10,000 m, marathon) had an average height of 169.2 cm and an average weight of 56.3 kg [T & F News Panel, 2006], significantly shorter and less massive than the average US male (20-39 age group), currently about 176.4 cm and 84.7 kg [McDowell, et al., 2005]. Short-distance runners (sprinters), who depend upon a brief, explosive burst of energy, are physically larger, more massive and muscular than endurance runners [Khosla, 1978; Weyand and Davis, 2005]. A short sprint requires more muscle to generate force and additional tendon and bone to transmit it safely [Weyand and Davis, 2005]. Among track and field athletes, the lowest muscle mass is found in distance runners, the highest in power throwers [Spenst, et al., 1993]. Body builders and weight lifters have more bone mass and strength. Endurance runners have less bone, weaker bones and are the least massive among runners [Frost, 1997; Weyand and

Davis, 2005]. This evidence seems incompatible with the idea that hominin robusticity was the result of postnatal, prolonged, endurance running.

An alternative explanation based on natural selection and differential reproduction will now be presented: hominin musculoskeletal robusticity was due to natural selection acting to improve throwing and striking, due to the reproductive benefits it provided.

The role of genetics. Some explanations have rejected a genetic element in hominin robusticity, but evidence of the skeletal trait in hominin children renders this doubtful. Klein [1999] thought it was compatible with genetic inheritance. This view is consistent with the presence of skeletal robusticity in all hominins of both sexes in the fossil record until 30,000 years ago. If a significant genetic component is involved, then an evolutionary explanation based on reproductive benefits is required.

Modern professional throwers and strikers are physically robust. As noted above, power throwers are large and powerfully built. High performance in throwing events (discus, hammer, shotput) is associated with muscular dynamic strength and peak power in both upper and lower limbs [Bourdin, et al., 2010]. Even handball throwers throw with greater velocity when they are muscular [Chelly, et al., 2010]. The correlation of robusticity with throwing prowess is well documented by analysis of height and weight statistics for athletes that are specialized in throwing and striking, epitomized by major league baseball players.

Gould [1996] was the first to study height and weight in major league baseball players. From the 1870's through 1909, the average height and weight of batters was 179.5 cm and 77.1 kg; pitchers, during the same period, averaged 179.5 cm and 78.0 kg. During the next four decades batters averaged 182.1 cm in height and 79.4 kg in weight while pitchers averaged 184.6 cm and 83.0 kg. From 1950 through the 1980s, batters averaged 184.6 cm and 86.2 kg; pitchers, 189.7 cm and 86.2 kg. Gould [1996] linked the increase in body size to an improvement in pitching and clubbing prowess.

All of these heights (from 179.5 cm to 189.7 cm) and weights from (77.1 kg to 86.2 kg) exceed the average height and weight of US males, age 20-39, which in the early 1960's were 176.5 cm and 75.3 kg. Between then and 2002 US males in this age group remained the same height but they gained more than 9 kg [Ogden, et al., 2005; McDowell, et al., 2005], an increase believed to be due primarily to adipose tissue. Even with this added bulk in the general population, major league clubbers and throwers are still more massive. (Note: the time-related increase in size of the professionals is due to "artificial" selection based on prowess).

A comparable body of statistics, restricted to the highest performing throwers and strikers in major league baseball yields a similar result [Baseball.com, 2009]. The sample is limited to pitchers and batters who won the Most Valuable Player Award beginning in 1911, and the Cy Young (pitching) Award which began in 1956. (Individuals who were multiple winners are entered only once in the following calculations).

Throwers. Pitchers who made their major league debuts from 1907 to 1941 and won these awards had an average height of 187.2 cm. This exceeds the mean height of recent (2002) US males by 10.7 cm. Their average weight of 86.3 kg exceeds the average weight of US males 20-39 years by 1.6 kg and the average weight of US males of that age in 1960-1962 by 10.8 kg. The average height and weight of elite major league baseball pitchers continued to rise. From 1975 to 1999 the average height was 192.3 cm and the average weight was 89.9 kg, which is 15.8 cm taller and 5.2 kg heavier than US males 20-39 years of age in 2002 (data from [Baseball.com, 2009]). In 2012, the starting pitchers on the New York Yankees team averaged 196.9 cm in stature and 111.4 kg in mass [Waldstein, 2012].

Clubbers. A similar result is obtained for Most Valuable Players who were honored primarily for their striking prowess. From 1911 to 1941 their average height was 184.6 cm, and their weight was 82.5 kg. They were 8.1 cm taller and 7.2 kg heavier than the average US male in 1960-1962. The size of these elite clubbers continued to increase in subsequent years. Those honored from 1975 to 1999 had an average stature of 187.2 cm and weight of 91.7 kg. This exceeded 2002 average heights and weights of US males, 20-39 years, by 10.7 cm and 7.0 kg. It is clear that elite throwers and club-swingers are larger than the average male in their age group.

Research shows that significant relationships exist among height, body mass, lean body mass, rotational power, rotational strength, lower body power, upper body power, torso rotational power, upper and lower body strength, angular hip velocity and angular shoulder velocity to linear bat velocity in adolescent, high-school and college baseball players [Szymanski, et al., 2010]. Those who are taller and have greater lean body mass are able to generate more force at impact than shorter players with less lean body mass [Szymanski, et al., 2010]. Those who swing their clubs with greater velocity are tall, muscular, powerful individuals.

Physical principles. The large stature and muscularity observed in elite throwers and club-swingers is consistent with fundamental physical principles. Greater height and limb-length associated with throwing and clubbing prowess reflects the greater acceleration that can be achieved with longer lever lengths. In the throwing or clubbing motion many segments of the body are brought into the movement in sequence, giving a whiplike action at the distal end of the lever system. "The essential mechanism of throwing is that momentum is transferred from the body to the object thrown...[It is] transferred from point to point in a flexible system like a whip" [Toyoshima and Miyashita, 1973, p. 86]. Greater muscular strength increases the force that can be applied to the missile or club, makes it possible to increase the velocity and kinetic energy of the throw or clubswing, and facilitates use of weapons with greater mass. The force is derived from muscular contraction. The more highly developed the muscle fibers, the more force (hence the muscular physique of elite throwers and clubbers), the greater mass of the weapon that can be used, the higher velocity to which it can be accelerated, and the more damage it can do when the momentum is transferred to a target.

This rule applies whether the momentum is transmitted indirectly in throwing (the hand no longer exerts force on the missile after its release) or directly in clubbing (when the hand(s) continue to exert force on the weapon during impact). In striking, the object absorbing the blow exerts a backforce.

This requires a strong prehensile action to avoid loss of the club, which is consistent with the evolutionary explanation of the hominin clubbing grip (Chapter 9). As the mass of the hand-held weapon increases, the velocity to which it can be accelerated decreases because of the increased inertia at the distal end of the forearm [Toyoshima and Miyashita, 1973; Cross, 2004]. Nevertheless, this velocity is greater in large, muscular individuals and (in throwing) the motion itself does not change: a force-velocity curve is obtained for a single, synergistic, musculoskeletal system [van den Tillar and Ettema, 2004].

The role of physique. The foregoing evidence indicates that male long-distance runners are smaller in height and weight than the average male of similar age. Their bones and muscles are less robust, despite the long hours they devote to training and racing. Their ectomorphic body type with low mass requires less energy for its transport. In contrast, elite male throwers and club-swingers are larger than the average male of similar age. Their large-sized mesomorphic body type is advantageous in this behavior. A robust, muscular body maximizes the ability to generate an explosive burst of energy. To increase muscle mass, baseball players undergo weight training and some have used steroids to promote muscular hypertrophy.

Trinkhaus and Shipman [1993] remark that the Neanderthals continued behavioral trends that were well established among their predecessors: "From the robust dimensions of their limb bones, which had massive shafts and large joints, to the pronounced bony crests and sturdy ridges where brawny muscles attached—on their necks, backs, shoulders, arms, hands, legs, and feet—the primary message bespoken by Neandertal anatomy is 'power' " [p.417]. I propose that this physique had been selected to enhance the power of throwing and club-swinging. The association between large size, muscularity, and robust bones with bipedal use of missiles and clubs supports the argument that natural selection acted to produce these body features because it improved prowess in these behaviors and thereby augmented reproductive success.

Hominin robusticity is detectable as early as 6.0 Mya. This implies an onset of natural selection for throwing and striking that is at or near the time of hominin origins. Subsequently, robusticity was maintained by natural selection for several million years. The entire hominin lineage preceding late *H. sapiens* apparently was engaged in a behavior in which brute force yielded reproductive benefits. Throwing and striking behavior is consistent with this evidence.

Different selective factors acted to optimize locomotor efficiency on the one hand and bipedal use of weapons on the other. Selection for efficient long-distance walking or running would have favored smaller hominins of lower body mass, like modern long-distance runners. Selection for powerful throwing and club-swinging would have promoted a larger body size with greater musculoskeletal robusticity. Since these behaviors seem to have evolved simultaneously (Chapter 14), the reproductive advantages provided by each of them must have constrained the other to some extent. Natural selection would have been determined by the net reproductive advantage associated with any genetic variation.

Bone and muscle are expensive tissues. Trinkaus [1986] noted the cost to Neanderthals of developing robust bones and muscles in children and maintaining them in adults. The expense is both energetic (calories) and nutritional, since most of the molecular components of these tissues are continually renewed. Selection for improved throwing and striking prowess that involved augmenting muscular strength and robust bones must have been sufficiently important for reproductive success to compensate for the additional dietary requirements and the detrimental effects of increased body mass on locomotor efficiency.

Why did robusticity decline? What accounts for the "gracilization" that occurred in our species beginning a relatively short time ago? [Ruff, et al., 1993; Trinkaus, 1997]. What were the reproductive benefits?

Kirschmann suggests that robusticity is correlated with mating competition. He concluded [1999, 6.6, 6.7] that food-sharing, lifelong pair-bonding, increased cooperation and decreased male fighting arose as part of culturally determined behaviors in *Homo sapiens* about 70,000 years ago. This innovative social behavior could have fostered natural selection that reduced physical robusticity and sexual size dimorphism, yielding nutritional and reproductive benefits by cost reduction.

Loss of robusticity is not predicted by the bipedal use of weapons proposal. Some other explanation is required. I suggest a hypothesis involving musculoskeletal robusticity, dietary requirements, and locomotor efficiency which skirts the social behavior issue. The hefty body type, which distinguished pre-*sapiens* hominins, required substantial nutritional intake to grow and maintain the expensive tissues that characterized it, and hampered bipedal locomotion because of the added costs of transporting the bulky mass (as noted above). Perhaps the evolution of a more gracile body form indicates that the balance of natural selection had shifted toward decreasing nutritional needs and improving the economy of locomotion. Did the diminution of muscularity lower the ability to exert force on a hand-held weapon, thereby reducing throwing and clubbing prowess? If so, the elite throwers and club-swingers of the modern era could represent a dwindling remnant of a bygone age of more powerful athleticism of this sort. Alternatively, prowess might have continued despite gracilization due to neural rewiring of the brain that enhanced such behavior [Young, 2010]. Maybe female choice was involved. Clearly, the decline of robusticity has yet to be satisfactorily explained.

Another unique human characteristic. Not only is the bipedal use of hand-held weapons a unique feature of hominin evolution that can be demonstrated in modern humans, the biomechanical means by which we perform this complicated and exceptionally rapid act is also unique. Alluded to in several earlier chapters and in this one, it is examined in detail in Chapter 13.

CHAPTER 13

Development and Maturation of Throwing and Clubbing

"A step in the direction of the ball, a forward rotation of the hips and spine, and the uncocking of the arms and wrists. In essence, this is the mature pattern for all striking skills" [Wickstrom, 1977, p. 163].

"...both striking and throwing represent tasks with similar purposes... that can be performed using similar biomechanical actions" [Langendorfer, 1987, p. 45].

[In children] *"The magnitude of the [gender] difference in throwing is remarkable and is out of proportion to those in any other physical measure"* [Espenschade, 1960, p. 432].

Introduction. Is throwing and striking behavior inherited or learned? Is it based upon an innate ontogenetic process or is it a cultural product dependent upon teaching or other societal factors? This question, recently examined for the first time [Young, 2009, 2010], will be further explored in this chapter. The conclusion is reached that throwing and club-swinging behavior in humans is dominated by a genetic element. Teaching, learning, practice and various other societal influences may play a role, but they are less significant than the influence of biological inheritance.

There is an inherited motor pattern that emerges shortly

after birth in all human children, yields a single type of bipedal throwing motion, is characterized by a gender disparity (boys have a significant advantage), develops with increasing age, involves remarkable complexity and rapidity, and in young adults reaches an exceptionally high level in a small number of elite athletes, most of whom are males. The same pattern is seen during the ontogeny and maturation of striking. The two behaviors share a sequence of similar biomechanical movements involving the entire body in a motion which begins in the lower extremities and generates a growing pulse of kinetic energy that is ultimately transmitted to a hand-held implement. The same throwing and striking motions develop without teaching in all ethnic groups and cultures. In short, all humans throw and swing clubs the same way as an outcome of a universal, human ontogenetic program. This evidence requires an *evolutionary explanation*.

Mayr [1961] called an evolutionary explanation the "ultimate cause" of a heritable behavior because it reflects the history of natural selection of the set of genes that underlie the behavior. The "proximate cause" (in an individual) is the interaction of these selected genes with environmental factors to yield mature individuals with an ability to throw a projectile or swing a club from a bipedal stance.

The evolutionary explanation to be proposed will be familiar to the reader who has proceeded this far: It is based on the concept that throwing and clubbing behavior was naturally selected because it yielded reproductive benefits to our ancient ancestors.

The development of throwing in young children. This research field was originated over seventy years ago by Wild [1938], whose pioneering investigations had a great and enduring impact on the field of science it created [Wickstrom 1977]. Wild filmed 32 right-handed children from Iowa between the ages of 2 and 12 years while they performed overhand throws. By analysis of the extensive movement and timing data she collected, Wild reduced throwing development to four age-related stages.

In Stage I, children 2-3 years old threw from an upright stance with a simple arm movement. The flexed arm was first moved high above the shoulder with trunk extended and the body facing forward. Then the trunk flexed forward as the elbow extended and the arm swung over the shoulder and down in front. In Stage II, ages 3.5-5 yrs, a horizontal rotation was introduced. The feet remained together in place, but now the body rotated right, then back to the left as the right arm extended and swung forward. Stage III (ages 5-6) added a step forward to the motion, although sometimes it was the right foot that moved forward with a right-handed throw. Stage IV was recorded in all boys 6.5 yrs and older but few girls. It involved a weight shift to the right foot as the trunk rotated rightward and the throwing arm was swung upward and back, followed by a step forward by the left foot. This forward weight shift was accompanied by forward trunk rotation, horizontal adduction of the throwing arm and extension at the elbow. The girls had fallen behind. Most of them displayed the body and foot movements, but had less advanced arm movements, and some had regressed to earlier stages.

In other early research, H. Halverson [1940] found that children first throw from a sitting posture before being able to stand, then throw bipedally when they are able to maintain upright balance. Boys outperform girls in throwing delivery, accuracy and distance by 3.5-5 years.

These pioneering reports showed that throwing behavior begins shortly after birth and then becomes more complex as it develops into a rudimentary version of the motion employed by mature athletes (Figure 11, p. 188). Boys outperform girls in maturity of the throwing motion and the distance (or velocity) of the throw.

Confirmation of early studies. The emergence of a standardized overarm throwing motion and a notable gender difference in its development proved to be a feature of child development wherever it was studied, including Japan [Sakurai and Miyashita 1983], Mexico [Malina and Buschang, 1985], Nigeria [Toriola and Igbokwe, 1986], New Guinea [Malina, et al., 1987], Senegal [Bénéfice and Malina, 1996; Bénéfice, et al., 1999], Tasmania [Cooley, et al., 1997], England [Marques-Bruna and Grimshaw, 1997], Brazil [Teixeira and Gasparetto, 2002], Germany [Ehl, et al., 2005], Australia [Thomas, et al., 2010], and many samples of children in the United States [Keogh, 1965; Wickstrom, 1977; Morris, et al., 1982; Malina, et al., 1987, and others cited below].

Very young children, under the age of 3, and preschoolers, ages 3-5 yrs from diverse cultures and ethnic groups are all able to throw. Even undernourished children, with reduced body size that adversely affects their throwing performance, show an age-related increase in skill and a gender disparity [Malina and Buschang, 1985; Malina, et al., 1987; Bénéfice and Malina, 1996; Bénéfice, et al., 1999]. They all throw with the same throwing motion, which develops in the same developmental sequence. There are no reports of unusual throwing styles.

Development of throwing skill is variable. Roberton [1977] initiated the analysis of different body components during childhood development of throwing. This approach revealed variation of individual pathways [Roberton and Langendorfer 1980]. It also disclosed refinements in the throwing motion that appear late in development, including rotation of hips before shoulders, shoulder motion before humerus, humerus before the forearm [Halverson, et al., 1982]. In some individuals, development stops before these late refinements are installed [Leme and Shambes, 1978; Atwater, 1979; Williams, et al., 1998]. This occurs more frequently in girls.

The developmental gender difference in throwing. From infancy through childhood the average performance of boys in throwing (velocity and distance) exceeds that of girls [Wickstrom, 1977]. This disparity surpasses in magnitude the gender difference in any other motor skill [Espenschade, 1960]. Gender differences detected at age 3 are already three times larger than in other motor behaviors [Roberton and Langendorfer, 1980; Thomas and French, 1985]. Keogh [1965] found that boys age 8 could throw farther than girls age 11. Boys in first grade threw faster and farther than girls in grade 4 [Rippee, et al., 1990] and boys in grade 2 had a more advanced throwing motion than girls in grade 5 [Butterfield and Loovis, 1993].

In a 1979 investigation of 13-year-old children from Wisconsin, boys threw on average 6.6 m/s faster than the girls [Halverson, et al., 1982]. In 1999, the same research team [Runion, et al., 2003] studied a similar sample from Ohio. The ball velocities and between-sex differences had not changed during 20 years. Next, the group applied the study protocol to 13-year-old German youths with different cultural influences [Ehl, et al., 2005]. (In Germany, many more youths play soccer than any throwing sport, a contrast to the emphasis on baseball and softball in the US). Although the US youths threw faster, the disparity between boys and girls in throwing velocity was identical in Germany and the US (6.9 m/sec)—comparable to the two US cohorts separated by 20 years.

Petranek and Barton [2011] studied a group of 38 girls (mean age 13.7 years) from a US softball league. Most of the girls (90%) practiced three or more times per week; many (83%) had been playing for 6-9 years [Petranek and Barton 2011]. Compared to girls of the same age with less experience from the US [Runion, et al., 2003] and Germany [Ehl, et al., 2005], they threw with greater velocity (2.8 m/s and 4.8 m/s faster, respectively), but the boys from the US and Germany still threw with more velocity than the more experienced softball players (4.1 m/s and 2.2 m/s, respectively) [Petranek and Barton, 2011]. The results are not directly comparable, due to differences in methods (in the early studies children were randomly selected from schools and threw tennis balls with much less mass than softballs), but they call attention to the continuing gender gap in throwing at about the time of puberty.

Throwing *accuracy* (distinct from velocity or distance thrown) has seldom been analyzed, but limited evidence suggests a male advantage in very young children. Keogh [1965] studied accuracy of the throw in 337 children ages 7-9 years. He recorded an improvement with age and a superiority of boys over girls at each age, even though boys threw from 1.5 m farther from the target. Wickstrom [1977] found a similar annual improvement in throwing accuracy among 960 children 6-9 years of age. Boys were more accurate than the girls at each age. Other studies have also found that boys outperform girls in tests of accuracy [Hicks, 1958; Thomas and French, 1985; Rippee, et al., 1990].

What causes childhood gender differences in throwing? There is no gender disparity in either static or dynamic balance [Keogh, 1965; Morris, et al., 1982; Thomas and French, 1985; Ulrich and Ulrich, 1985; Toriola and Igbokwe, 1986; Butterfield and Loovis, 1993, 1998]. Physical factors are not likely to be involved. Gender sameness prior to puberty is the rule in body type, composition, strength, limb lengths, height and weight [Keogh, 1965; Thomas and French, 1985; Bénéfice, et al., 1999; Thomas, 2000; K. Thomas, et al., 2001]. Nor is there any known physiological, anatomical or maturational basis to account for the gender difference [Nelson, et al., 1986, 1991; Butterfield and Loovis, 1993, 1998]. It is not until puberty that numerous changes occur which accentuate the pre-existing male advantage (Chapter 5).

The notion that societal factors are important has often been proposed. Gender inequalities might be due to the different treatment society has for boys and girls, induced by parents, peers, teachers and coaches, rearing factors, cultural expectations, experiential differences, motivation, effort, encouragement and opportunities for training and skill development [Espenschade, 1960; Sakurai and Miyashita, 1983; Ulrich and Ulrich, 1985; Thomas and French, 1985; Thomas and Thomas, 1988; Nelson, et al., 1991; Thomas, et al., 1994; Butterfield and Loovis, 1998; K. Thomas, et al., 2001]. Practice is considered essential to improving complex motor skills. In several studies boys reported more throwing than did girls [Halverson et al., 1982; Butterfield and Loovis, 1998; Runion et al., 2003; Ehl, et al., 2005], yet in others, girls participated in youth sports at a rate similar to boys [Butterfield and Loovis, 1993] and more advanced throwing patterns occurred in children of both genders who participated in throwing sports [Butterfield, et al., 2003].

Instruction seems to make little difference in outcome scores such as distance or release velocity [Dusenberry, 1952]. Halverson and colleagues [1977] studied children in kindergarten. Eight weeks of guided practice in the overhand throw did not significantly change ball velocities of those who received it compared to those without it [Halverson and Roberton, 1979]. Similarly, when young children were trained in throwing for 6 weeks, girls learned to mimic some aspects of a more advanced motion but without increasing the distance they could throw [Thomas, et al., 1994]. Children who completed a teaching program in kindergarten showed no evidence of learning from instruction when re-examined at the close of first and second grade [Roberton, et al., 1979].

Do young girls trained in throwing "catch up" with the boys? Thomas and coworkers [1994], who reviewed investigations that dealt with this question in children ages 5-8 yrs, found there was no catch-up effect for girls. Gender differences appear to be resistant to instructional training in young children [Thomas, 2000], although practice likely assists throwing development [Langendorfer and Roberton 2002]. Throwing at later ages may be improved by practice, strength and flexibility training, as well as slight adjustments of mechanical factors under the guidance of knowledgeable instructors [Stodden, et al., 2005; Escamilla, et al., 2010]. As much as 5% of the velocity difference between elite Korean and American adult throwers might be due to such factors [Escamilla, et al., 2002]. However, evidence is lacking that such interventions affect the development of this skill in young children.

Two motions or one? Thomas and Marzke's study of two adult throwers [1992] implied that there might be two distinct, gender-related, throwing motions [K. Thomas, et al., 2001]. An alternative view, that in children there is a single motion which does not develop as rapidly or completely in girls, is supported by much evidence. Girls appear to follow the same sequence of development but tend to lag behind boys in its progress [Halverson, et al., 1982; Nelson, et al., 1986, 1991; Roberton and Langendorfer, 1980]. In one investigation, girls' throwing development trailed that of boys, but increased at the same rate [Butterfield, et al., 2003]. In most studies, however, girls fall farther behind each year [Espenschade, 1960; Keogh, 1965; Roberton, et al., 1979; Halverson, et al., 1982; Seefeldt and Haubenstricker, 1982; Thomas and French, 1985; Roberton and Konczak, 2001; Teixeira and Gasparetto, 2002] and their progress may terminate earlier [Sakurai and Miyashita, 1983; Rippee, et al., 1990; Nelson, et al., 1991; Butterfield and Loovis, 1998; Roberton and Konczak, 2001].

The throwing motion that develops in children is a precursor of the one used by mature athletes. Investigation of throwing development by analysis of the movements of body components has been integrated with analysis of throwing biomechanics in adult athletes by a team of researchers from both fields who studied 34 boys and 15 girls, 3-15 years of age [Stodden, et al., 2006a, b]. Kinematic variables and developmental movement patterns (step, trunk, humerus and forearm actions) were positively correlated. Kinematics correctly predicted developmental levels 85% to 96% of the time. As developmental levels advanced, body movements more closely resembled those characteristics of elite throwers [Stodden, et al., 2006a, b]. By combining the genders, the researchers implicitly supported the view that boys and girls follow the same developmental program, while the tabulated results (in which girls predominated at lower developmental levels and boys at higher ones) illustrated the gender disparity.

Some children, mostly boys, complete the full developmental sequence at an early age (9-10 yrs). In studies of US throwers, Cosgarea, et al. [1993] analyzed male baseball pitchers in four groups (9-12 yrs, 13-16 yrs, collegiate and professional). The kinematics of throwing were similar in all groups, although ball velocity increased with level of competition, likely due to increasing size and muscle strength. Fleisig and coworkers [1999] studied groups of male baseball pitchers from youth (10-15 yrs), high school, college and professional levels. Again, velocity of the throw increased at each stage, but there was no difference in the throwing motion. Subsequent studies of male pitchers ages 9-12 yrs [Fleisig et al., 2006] also support these conclusions. The velocity of the throw issuing from the same motion increases as male throwers gain size and muscularity during puberty.

The throwing motion in adult athletes. The following summary of the motion used by adult, skilled throwers is based importantly on research at the American Sports Medicine Institute under the leadership of Glenn Fleisig. Unless otherwise indicated, the description refers to baseball pitching.

In expert throwers, movement begins with a runup (javelin) or windup (baseball pitching). Both actions add to the forward momentum of the throw. The right-handed baseball pitch (a mirror image of the left-handed version) begins with the head facing the target, body turned to the right and leaning rearward, left leg raised, throwing arm directed backward with elbow extended (Figure 11). The partially flexed right leg then suddenly extends, thrusting the foot backward and downward into the ground, producing a reaction force and a forward weight shift. The left knee extends and the thigh rotates externally so that the foot points toward the target. As the foot strikes the ground, the knee flexes slightly, then extends, pushing backward, converting forward motion to rotatory motion around the long axis of the body [Mero, et al., 1994; Escamilla, et al., 2002].

Backward force against the left side of the pelvis causes the right side (thrust forward by the right leg) to accelerate, reinforced by the medial rotators of the hip joint. Sudden forward rotation of the right side of the hip produces a powerful stretch on the trunk. Now hip rotation decelerates and the trunk begins to rotate forward, aided by contraction of back and abdominal muscles [Watkins, et al., 1989; Hirashima, et al., 2002]. The left arm flexes at the elbow and moves closer to the body, accelerating trunk rotation [Ishida and Hirano, 2004]. Suddenly, hip rotation is blocked facing the target, arrested by the external rotators at the hip joint. This transfers angular momentum to the thorax, the right side of which swings forward with increasing velocity. When the shoulders near a position facing the target, rotation of the left side is blocked, producing an acceleration of the right side.

The arm, abducted and extended, now undergoes elbow flexion to about 90°. After the right shoulder has attained maximum velocity, movement of the right side of the torso decelerates, and the humerus rotates externally so that the hand gripping the ball is directed backwards. Just before maximum external rotation is attained (about 180°), elbow extension begins, then a powerful torque of internal rotation is applied to the humerus which whips the forearm forward [Dillman et al., 1993; Werner, et al., 1993; Matsuo, et al., 2001; van den Tillar and Ettema, 2004]. The hand lags behind, extending the wrist joint. Forward motion of the upper arm decelerates, accelerating the forearm and hand. When internal rotation of the humerus

CHAPTER 13: DEVELOPMENT & MATURATION OF THROWING AND CLUBBING

Figure 11. Illustration of the throwing motion in an adult, skilled thrower (above) and a four-year-old boy, drawn to the same height [from Young, 2009, 2010]. Without prior instruction or training, the child uses a rudimentary version of the same throwing motion used by the adult athlete. The windup, step with weight shift (usually first seen in boys 5-6 yrs old), hip and torso rotation, forearm lag, elbow extension, overarm release and follow-through are all present. Throwing behavior appears in all normal children very early in life without teaching. It is inherited, but its development is variable. Only a small proportion of children, primarily males, advance to the top level of adult throwing capacity.

---

reaches about 90°, the wrist flexes from extension to neutral, imparting additional thrust. The thumb drops away and the fingers extend at the moment of release [Tullos and King, 1973; Sing, 1984; MacWilliams, et al., 1998; Wilk, et al., 2000; Matsuo, et al., 2001; Stodden, et al., 2001; Escamilla, et al., 2002].

During the acceleration phase of the throw, when the

hand-held missile begins its forward motion until it is released, the humerus is held in a position of 90° abduction, in line with the plane of the shoulders. It rotates medially about its long axis at an extremely high rate, but its orientation in the glenoid fossa does not change. The alignment of the humerus with the shoulders and with the forearm and fingers at missile release assures the greatest possible biomechanical lever-arm action for transfer of the maximum amount of energy to the hand-held weapon. Whether a missile is thrown or a club is swung "overarm", "sidearm" or somewhere in between, the upper body is always tilted to maintain the humerus in a lateral orientation in line with the shoulders [Escamilla, et al., 1998; Escamilla and Andrews, 2009] (Figure 12).

Escamilla and colleagues [2001] investigated 21 kinetic parameters in Olympic-level baseball pitchers from eight countries. The throwing motion was similar in all of them. Coaching, training and anthropometric features had no effect. The same result was obtained in a comparison of professional baseball pitchers from the US and Korea [Escamilla, et al., 2002] and in novice, club and elite javelin throwers [Bartlett, et al., 1996]. The throwing motion is basically the same in throwers of various standards and from different countries. The throwing motion which develops in all children progressively approximates the motion of all adult throwers world-wide.

The rapidity of the throwing motion. Throwing is an explosive act. In baseball pitchers, after the windup, the time between front foot contact and ball release (Figure 11) is 145 msec [Stodden, et al., 2005]. In a javelin throw the duration from final foot contact to release is similar [Mero, et al., 1994; Morriss and Bartlett, 1996]. The acceleration phase lasts only 30-50 ms [Wilk, et al., 2000; Werner, et al., 2001; Stodden, et al., 2005]. During this exceedingly brief interval, numerous independent muscle groups are successively activated and deactivated in a precisely coordinated manner under the control of the central nervous system (Chapter 15). This sequence is one of the most rapid human motions ever documented [Fleisig, et al., 1995, 1996]. It seems likely its rapidity is a unique modern human behavioral characteristic.

CHAPTER 13: DEVELOPMENT & MATURATION OF THROWING AND CLUBBING

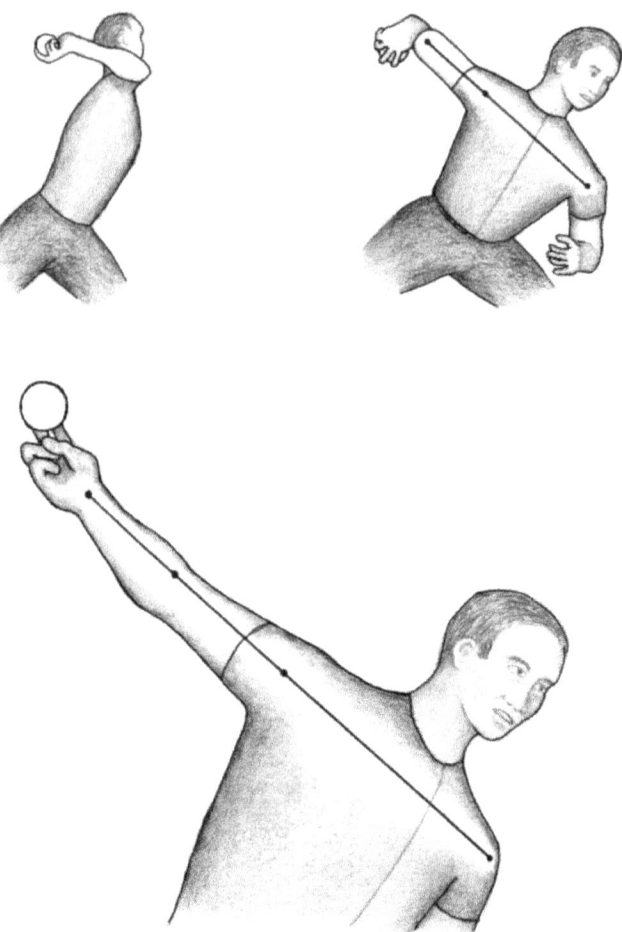

Figure 12. During the transition from *Australopithecus* to *Homo* the shoulder joint changed from an upward to a lateral orientation. Evidently, natural selection acted to improve a behavior in which the humerus was oriented *laterally*, abducted at right angles to the vertical axis of the upper body and aligned with the plane of the shoulders. The loss of vertical orientation can be explained as a shift from an arboreal adaptation to life on the ground. The terrestrial behavior which was enhanced in the horizontal orientation of the shoulder joint can be attributed to selection for improved throwing and club-swinging which yielded reproductive benefits (Chapter 6, 7). *(Caption continued on the next page)*

In a throw (illustrated), starting when the hand-held missile begins its forward acceleration (top images) until it is released, the humerus is maintained in a position of 90° abduction, in line with the plane of the shoulders. It rotates medially about its long axis at a rapid rate, but its orientation in the glenoid fossa does not change. Alignment of the humerus with the shoulders, forearm and fingers at missile release (lower image) augments biomechanical lever-arm-action for maximal transfer of energy to the hand-held weapon. Whether a missile is thrown or a club is swung "overarm", "sidearm" or in some intermediate position, the upper body is always tilted to maintain the humerus in a lateral orientation in line with the shoulders.

Injury-prevention mechanisms. The throwing motion is so swift and violent that innate mechanisms are incorporated into the throw to minimize damage. As the arm accelerates, internal rotation of the humerus occurs with such high velocity that a compensatory axial compression which drives the humerus into the shoulder joint is exerted, preventing distraction of the arm and doubling the stress that can be applied to the humerus without fracture [Gainor, et al., 1980; Fleisig, et al., 1995; Escamilla and Andrews, 2009]. Innate mechanisms also protect the elbow. During acceleration, as shoulder muscles bring the humerus forward, the forearm and hand lag behind, putting stress on the medial side of the elbow joint. This is counteracted by an inward torque applied by forearm muscles to protect the ulnar collateral ligament [Tullos and King, 1973; Werner, et al., 1993; Fleisig, et al., 1995]. Then, at ball release, a powerful proximal force is produced at the elbow joint by flexor muscles, which acts to prevent distraction from centrifugal force acting on the forearm [Fleisig, et al., 1995; Loftice, et al., 2004].

The importance of stride length. In Chapter 7, I proposed that the increase in leg length and muscularity that characterized the transformation of *Australopithecus* to *Homo* may have been influenced significantly by natural selection for improved throwing and club-swinging. The advantages of longer, more muscular legs in these behaviors were addressed in Chapter 12: A long, powerful stride by robust and muscular legs can impart greater velocity to a hand-held weapon. In the following, the focus will be on the importance of stride length in throwing.

The effect is already demonstrable in young children. Roberton [1977] found that when a forward stride was added to the developing throwing motion it enhanced trunk rotation that transmits energy from the pelvis to the shoulders. In older children (13 years) Roberton and Konczak [2001] observed that a longer stride length was associated with throws of higher velocities. The longer stride, they noted, lengthens the time for push-off from the rear foot.

In adult baseball pitchers, wrist velocity correlates highly with increased leg drive [MacWilliams, et al., 1998]. The forward thrust by the rear foot initiates the forward momentum of the entire body. More kinetic energy is produced by greater magnitude of the thrust in the direction of the pitch. The longer the stride length, the longer the acceleration path through which the thrust of the rear leg can act. Thus, a lengthened stride augments forward linear velocity [Escamilla, et al., 1998] which is then converted to rotational velocity of the pelvis. In addition to thrust from the rear leg, gravity from the raised stride leg may act to increase the stride [Fleisig, et al., 1996]. The vertical vector of the push-off generates potential energy, which can be transformed into kinetic energy later in the throw. The landing leg anchors the transformation of forward and vertical momentum into rotational components [MacWilliams, et al., 1998]. Greater stride length and faster pelvis and upper torso rotation are associated with advanced developmental levels of stepping, trunk action, and faster throwing velocity [Toyoshima, et al., 1974; Kibler, 1995; Stodden, et al., 2006a].

In skilled male throwers, stride length is about 75% of body height [Dillman, et al., 1993; Fleisig, et al., 1996; Wilk, et al., 2000]. Some step towards the target a distance of up to 84% of their standing height [Escamilla, et al., 1998]. In skilled women throwers, stride length is 60% of standing height, whereas in average women throwers it is only 33% [Atwater, 1979]. Part of the gender disparity may be due to a difference in leg length and strength.

The development of striking in young children. Evidence of club-swinging behavior in children is consistent with that of throwing, including the gender difference. Sidearm striking in children under the age of 30 months resembles the early overarm throwing motion—a simple, forward arm thrust [Wickstrom, 1977]. When the stage of trunk rotation emerges, it develops with the same sequence as in throwing. It begins in the upper spine, then the pelvis and upper spine (thorax) begin to rotate as a unit, then pelvis rotation precedes the rotation of the spine [Halverson, et al., 1973]. Langendorfer [1987] found that when boys ranging in age from 1.3 to 10.3 years used a racquet to strike a ball suspended above head height, development of movement sequences in the pelvis, spine, trunk, humerus, and elbow were all comparable to those recorded in throwing development. After age 6 a weight shift and forward stride was commonly observed. The result by age 8 was an overhand striking motion essentially the same as that used in the mature throwing motion (Figure 13).

Overhead striking and throwing are performed using similar biomechanical actions [Langendorfer,1987]. Striking with a sidearm motion is also very similar (Figure 14).

Gender differences in development of striking. Seefeldt and Haubenstricker [1982] devised stages for the development in children of striking a ball in a sidearm swing with a bat. It begins with the feet stationary, arms rotating forward in a chopping motion. It culminates in a motion that incorporates a step forward, followed by rotation of hips, spine, and shoulders, elbow extension and hand pronation at ball contact. Boys passed through the intermediary stages more rapidly than girls. Gender differences were apparent by 2.7 yrs of age and the disparity increased in older children [Seefeldt and Haubenstricker, 1982]. Ulrich and Ulrich [1985], using a similar approach, observed that at 3-5 years of age girls lagged behind boys, comparable to the disparity in throwing in the same children. Development of sidearm striking in older children (ages 4-14 yrs) also was similar to overhand throwing in both its developmental sequence and gender disparity [Loovis and

CHAPTER 13: DEVELOPMENT & MATURATION OF THROWING AND CLUBBING

Figure 13. In this illustration of overhead striking by an 8-year-old boy [from Young, 2009, adapted from Langendorfer, 1987], it can be seen that apart from differences in handgrip and wrist movement the overhead striking motion is similar to the overhand throwing motion depicted in Figure 11.

Butterfield, 1995; Butterfield and Loovis, 1998]. The earliest stage was confined to arm action. Later developments, such as a forward step, a weight shift to the front foot and separation of the rotation of pelvis and trunk were displayed by many boys in kindergarten. In girls, development was slower. Tasmanian schoolgirls also scored lower than boys on tests of striking skill [Cooley, 1997].

The club-swinging motion in adult athletes. The mature striking motion is similar in many respects to the throwing motion [Young, 2009] (compare Figures 11 and 14). In the right-handed swing of a baseball bat, the batter stands with eyes on the target, body facing at right angles to it. Both hands grip the club and the arms extend it backward. The left heel is lifted as the left knee turns inward, redistributing weight to the right foot, which is then thrust into the ground, shifting weight forward. The left foot stays in place or strides forward as the thigh rotates outward. When the foot plants, the leg then extends at the knee, pushing the left hip backward, while the right side of the pelvis rotates forward, creating rotation around

CHAPTER 13: DEVELOPMENT & MATURATION OF THROWING AND CLUBBING

Figure 14. Drawing of the sidearm striking motion in an adult, skilled athlete (above) and a 4-year-old boy, drawn to the same height [adapted from Wickstrom, 1977]. The similarity of the motion in the child and the adult is part of the evidence that the complex throwing and club-swinging behaviors are due primarily to genetic inheritance.

the vertical axis. Next, as hip rotation is blocked, the shoulders rotate forward, leaving the arms behind, radially deviating the wrists and bat. As the shoulders turn towards the target, their movement decelerates rapidly as shoulder and arm muscles swing the upper limbs forward. The elbows extend, aligning arms and bat as the wrists move ulnarly, stabilized at impact to deliver maximum kinetic energy [Jorgensen, 1994; Welch, et al., 1995]. Because the bat is not released, the hitter counteracts the forces propelling the bat swiftly forward by leaning backward to maintain upright balance [Race, 1961]. The baseball club swing, like the throw, is very rapid. The time between onset of forward motion until ball contact is about 0.58 sec [Welch, et al., 1995].

The striking motion that develops in children is a precursor of the one used by mature athletes. This is evident in young children (Figure 14). Escamilla and coworkers [2009] studied 12 skilled youth (12-17 yr) and 12 skilled adult baseball players. The adults were on average 7.5 years older, 35-40% heavier and 5% taller, and they used bats that were 15% heavier and 5% longer. The larger and stronger adult hitters were able to generate greater angular velocities and bat speed compared with youth hitters, but analysis of linear and angular displacements indicated that hitting mechanics were similar in both groups. Earlier, Fleisig and colleagues [1999] had shown that in throws by skilled youths and adults most such parameters were similar between the two groups. The same pattern of body movement and kinetic energy transfer from larger, slower-moving segments to smaller, faster-moving segments was found in both youths and adults in clubbing as in throwing.

The kinetic energy linkage mechanism in throwing and striking. Both motions use similar sequences of muscular contractions which generate kinetic energy that is transferred through the body from the lower limbs to the hand-held implement. The two motions differ in wrist movements and handgrips associated with throwing the missile or swinging the club and in angular tilt of the torso to maintain balance after release of a projectile or while gripping a rapidly swinging club.

In both acts the forces of each contracting muscle group supplement those generated by muscles acting earlier and contribute to the acceleration of the next segment. Velocity can be raised by lengthening the acceleration path through which force is applied. This is the function of the windup, which coils the body before uncoiling it, activates stretch reflexes, optimizes length-tension conditions and stores recoverable elastic energy in tendons and ligaments [Matsuo, et al. 2001; Escamilla, et al., 2002]. The generation of kinetic energy begins with the thrust of the rear foot into the ground, creating a ground reaction force and initiating a weight shift towards the target. Lifting the front leg stores potential energy; extending it adds kinetic energy in the anterior direction. When the front foot plants and the knee extends, it produces a force couple—rear leg thrusting forward, front leg thrusting backward—accelerating the rotation of the pelvis, translating linear momentum into angular momentum. Then hip rotation is blocked, transferring momentum to the shoulders which, when their movement is terminated, pass the accumulating energy along the arm and hand to the sphere or club. As a general rule, each segment delays its motion until the maximum amount of energy can be transferred to it, linking the transfer to parts of smaller mass moving at higher velocities [Race, 1961; Welch, et al., 1995; Bartlett, et al., 1996; Fleisig, et al., 1996; Barrentine, et al., 1998; Wilk, et al., 2000; Marshall and Elliot, 2000; Kibler and McMullin, 2003; Hore, et al., 2005a; Stodden, et al., 2005; Young, 2009]. The fine structure of this innate mechanism is exceedingly complex (Chapter 15).

Gender differences in mature throwing and striking. Adult females can throw and swing clubs with skill far exceeding that of individuals of either gender in any other species of animal. As a rule, however, they cannot match the prowess of human males at the elite level. World records indicate that women throw with about 60% of the power of men. Among Olympic javelin throwers, despite similar throwing motions, males outperform females [Mero et al., 1994]. In the overhand tennis serve at the 2000 Olympics the ball velocity of females was about 80% of that in males, although the kinematics of elite females are almost identical to those of elite males [Fleisig et al., 2003]. A gender disparity is also observed in professional golfers. Using a similar club-swinging motion, females drive the ball from the tee an average distance 15% less than that achieved by males [Zheng et al., 2008].

The throwing motions of eleven elite baseball pitchers of each gender were compared by Chu and coworkers [2009]. Their pitching kinematics were very similar, but the release velocity in females was 74% of that of the males (26.8 m/sec vs. 36.3 m/sec). Because the throwing motion differed so little between the genders, the researchers suggest that physical factors such as the greater height, body mass and muscularity of the males may have been responsible. (Males were about 10% larger in stature and mass). If the females had shorter arms, this would reduce the radius of rotation and might reduce ball velocity. They lifted their stride knee less and their forward stride was not as long, possibly due to less lower-body strength and shorter legs [Chu, et al., 2009].

During the acceleration phase of the throw, females generated 31% lower peak elbow extension velocity, which might be a sign of less triceps strength. The peak velocity of knee extension during the stride was also less and the knee was more flexed at ball release. This could have been due to lower strength in the quadriceps muscles. The time from stride foot contact to ball release was also greater in females, possibly also the result of greater male muscularity. Muscular forces that act to protect the arm during the acceleration phase were reduced in the females [Chu, et al., 2009].

This analysis of gender differences in adult elite throwers is consistent with the conclusion that both genders use the same throwing motion, but greater stature, mass and muscularity in males contributes to their greater throwing velocity. There may be other participating factors, but it is probable that gender disparities in height, weight, strength and power underlie the adult male advantage [Kirschmann, 1999; Ransdell and Wells, 1999; Zheng et al., 2008; Chu, et al., 2009].

The ontogeny of human throwing: Evidence of a genetic basis. Is human throwing and striking behavior inborn or acquired? Factors such as teaching, learning, practice and societal influences may play a role, but they seem to have a relatively insignificant effect compared to the genetic component. If throwing and club-swinging are primarily inherited behaviors, they must in some measure be coded in the DNA. If true, that would require an *evolutionary* explanation of why these genetic elements had been naturally selected from the ancestral stock.

Accumulated evidence of throwing development shows that this behavior appears before children can stand up. It subsequently develops in a characteristic manner, without instruction, as it unfolds to higher levels of prowess. In some children the developmental program proceeds rapidly and is completed before puberty. In others a rudimentary throwing motion persists throughout life. A remarkable gender difference favoring males is evident by three years of age and increases thereafter. Girls, compared to boys, seem to develop this motor

CHAPTER 13: DEVELOPMENT & MATURATION OF THROWING AND CLUBBING

behavior at a slower pace and may cease development at a lower skill level. The club-swinging motion develops concurrently, at the same pace, in the same pattern, and with the same gender difference.

The throwing and clubbing motions that develop in each child are rudimentary forms of the one used by skilled adult athletes. Throwers from every part of the globe use the same motor program, even where throwing is no longer either useful or practical. All humans inherit a tendency to develop the most mechanically effective motion for these behaviors (although few develop the best possible motions while also becoming tall and muscular). These action patterns are so complex and rapid they seem refractive to learning at any age. How could one learn (or teach) all the subtle adjustments of muscle contractions that prevent self-injury in the violently moving upper extremity during a forceful throw? How could one learn the multitude of coordinated body movements in the right sequence when they are part of a motion lasting less than a second? There must be a strong genetic element. No other species of animal has evolved a specialized throwing behavior. We are the only ones now, and probably always have been.

In summary, the conclusion that throwing has a strong genetic element is based primarily on the following evidence: (1) Throwing behavior occurs in young children of all cultures and ethnic groups that have been studied. (2) It begins very early in childhood, before vertical balance is possible. (3) When upright stance is achieved, throwing becomes a bipedal act. (4) Instruction seems to have negligible effect on its early development. (5) Its complexity exceeds learning capacity in young children. (6) This motor behavior is so complicated and so rapid that it seems incompatible with a basis in learning at any age. (7) A standardized throwing motion emerges which increasingly comes to resemble the one used by mature athletes. (8) In some children the ontogenetic program proceeds rapidly and ceases before puberty. In others progress is slower and may terminate at an early stage. (9) Girls often acquire this motor behavior at a slower pace and complete development at a lower skill level. (10) A gender disparity favoring boys is evident as early as 3 years of age and subsequently increases. (11) This difference is not reduced by teaching. (12) The club-swinging motion that develops concurrently with the throwing motion has the same general features and gender difference. (13) The gender disparity persists in adults, enhanced by anatomical and physiological changes that occur in males at puberty. (14) Elite throwers everywhere use the same throwing motion. (15) This motion appears to be the most effective technique for throwing, given the structure of the human body. (16) A forceful throwing and clubbing motion, which lasts less than a second, is accompanied by neuromuscular mechanisms which protect against injury from the powerful forces that accompany it.

An evolutionary explanation. The innate bipedal throwing and striking behavior that emerges during human ontogeny represents a unique outcome of hominin evolution. Individuals who excelled at these behaviors must have been naturally selected because their prowess in throwing and club-swinging enabled them to produce more offspring, causing the frequency of their genes in the breeding population to increase. I suggest that our ancient ancestors enhanced their reproductive success by throwing stones and swinging clubs to gain an advantage over their competitors. Practice, and guidance by informed instructors may have a positive influence on throwing and striking; poor nutrition can have a negative effect. These factors may influence the development of the inherited motor pattern.

Evolution of gender differences in throwing and striking. Evolutionary change results from natural selection of heritable variations that enhance reproductive success. The two genders have different perspectives on that outcome (Chapter 3). Males require access to fertile females, which are in short supply; fighting ability can provide an advantage in this competition. In females, fulfillment of nutritional and caloric needs for themselves and their offspring is the paramount issue. Throughout the hominin lineage, males were the fighters. They fought amongst themselves, against predators, and in combat with prey animals for food. Thomas and Marzke [1992] suggested that gender disparities in throwing development have an evolutionary origin, if throwing affected survival in males but not females, and if selection for this skill had taken place. Males were more likely to be hunters and warriors, resulting in greater selection of those who threw effectively [Thomas, 2000; K. Thomas, et al., 2001]. Performance in the javelin throw is higher in men than in women, even though their spears are heavier. Kirschmann [1999. 3.3.1.] wrote: "For such a grave difference in capability...no other explanation occurs to me than a narrow specialization of just men for throwing in the course of evolution."

The gender disparity presumably yielded greater reproductive advantages for both sexes. Females who devoted more time and energy to gestation, lactation, and nurturing children (and less to belligerent use of weapons) were probably more successful in transmitting their genes to future generations. On occasion, they might have used weapons to protect themselves and their children, or to increase access to disputed food resources, but among modern primates, including humans, males are more active in fighting and hunting. There is no reason to suspect that ancient hominin males were an exception. Females may have less interest in these activities because they are pregnant or lactating, or because of the risk to dependent offspring and reproductive costs associated with pursuing dangerous, mobile prey [Zihlman, 1978; Tooby and DeVore, 1986; Manson and Wrangham, 1991; Kaplan and Hill, 1992].

Natural selection cannot act directly on reproduction until individuals become reproductively competent, after puberty. How can the onset of throwing and striking in early childhood be explained? The precocious start for throwing in children suggests that it takes a long time to create the physical and neural underpinnings of the mature throwing and striking apparatus. It cannot simply be switched on at puberty because it is much too complex. Children, particularly boys, who developed the propensity and skill for throwing and striking at a young age would more likely survive until puberty and succeed in the competition for mates afterwards. The onset of such behaviors early in life would provide a longer duration for their development, supplying advantages later in life when they affected reproductive success [Young, 2009].

In modern humans, genetic factors predominate in regulating the development of throwing and striking. These behaviors are relics of events that occurred long ago during the evolution of our ancestors. We have inherited genes that were selected to enhance combat, especially in males, during many thousands of generations beginning millions of years ago. Throwing and striking are of little practical use in the present era, but they evidently played a major role early in the human lineage, perhaps from its very beginning, when our ancient progenitors enhanced their reproductive prospects by throwing stones and swinging clubs.

CHAPTER 14

Bipedal Locomotion

"The commonest explanation for the origin of human upright posture and gait relates these to tool and weapon using, and to offense and defense" [Hewes, 1961, p. 692].

"...the origin of a bipedal form of locomotion was so fundamental a change, so replete with profound evolutionary potential, that we should recognize the roots of our humanity where they really are" [R. Leakey and Lewin, 1992, p. 82].

"The evolution of hominid bipedalism is recognized as a crucial element in the hominization process. However, despite a voluminous literature, our ignorance concerning bipedalization is almost complete" [Rose, 1991, p. 38].

Introduction. The earliest part of the hominin fossil record clearly reveals some sort of behavior performed in an upright, bipedal posture. The evidence is in the skull, the pelvis, the legs and feet. *What was that behavior?* Currently there is near-unanimous agreement that it was *locomotion*, and that the origin of our lineage is linked to the onset of bipedal walking.

Several 19th and early 20th century writers following Darwin [1871] called attention to our distinctive upright gait

[Landau, 1991]. In the modern era, this became a central focus of human evolutionary studies. After Washburn [1950] asserted that bipedal locomotion was the first human adaptation, this idea gained such strong acceptance that bipedalism became a synonym for upright locomotion. It is now widely believed to represent the primordial hominin behavior and the initiating event in human evolution [Bartholomew and Birdsell, 1953; Hewes, 1961; Du Brul, 1962; Zihlman, 1978; Yamazaki, et al., 1979; Lovejoy, 1988; Corballis, 1991; Coppens, 1991; Langdon, et al., 1991; R. Leakey and Lewin, 1992; Tattersall, 1995; Wrangham and Peterson, 1996; Fleagle, 1999; Richmond and Strait, 2000; Wrangham, 2001; Richmond, et al., 2001; Videan and McGrew, 2001; D'Aout, et al., 2002; Ward, 2002; Kingdon, 2003; Stanford, 2003; Schmitt, 2003; Hilton and Meldrum, 2004; Crompton and Günther, 2004; Hart and Sussman, 2005; Thorpe, et al., 2007a; Richmond and Jungers, 2008; Parker and Jaffe, 2008; Pontzer, et al., 2009; Carvalho, et al., 2012].

If correct, this means the onset of bipedal locomotion signals the moment when our lineage diverged from its ancestors and set off in a new direction. It also indicates that an evolutionary explanation of this event is crucial to the understanding of hominin origins.

Nevertheless, after decades of speculation, the query of why hominins began to walk upright is still unanswered. It is widely acknowledged that no compelling explanation for bipedal locomotion has yet been proposed [Robinson, 1963; Rose, 1991; Langdon, et al., 1991; Stringer and McKie, 1996; Fleagle, 1999; Stanford, 1999; Kramer and Eck, 2000; Wrangham, 2001; D'Aout, et al., 2002; Kingdon, 2003; Steudel-Numbers, 2003; Schmitt, 2003; Preuschoft, 2004; O'Higgins and Elton, 2007; Sockol, et al., 2007; Sylvester and Kramer, 2008; Parker and Jaffe, 2008; Pontzer, et al., 2009; Carvalho, et al., 2012; Tattersall, 2012]. Why our unique two-legged gait evolved is an enigma. It is a fundamental question in the study of human evolution at the present time, but the answer remains elusive.

What were the bipedal proclivities of the hominin ancestor? Because the crucial fossils from the time of the hominin/chimpanzee divergence (6-7 Mya) are missing from the record (Chapter 2), paleoanthropologists have had to address the question indirectly. Two general approaches have been used.

The hypothetical ancestor. This method involves constructing a model of the last common ancestor's locomotion based on study of living arboreal apes that are adapted to clambering in trees, but also spend time on the ground. Tuttle proposed that traveling in arboreal contexts might have predisposed our hominin ancestors for bipedal locomotion by adapting them to vertical climbing on tree trunks or vines and bipedal travel on horizontal branches [1974, 1981, Tuttle, et al., 1991]. The concept that hominins were "preadapted" to terrestrial bipedal locomotion by upright adaptation to an arboreal setting gained wide support [Stern, 1977; Prost, 1980; Fleagle, et al., 1981; Rose, 1991; Senut, 1991; Richmond, et al., 2001]. In Prost's [1980] view, human bipedal gait arose from specialized climbing of vertical tree trunks; when hominins adopted terrestrial life they already had bipedal capacity. Fleagle and coauthors [1981] offer support for this idea. Senut [2006a, 2007] observed that fossil evidence indicated that early hominins lived in wooded habitats and their hominoid ancestors had been arboreal apes during the Miocene (~ 23-5 Mya). Some of these apes may have traveled bipedally on the ground, but not in the modern manner [Pickford, et al., 2002; Senut, 2006b; Nakatsukasa, et al., 2007; Harrison, 2010; Wood and Harrison, 2011]. Crompton and coauthors [2008] concluded that the last common panin/hominin ancestor was a mainly arboreal, orthograde, short-legged, long-armed ape capable of hand-assisted bipedality in trees, in a manner like that observed in modern orangutans. Hand-assisted bipedal locomotion in trees could have preadapted the ancestors of early hominins for habitual terrestrial bipedalism [Thorpe, et al., 2007a, b; Crompton, et al., 2008, 2010 and see below].

The modern chimpanzee model. Chimpanzee bipedal walking has been studied in wild and captive animals. These investigations show that upright walking in chimpanzees bears little resemblance to the way modern humans walk. Thus it is a questionable model that may even obscure understanding [Senut, 2007; Crompton, et al., 2008; Lovejoy, et al., 2009b, g].

Chimpanzees rarely use upright gait, and then only in short bursts [Hunt, 1992, 1996; Doran, 1992; Stanford, 2003]. They are not adapted for it [Hunt, 1992; Richmond, et al., 2001].

Their common gait on the ground is quadrupedalism, supported by the knuckles of the hands and the plantar surface of the feet. Hip and knee joints are flexed and the pelvis is oriented horizontally, with the iliac blade well in front of the hip joint. In this configuration the gluteus medius and minimus are able to rotate the femur which can be extended beyond the vertical at the hip joint by the hamstring and gluteus maximus muscles to produce forward propulsion. However, when the animal stands upright, the pelvis is oriented vertically, producing instability because the long iliac blade raises the sacroiliac joint high above the hip joint [Robinson, 1972]. The hamstrings and gluteus maximus cannot fully extend the femur and the trunk is bent forwards [D'Aout, et al., 2002]. To compensate for the tilt, the knees have to be flexed, which requires strong contraction of hip and knee extensors to prevent collapse [Tuttle, et al., 1979; Kimura, et al., 1983]. When the femur is maximally extended, rotation by the gluteus medius and minimus is lost [Zihlman and Brunker, 1979], limiting the ability to swing forward the opposite side of the pelvis. The abducted femur and angle of the knee joint position the foot outwardly [Jenkins, 1972]. Mobility of hip, knee, ankle and foot joints adds instability [Robinson, 1972]. The hallux, deployed sideways, cannot exert forward thrust, and the other toes, long and curved, cannot be extended fully [Elftman and Manter, 1935; Robinson, 1972]. As weight shifts from heel to toes, the foot collapses at the midtarsal joint, wasting energy [Elftman, 1944].

Chimpanzees walk upright with short steps, feet wide apart, and minimal hip rotation [Zihlman and Brunker, 1979]. Leg, hip, and torso tend to swing forward as a unit on the unsupported side. The upper body leans markedly toward the supported side to bring the center of mass over the support leg [Elftman, 1944; Tuttle, et al., 1979]. Asymmetric motions of the arms are required to maintain balance [Tardieu, et al., 1993] and there is great variability in the gait. Chimpanzee bipedalism (1-1.7 m/sec) [Tardieu, et al., 1993; Kimura, 1996] is about ten times slower than the estimated top speed of quadrupedalism [Kortlandt, 1980].

Which model works best? Did upright walking in the modern manner arise in the trees before hominins emerged from their ape ancestry? This cannot be the full story because the fossil record, described below, shows that adaptation for bipedal locomotion continued for several million years *after* onset of the hominin lineage. Did hominin bipedal gait arise in an arboreal ancestor that commonly used quadrupedalism on the ground? This also seems questionable, based on the chimpanzee model. Explication of the evolution of human upright walking in either case will require the identification of a behavior that provided profound reproductive advantages to produce such an eccentric mode of locomotion.

Modern human bipedal locomotion. Modern humans are clearly adapted for bipedal locomotion. The behavior is instinctive and appears within two years of birth. Energy expended per unit distance traversed is minimized [Saunders, et al., 1953; Ralston, 1976]. Upright stance requires only slight, intermittent muscle action to maintain balance, due to extensive skeletal remodeling [Elftman and Manter, 1935; Weidenreich, 1947; Tuttle, et al., 1979; Reeser, et al., 1983]. The heels rest on the ground; the legs are extended at hip and knee joints; the vertebral column is curved alternately forward and back; the skull is balanced on top of the spine, and the center of mass has descended to the level of the hip joint in a plane passing through the foramen magnum, hip, knee and ankle joints [Robinson, 1972; Tardieu, 1992]. Normal walking speed is about 1.25 m/sec [Ralston, 1976; Steudel, 1996], comparable to chimpanzee bipedal walking, but top running speed is about 25% slower than full speed chimpanzee quadrupedal running

Human bipedal walking is steady, rhythmic, and bilaterally symmetrical, with each step a replica of preceding ones (Figure 15). One leg swings forward, accepts the weight of the body, then produces forward acceleration as it extends, thrusting into the ground behind the hip joint while the opposite leg swings forward. Vertical and transverse displacements are small and in phase with one another, reducing energy expenditure. Efficiency is enhanced by slight rotations of the thorax and hips around the vertical axis [Zihlman and Brunker, 1979; Tardieu, et al., 1993]. The pelvis rotates 4° on the support leg to position the swinging foot in the midline while the femur on the swing side rotates externally to point the foot straight ahead [Saunders, et al., 1953; Zihlman, 1978]

Figure 15. In this comparison of walking and throwing, the different stages of the throwing motion have been artificially extended horizontally, rather than superimposed. The behaviors are both bipedal, vertically oriented, and dynamically balanced upon strong legs and springy feet for propulsion. Both involve rotation of the femur, pelvis and torso and coordinated arm and leg movements. However, throwing involves a sequential pattern of muscular contractions adapted to maximize the amount of energy transferred to a hand-held missile and control its trajectory in a single explosive act while the body remains in one place. In contrast, bipedal locomotion consists of a rhythmic, repetitive pattern adapted to minimize energy expenditure while transporting the body to another place. Walking follows a simple, bilaterally symmetrical pattern, whereas in throwing, movements of right and left sides differ.

As the pelvis rotates, the thorax swivels in the opposite direction to stabilize the head and shoulders. Rotational momentum around the vertical axis due to movement of the swing leg is counteracted by the forward motion of the opposite arm [Elftman, 1939; Zihlman, 1978; Tardieu, 1992].

Adaptive changes in the human lower thorax and pelvic region facilitate bipedal gait. Compared to living apes, vertebral bodies and intervertebral discs are larger and the femoral head and acetabulum of the hip joint is expanded, distributing the stress of weight-bearing over a larger surface. The ilium is reduced in height and increased posteriorly, lowering the center of mass and shifting it backward over the hip joint [Robinson, 1972]. Widening the sacrum pushed the ilia apart and reoriented them mediolaterally, permitting the hamstring and gluteus maximus muscles to extend the leg past the vertical [Robinson, 1972; Kimura, et al., 1977] and converting the gluteus medius and minimus muscles to rotators and abductors at the hip joint. This solved the problem of lateral balance insofar as abduction on the side of the supporting leg prevents the upper body from tilting to the other side while the rotator function swings the opposite side of the pelvis forward on the fixed femur [Zihlman and Hunter, 1972; Lovejoy, 1988].

Medial angulation of the femora placed the knees and feet close to the midline. Stability was increased in knee and ankle joints [Latimer, et al., 1987] and in the foot, which lost muscle and its capacity to grip, gaining bone, fascia and ligaments for stability [Zihlman and Brunker, 1979] along with longitudinal and transverse arches for resiliency [Hicks, 1954; Reeser, et al., 1983]. In walking, when body weight is shifted from the heel to the first metatarsal [Reeser, et al., 1983; Stern and Susman, 1983], the transverse tarsal joint locks, rather than collapsing. Simultaneously, ligamentous arches undergo tension, storing elastic strain energy and a unique plantar aponeurosis is stretched, producing tension which is released at toe-off from the hallux, adding to the propulsion generated by calf muscles [Hicks, 1954; Bojsen-Møller, 1979; Lewis, 1989].

Paleontological evidence of hominin bipedalism. In specimens of the earliest putative hominin, *Sahelanthropus tchadensis* from Chad, dated at 7 Mya (near the predicted branch point of the hominin line) the anterior position of the foramen magnum suggests adaptation to some kind of upright behavior [Brunet, et al., 2002; Zollikofer, et al., 2005]. Three femoral fragments assigned to *Orrorin tugenensis* from Tugen Hills, Kenya, (6.0 Mya), indicate the likelihood of bipedality

[Senut, et al., 2001; Pickford, et al., 2002; Galik, et al., 2004; Richmond and Jungers, 2008].The proximal femur is said to resemble those of humans more than those of *Australopithecus* or African apes.

Specimens of *Ardipithecus kadabba* from Ethiopia (~5.8-5.5 Mya) [White, et al., 2006] include a proximal pedal phalanx with a dorsal orientation of the base that may indicate bipedal behavior [Haile-Selassie 2001]. A similar specimen from *Ar. ramidus* (4.4 Mya) retains a broad proximal facet, oriented dorsally [Semaw, et al., 2005]. In braincase fragments of this hominin the foramen magnum is situated somewhat anteriorly [White, et al., 1994].

Discovery of an important collection of *Ar. ramidus* fossils reported in 2009 has contributed significantly to knowledge of bipedalism in this ancient hominin. Dated at 4.4 Mya, the specimens were found at Aramis, in eastern Ethiopia. A near-complete but damaged left hip, a portion of the right ilium and a distal sacral fragment were recovered. The fragmented pelvis, after reconstruction by a tomographic modeling program [Lovejoy, et al., 2009g], indicated "habitual" upright walking [Lovejoy, et al., 2009b] or "facultative" bipedal gait more rudimentary than that of *Australopithecus* [White, et al., 2009]. The pelvis has short, broad, laterally flared ilia [Lovejoy, et al., 2009b, 2009f]. This, and a growth site for the anterior inferior iliac spine, suggest that gluteal muscles may have been positioned so that *Ardipithecus* could walk without laterally shifting the center of mass [Lovejoy, et al., 2009f]. Lordosis of the lumbar spine might have been present because this part of the vertebral column was not trapped within the pelvis [Lovejoy, et al., 2009g; Lovejoy, et al., 2009h, Fig. S 10]. However, the ischium is like that of African apes in other features [Lovejoy, et al., 2009g, 2009h], and the ischial surface suggests that the hamstring muscles are disposed in an ape-like manner [Lovejoy, et al., 2009b].

Ar. ramidus is the only known hominin with an abducted great toe, like that of chimpanzees [Lovejoy, et al., 2009j]. This is evidence that it was arboreal and its upright gait was rudimentary [Lovejoy, et al., 2009f, 2009i]. The talocrural joint angle lies within the range of quadrupedal primates [Lovejoy, et al., 2009j]. A pedal os peroneum may have increased the ability to adduct the hallux and plantarflex the foot [Lovejoy, et al., 2009i, j]. Also, an expanded second metatarsal base suggests an absence of midtarsal laxity [White, et al., 2009; Lovejoy, et al., 2009j]. In *Ar. ramidus* a dorsal cant of the pedal phalanges supports dorsiflexion of the pedal phalanges at toe-off, but less so than in modern humans [Lovejoy, et al., 2009k]. The phalanges are significantly curved [Lovejoy, et al., 2009h; Lovejoy, et al., 2009k]. Like curvature of the hand phalanges, this trait indicates an adaptation to life in the trees (Chapters 2, 9). Metatarsals II and III are also curved [Lovejoy, et al., 2009j]. The evidence for bipedalism in the foot of *Ar. ramidus*, like that from the pelvis, is mixed with signs of an adaptation to arborealism [Lovejoy, et al., 2009i; White, et al., 2009].

In *Australopithecus anamensis* from Kanapoi, Kenya (4.2-3.8 Mya) tibia, knee and ankle joints suggest improved bipedality [M. G. Leakey and Walker, 1997; M. Leakey, et al., 1998]. The tibia, similar to those of *A. afarensis* from Hadar, is adapted for bipedal locomotion. The diaphysis is oriented orthogonally to the talocrural joint so that the knee was placed directly over the ankle during the support phase [Ward, et al., 1999b, 2001]. The tibial metaphyses are expanded in all dimensions, increasing the osseous shock-absorbing mechanism during bipedalism [Ward, et al., 2001].

A. afarensis is known from ~3.5-3.0 Mya at Hadar and Maka in Ethiopia and at Laetoli, Tanzania. The pelvis, spinal column and foot are those of an erect creature [Stern and Susman 1983; Kimbel and Delezene, 2009]. A greater volume of bone in the calcaneus shows an adaptive response to ground reaction at heel strike [Latimer and Lovejoy, 1989]. Footprints from Laetoli [White 1980] and foot bones of *A. afarensis* reveal that the hallux was adducted [Susman 1983]. As in humans, it is robust and has an enlarged head [Latimer, et al., 1982; Kimbel and Delezene, 2009]. The pedal phalanges are shorter, relative to body size, than in extant apes, but the proximal phalanges remain curved [Latimer, et al., 1982; Langdon, 1985; McHenry 1986]. Some intermediate phalanges also are curved [Latimer, et al., 1982]. It has been suggested that this curved foot is incompatible with the footprint trail at Laetoli [Harcourt-Smith and Aiello, 2004; Kimbel and Delezene, 2009].

Hailie-Selassie and colleagues [2012] have reported a

partial foot skeleton from a hominin apparently contemporaneous with *A. afarensis* (~3.4 Mya) and from the same Afar region of Ethiopia, but as far as its big toe opposability is concerned, the foot more closely resembles *Ar. ramidus*, with a grasping capacity suitable for arboreal settings.

In modern humans, the heads of the metatarsals have rotated in relation to their bases so as to lie squarely on the ground. There is very little torsion in the first metatarsal but progressively more from metatarsals II-V. In apes the head of metatarsal I is oriented toward the other metatarsals with the second to fourth oriented toward the first [Zipfel, et al., 2009]. This positions the hallux in opposition to the lateral toes during grasping [Susman and de Ruiter, 2004]. An *A. afarensis* fourth metatarsal (from Hadar, AL 333-160) exhibits torsion of the head relative to the base, indicating transverse and longitudinal arches. It also implies the absence of an ape-like midtarsal break and a more rigid foot, providing a mechanical advantage during the propulsive phase of gait [Ward, et al., 2011]. The articular surface of the metatarsal head is extended dorsally, signaling dorsiflexion at the metatarsophalangeal joint during toe-off [Kimbel and Delezene, 2009; Ward, et al., 2011]. A proximal pedal phalanx, attributed to *A. garhi* (2.5 Mya) from Ethiopia is similar to that of *A. afarensis* [Asfaw, et al., 1999].

Hominin foot bones from South Africa (StW 573) [Clarke and Tobias, 1995], possibly as old as 3.5 Mya, may be from an early form of *A. africanus*. Four articulating bones have human features in the hindfoot and apelike traits in the forefoot. The hallux is appreciably abducted and strongly mobile, as in apes. Evidently, the hominin was bipedal on the ground, but included arboreal climbing in its locomotor repertoire.

Postcranially, *A. africanus* (3.2-1.8 Mya) was similar to *A. afarensis*, although the upper and lower limbs suggest a greater dedication to arboreality in *A. africanus* [McHenry and Berger, 1998]. Most structural aspects of efficient bipedal behavior were present [Robinson, 1972]. A complete hominin fifth metatarsal (StW 114/115) from Sterkfontein (an adult of uncertain taxon) estimated at 2.5-1.5 Mya has primitive features, such as sagittal curvature of the shaft, but the general morphology is human-like, with medial torsion, an upturned set of the metatarsal head, extension of the phalangeal articular surface onto the dorsum of the bone, and suggestions of longitudinal and transverse arches [Zipfel, et al., 2009].

A first (hallucal) metatarsal head (SK 1813) from Swartkrans unassigned to taxon or age, displays more axial torsion than seen in modern humans, but less than in chimpanzees [Susman and de Ruiter, 2004]. The short, stout form and basal morphology imply well-developed plantar ligaments. There is a modern range of dorsiflexion at the metatarsal-phalangeal joint, but it does not close-pack in this position as it does in modern humans [Susman and de Ruiter, 2004]. *A. robustus* (*Paranthropus*) hominins from East and South Africa (2-1 Mya) were becoming more adept bipeds, as shown by evidence of a plantar aponeurosis [Susman and Stern, 1991], but because the hallucal metatarsal-phalangeal joint failed to close-pack in dorsiflexion, a modern toe-off mechanism was absent [Susman and Brain, 1988; Susman and de Ruiter, 2004].

A./H. habilis from East Africa (~2-1.5 Mya) had legs of a habitual biped and feet which had become stable, arched, resilient platforms [Susman and Stern, 1982; Susman, 1983], although it retained the short legs and long arms of the arboreal ancestor [Richmond, et al., 2002].

A. sediba from Malapa, South Africa (2.0 Mya) had numerous features of the hip, knee and ankle that indicate a dedicated biped [Berger, et al., 2010a]. The distal ends of the femur and tibia resemble those of *A. africanus*. However, while several features in the pelvis are like those from Koobi Fora and Olduvai (possibly early *Homo*), the calcaneous is markedly primitive [Berger, et al., 2010a]. *A. sediba* walked with an extended leg, but with a uniquely exaggerated pronation of the foot at toe-off [DeSilva, et al., 2013], possibly due to an adaptation to climbing trees [Lieberman, 2012a].

At the Dmanisi, Georgia, hominin site (1.8 Mya) among skeletal remains attributed to early *H. erectus*, the legs are longer relative to the arms than in *Australopithecus* and *A/H. habilis*, and relative to body mass, longer than those of African apes [Pontzer, et al., 2010]. Robust metatarsals with straight shafts have been recovered from the 1st, 3rd, 4th and 5th rays. Their lengths overlap the low end of modern human variation. Metatarsals III and IV have a high degree of torsion, compatible with a transverse arch. The wide base of the first metatarsal suggests a plantar ligament and a longitudinal arch according to Lordkipanidze and coworkers [2007]. The heads of the first metatarsals are relatively narrow, but the proximal shaft is expanded, and the phalangeal articular surface curves back onto the dorsal aspect of the shaft. Metatarsals III-V are longer than in modern humans. The talus suggests a human-like ankle joint, and the adducted, robust, hallucal metatarsal indicates full commitment to terrestrial bipedalism [Pontzer, et al., 2010]. Combining femoral and tibial torsion in the Dmanisi hind limb yields a net −7° rotation (intoeing) distinct from that in modern humans (+20°). Calculations indicate the hominins were intermediate between *Australopithecus* and *Homo* in locomotor economy. These considerations imply that locomotor evolution was still incomplete 1.8 My ago [Pontzer, et al., 2010].

Hominin footprints dated at ~1.5 Mya at Ileret in Kenya provide evidence of a foot with a hallux adducted to 14° from the midline. (It is 8° in modern humans and 27° in the 3.5 Mya Laetoli prints) [Bennett, et al., 2009]. A medial longitudinal arch is present. The size of the footprints is consistent with stature and body mass estimates for *H. erectus*. The heel contacts the ground first, followed by weight transfer to the lateral side of the foot, then to the ball of the foot with peak pressure under the medial metatarsal heads, ending with toe-off from the hallux, all in the modern manner [Bennett, et al., 2009].

The fossil record reveals that some form of bipedalism, began at or near the origin of the hominin lineage. Evolution of bipedal locomotion (which may have begun *after* bipedal use of weapons—see below) continued until the appearance of African *Homo erectus*, a duration of more than five million years. Locomotor diversity among early hominins raises the possibility that bipedal walking could have evolved independently in different hominin lineages [Harcourt-Smith and Aiello, 2004; Haile-Selassie, et al., 2012; Lieberman, 2012a], but this does not require that it *began* independently several times. The more parsimonious view is that after a singular origin, bipedal gait became increasingly refined in separate lineages as more efficient versions of the same adaptation appeared independently [Berger, 2000; Begun, 2007].

Hypotheses to explain the origin of hominin bipedal locomotion. For many years, beginning with Darwin and continuing to the present day, scholars have sought to find a way to explain how bipedal locomotion could have arisen. Hypotheses abound. Here are some of them.

The sentinel hypothesis. Dart and Craig [1959] and Ravey [1978] proposed that there would be advantages to an ape that could search for food and enemies in tall grass on the savanna by standing upright.

The carrying hypothesis. In many permutations, this idea has had an enduring history. Hewes [1961, 1973] suggested that hominins became bipeds because the use of animal parts for food required their transport. Bipedal gait freed the hands for carrying the food "home". In Lovejoy's opinion [1981, 1988, 2009a, b] it was plant foods carried by males back to their mates and children that led to bipedal locomotion. Sinclair and coauthors [1986] suggested it was babies that were carried while hominins tracked migrating ungulates in order to be the first scavengers when animals died. Fisher [1992] proposed that hominins had to walk erect to carry food to a place where it could be eaten unmolested. Wrangham and Peterson [1996] considered it was roots carried from where they were found to a safer place where they were consumed that led to bipedal gait. Kirschmann [1999] thought it was carrying throwing-stones that required upright walking. Observations of chimpanzees carrying nuts and tools have been interpreted as support for Hewes' 1961 food-carrying hypothesis [Carvalho, et al., 2012].

The postural feeding hypothesis. Several authors have suggested a link between feeding stance and the origins of bipedal gait. Upright walking might have evolved from a bipedal feeding posture used for harvesting fruits [Tuttle, 1967, 1981]. This would eliminate the need to raise the body from quadrupedal stance [Wrangham, 1980]. Jolly [1970] thought bipedalism could have arisen from a behavior like that of Gelada baboons, who forage in a sitting position, shuffling on their buttocks for short distances. Rose [1984] and Kingdon [2003] emphasized the importance of bipedal travel for moving between closely clumped food sources. Perhaps a blend of shuffling and tree-hanging bipedalism began as a feeding posture which preadapted hominin ancestors for bipedal gait

[Hunt, 1994, 1996]. Stern [2000] supports this hypothesis, but requires an "unknown event" to make it work.

The tool-Use hypothesis. Bartholomew and Birdsell [1953] proposed that bipedal gait resulted from the advantage gained by freeing the forelimbs for use of tools. Eimerl and DeVore [1965] thought the two behaviors were connected in a feedback circuit. The tools might have been used for defense against predators. Washburn [1967] also believed the adaptive success of tool use (sticks or stones for defense) promoted locomotor evolution, as did Marzke [1992b]. Lovejoy [2009b] observed that the role of tools has "long been a tempting explanation for upright walking... however, it is now known that habitual bipedality evolved millions of years before any evidence of stone tools" (p. 74e5) (i.e., no hand-made stone tools = no stone tools; see Chapter 4). A case will be made below that bipedal locomotion arose from the use of naturally occurring objects for tools deployed bipedally as *weapons*.

The display hypothesis. Chimpanzees and gorillas move while upright for short distances during display behavior. Jablonski and Chaplin [1993] suggested that threatening bipedal display coupled with appeasement behavior fostered peaceful conflict resolution, leading to increased survival of progeny resulting in dedicated bipedal locomotion. Alternative display hypotheses are discussed by Dawkins [2004]. One is that hominin males first rose on their hind legs to show their genitalia, whereas females did it to conceal theirs; the outcome was bipedal walking. Dawkins's own idea is that vertical walking first arose capriciously as a "gimmick" which gained sexual attractiveness by the sheer virtuosity of its precarious posture. Those who were best at it attracted the most mates and left the most children.

The energetic hypothesis. This concept, first proposed by Rodman and McHenry [1980], has elicited much discussion. It was based on the calculation that modern chimpanzee quadrupedalism is less efficient than modern human bipedal gait, but did not specify how this would apply to the onset of bipedal locomotion. Steudel (1996) concluded that increased energetic efficiency would not have accrued to early bipeds. Selection for more economical locomotion does not have to involve bipedality and usually does not. This suggests that other factors must have been important in leading our ancestors to adopt bipedality [Steudel-Numbers, 2001, 2003]. Schmitt [2003] concurred that it was doubtful bipedal gait in the earliest hominins would have provided significant energy savings. Japanese macaques, trained to walk bipedally, use a longer stride, lower stride frequency, more extended hindlimbs and greater lumbar lordosis than untrained macaques, but their energy expenditure is still 20% higher than in quadrupedalism [Nakatsukasa, 2004]. Pontzer, et al. [2009] asserted that relatively small changes could have lowered the cost of bipedalism in a chimpanzee-like ape, but calculated that a bipedal chimpanzee spends 30% more energy when it walks on two legs compared to four. (For further discussion of this hypothesis, see Chapter 7).

The aquatic hypothesis. Morgan [1972, 1990, 1995, 1997] developed this explanation, which postulates an aquatic stage of human ancestry that began when arboreal primates entered the water to escape predators and returned to shore only for sleep or forage. Adapting to aquatic life, they evolved bipedalism for walking with their heads above water. Eventually, they returned to living on the land supported on two legs and became terrestrial animals again.

The shore dweller hypothesis. Niemitz [2002, 2006, 2010] changed the emphasis from aquatic to amphibious and adopted a multidisciplinary approach. His scenario begins with a generalized, omnivorous, arboreal ape, quadrupedal when on the ground. The origin of bipedal locomotion is attributed to wading in shallow lakes or rivers in search of food. The ability to obtain energy sources and animal protein (such as snails, fish and crabs) by wading in water would be the major advantage of such behavior, which was a transient phase in our ancestry [Niemitz, 2002]. There are fewer predators in shallow water than on land. In wading, buoyancy reduces the load on joint surfaces. It also triggers upright posture, forces the individual to maintain this position and to walk bipedally, thereby aiding the transition from quadrupedalism to bipedalism. During the transition, our ancestors may have walked bipedally in the water, then quadrupedally across dry ground to nearby forests where they climbed trees to harvest food and sleep at night. As they gradually acquired longer and more extendable hind limbs due to selection for wading, it ultimately became advantageous to keep walking upright when on land. When the legs became so long they hampered fast quadrupedal locomotion, selection for habitual upright walking and running could start [Niemitz, 2010].

The orthograde clambering hypothesis. Crompton and colleagues [2003] proposed that upright arboreal scrambling was subsequently exapted for terrestrial bipedalism. Modern orangutans suggest how this could have occurred. In these apes the most frequent locomotor mode is an upright clamber in which body weight is borne predominantly by forelimb suspension [Thorpe and Crompton, 2006]. However, in thin branches high in tall trees, orangutans sometimes adopt a standing posture, gripping small branches with their feet with extended hip and knee joints, while grasping other branches with one or both hands for stability [Thorpe, et al., 2009]. This hand-aided bipedalism enables them to reach fruit on terminal branches. On the ground, however, they have too much unstable mass above the hip to sustain unassisted bipedalism, so they use quadrupedalism, bearing weight on their proximal manual phalanges, or the side of the hand [Schultz, 1969; Crompton, et al., 2010]. Their feet, adapted to arboreal gripping, are ill-formed for walking [Crompton, et al., 2008]. If a Miocene hominoid became adapted for arboreal hand-assisted bipedalism, it might have led to bipedal locomotion on the ground [Thorpe, et al., 2007a, b; Crompton, et al., 2008, 2010]. Although hampered by a top-heavy physique, grasping feet, and joint mobility for clambering [Thorpe, et al., 1999; Payne, et al., 2006], they could have been sufficiently preadapted to hand-assisted locomotion in the trees to evolve unassisted bipedalism on the ground, perhaps driven by fragmentation of woodlands [Crompton, et al., 2008]. This would bypass a quadrupedal transition and a bent-hip-bent-knee stage [Crompton and Thorpe, 2007; Crompton, et al., 1998].

The genetic mutation hypothesis. Bowers [2006] proposed that an alteration in the control region of certain Hox D genes may be responsible for the sudden appearance of bipedality. Filler [2007] similarly proposed that a mutation in Hox genes could produce major shifts like bipedal gait. He specified that this occurred in an ape ancestor 21 Mya.

Combining hypotheses. A frequent approach has been to cobble together existing hypotheses. Sigmon [1971] felt it was incorrect to isolate one reason as being the most important. Robinson's [1972] view was similar. Rose [1984] concurred that there had been many causes, and it was their increasing number that led to selection for upright gait. Combining hypotheses has become commonplace [Marzke, 1992b; Jablonski and Chaplin, 1993; Tattersall, 1995, 1998; Aiello, 1996; Hart and Sussman, 2005]. However, a simple explanation is generally preferable to a complex one and it is debatable if hypotheses which fail to convince on their own are more persuasive when combined.

A major obstacle in the path to an explanation of bipedal gait is that the first hominins to assume this upright mode must have been awkward, inefficient, and more vulnerable to predation [Wrangham, 2001]. Verticality changed requirements for suspension of viscera, support of the fetus, and homeostasis of blood and lymphatic pressure. Practically every postcranial bone and muscle group was remodeled to achieve current levels of bipedal efficiency [Weidenreich, 1947]. Yet even today, numerous ailments associated with upright behavior continue to beset the human body, including back strains, spondylolisthesis, visceroptosis, prolapsed uterus, osteoarthritic hips, cartilage tears of the knee, flat feet, hammer toes, herniated, slipped or shrunken intervertebral disks, varicose veins, hemorrhoids and inguinal or femoral hernias [Straus, 1962; Napier, 1970]. Traveling on two legs is by its nature more unstable than traveling on four. These disadvantages must have been much worse at the onset of bipedal locomotion and through the millenia that followed [Morgan, 1990; R. Leakey and Lewin, 1992; Wrangham and Peterson, 1996].

Two forms of bipedalism. Bipedalism is defined as upright activity performed while supported by the lower limbs. In human evolution studies it has become synonymous with *bipedal locomotion,* to the exclusion of another upright behavior, *the bipedal use of weapons.* Either or both of these activities could account for the evidence of "bipedalism" in early fossil hominins.

Because evolution depends on differential reproductive success, a compelling hypothesis of the origin and evolution of hominin bipedal locomotion should include a description of why at the onset of this behavior there was an immediate improvement in reproductive success, and why this advantage would continue for millions of years while modern human bipedal locomotion evolved. The following explanation proposes that *bipedal use of hand-held weapons* yielded immediate and persisting reproductive advantages which resulted in prolonged natural selection for improved bipedal *stance* that eventually led to increasing use of bipedal *locomotion*.

Bipedal use of weapons and bipedal locomotion. The idea that these two behaviors are linked is not new, but it has only lately been discussed in modern evolutionary terms [Young, 2003, 2009, 2010]. It provides a way to avoid the stubborn issue of how a rudimentary form of upright locomotion could have enhanced reproductive success in the first place. The reproductive benefits stemming from bipedal use of weapons would have provided immediate benefits to the first apes that exploited such behavior and would have endured because it was applicable to fundamental biological needs, prolonging natural selection that would increase effectiveness of that bipedal behavior. Habitual bipedal locomotion could have emerged later, as it was gradually facilitated by adaptive changes in anatomy and physiology that were the result of natural selection for improved bipedal throwing and striking.

This concept, like so many others, can be traced back to Darwin [1871], who conjectured that weapons would have enabled hominins to defend themselves, hunt and fight, and that "a man must stand firmly on his feet" to perform these acts. Bipedal throwing and clubbing remains on the list of possible non-locomotor behaviors that might have given an impetus to bipedal gait: Dart [1949a, 1953, Dart and Craig 1959] agreed that the prime benefit of freeing the hands from locomotion was their use for wielding clubs and missiles. Hewes [1961] cited wide support for this idea. Fifer [1987] and Knüsel [1992] concurred that throwing stones played a role. In 1991, when Rose listed twenty competing hypotheses to explain vertical walking, one of them was "tool throwing". Tuttle and coauthors [1991] thought the weapon-use hypothesis still had appeal. Dunsworth, et al. [2003] mentioned a possible link between throwing and bipedal locomotion. Preuschoft [2004] reaffirmed that throwing was still an option in this regard. Carrier [2004] thought an ability to support the body on the hindlimbs in *Australopithecus* would free the forelimbs for fighting, but did not mention hand-held weapons. Kirschmann [1999] related throwing to upright walking as follows: Throwing facilitated the transition from quadrupedalism to bipedalism due to many advantages arising from use of throwing stones, of which the foremost was defense against predators. Ability to use hand-held weapons increased the fitness of early hominins and thus influenced their evolution. When crossing open country, they would have carried multiple stones as weapons, obligating them to walk upright. Vertical orientation of the upper body for throwing may have contributed to the evolution of vertical gait while increased flexion-extension in the wrist joint to improve throwing would have hampered knuckle-walking.

In 2003 I reasserted the idea that selection for throwing and striking with hand-held weapons might have led to habitual bipedal locomotion and showed how this concept can account for the evolution of the human hand. Parker and Jaffe [2008] concluded that bipedal throwing and clubbing was the best available alternative for explaining the origins of bipedal locomotion. Subsequently, the idea was examined in more detail [Young, 2009, 2010]. This chapter expands, elaborates and updates those two publications.

Similarities of the two bipedal behaviors. What features of bipedal throwing and striking are shared with bipedal locomotion? Both are inherited (they are part of the human ontogenetic program) and both are bipedal motor behaviors performed with the body oriented vertically. Strong, robust legs and stable, springy feet which supply propulsive force are essential to both. Each behavior involves rotation of the femur, pelvis and torso, and coordinated arm and leg movements (Figure 15, p. 208). Both depend upon dynamic vertical balance—the ability to remain upright while moving the body.

Differences between the two bipedal behaviors. Despite these similarities, there are some fundamental differences. The throwing and clubbing motions involve a sequential

CHAPTER 14: BIPEDAL LOCOMOTION

Figure 16. In this drawing three stages of the throwing motion are superimposed. In distinction to bipedal walking, (with its rhythmic, low-energy, repetitive movement) bipedal throwing involves an exceedingly rapid and complex series of coordinated movements which generate a large quantity of kinetic energy that is transferred to a projectile. The entire motion, including the windup, acceleration, and follow-through is completed in less than two seconds. There is no locomotion involved, but rather a propulsive forward stride that is suddenly interrupted at foot strike to exert a rotation around the vertical axis of the body. It is an explosive movement. The body must grasp another missile and rewind the system before another throw can be made.

---

pattern of muscular contractions adapted to maximize the amount of energy transmitted to the weapon and control its trajectory in a single explosive act while the body remains in one place (Figure 16). In contrast, bipedal locomotion consists of a rhythmic, repetitive pattern of muscular contractions adapted to minimize energy expenditure while moving the body to another place. One behavior suddenly generates and immediately expends a large quantum of energy; the other uses small amounts of energy over an extended duration. Thus, the energetics of the two behaviors are drastically different.

The complexity of the throwing and clubbing motions greatly exceeds that of bipedal gait. Walking follows a simple, bilaterally symmetrical pattern, whereas in throwing and clubbing, movements of the left and right side of the body are different (Figures 15, 16). They are asymmetric, as noted in Chapter 10. They are also much more rapid and violent than in bipedal gait and the rotations are greater. During a strong throw, hip rotation is 11 times larger (90° vs 8°) and enormously faster (700°/sec vs about 8°/sec) than in walking [Escamilla, et al., 1998; Wilk, et al., 2000]. There is minimal rotation of the torso during walking, but it may exceed 900°/sec during a club-swing [Welch, et al., 1995], and 1,200°/sec during a throw. In walking, elbow extension and humeral rotation are not required. During a throw, elbow extension occurs at 2,500°/sec and humeral internal rotation may exceed 7,000°/sec [Fleisig, et al., 1995, 1996; Escamilla, et al., 1998]. Both behaviors require vertical equilibrium, but the rapid, violent, thrusting, rotating accelerations and decelerations of throwing and striking place much greater demands on dynamic bipedal balance.

Walking depends mainly upon the lower part of the body. The arm-swing (minor flexion and extension at the shoulder joint) smooths the motion slightly [Elftman, 1939; Zihlman, 1978; Tardieu, 1992] and gains importance in running [Hinrichs, 1999], but the role of the arms in bipedal walking is relatively insignificant [Lovejoy, 1988; Fleagle, 1999] and can be considered largely independent of the evolutionary transition to bipedal locomotion [Latimer, 1991]. The upper body is so unimportant in the act of walking that it is commonly ignored in biomechanical studies of this behavior: head, arms and torso are simply represented as a single rigid body that articulates with the pelvis [Kepple, et al., 1997; Anderson and Pandy, 2003]. In contrast, movements at all upper limb joints contribute to the throwing and club-swinging motions, and the hand, irrelevant to bipedal locomotion, is the *sine qua non* of throwing and club-swinging.

Two different bipedal behaviors evolved simultaneously. Bipedal use of hand-held weapons and bipedal locomotion are two different behaviors subserving two different functions. Nevertheless, fossil evidence shows that for most of their evolutionary history they evolved simultaneously, beginning near the origins of the hominin lineage and continuing for millions of years afterwards. Natural selection preserves inherited variations that promote *net* reproductive success. When two behaviors have similar requirements, they can be promoted synergistically. When they have conflicting requirements, they constrain each other. For optimum effectiveness, bipedal use of weapons and bipedal locomotion require two different body types—muscular and robust for prowess with weapons, smaller and less massive for locomotion (Chapter 12). This indicates that mutual constraint on body type would have occurred during their evolution.

Which behavior came first during hominin evolution? Currently, the fossil record does not provide the basis for an answer. Use of spherical and cylindrical tools, diminution of canines, and some kind of bipedal behavior can be traced as far back as the earliest hominin hand specimens, canine teeth, pelves and lower limb bones. The bipedal behavior might have been use of hand-held weapons or it might have been locomotion. Insofar as the paleontological record can be read, the evolution of these two bipedal behaviors could have coexisted for five million years. Although an answer to the question of which had chronological precedence may be premature, probabilities can be discussed.

What if bipedal use of weapons came first? The major advantages of this sequence are threefold. First, the obstacle to explaining the origin of *bipedality* is hurdled. Important reproductive advantages would accrue immediately with the onset of hand-held weapons dispatched from a bipedal stance (Chapter 3). Second, natural selection that acted to promote this type of bipedalism, by improving bipedal stance and upright dynamic balance on stable, powerful lower limbs and springy feet would promote a transition to more frequent use of bipedal locomotion. Third, the early onset of changes in the hand and canine teeth are readily explicable by natural selection for the use of hand-held weapons, but not bipedal locomotion.

This proposal is expanded in the following scenario: The last common ancestors of hominins and chimpanzees were arboreal apes who spent part of the day on the ground, where they probably traveled quadrupedally due to the added stability of four limbs versus two. Their orthograde body plan, adapted for vertical climbing and feeding, permitted short bouts of bipedal terrestrial walking, but this was hampered by grasping feet, small, compliant lower limbs, top-heavy center of mass and unstable vertical balance. Bipedal use of hand-held weapons provided the reproductive benefits that drove natural selection for improved bipedal stance when bipedal locomotion was rare, rudimentary and precarious.

A male dominance hierarchy established by threat and physical force (such as grabbing and biting) in the panin/hominin ancestor seems likely. In such circumstances, about seven million years ago, it is proposed that an aggressive male began to throw rocks at adversaries with the intent to injure and dominate them. This signals the advent of the hominin lineage and the use of tools as weapons. The advantages it provided in the competition for scarce resources assured that other males would take up the behavior. Sticks, bones and rocks were available in the natural environment for use as clubs. Stones could be used as throwing weapons. Fractured bones might serve as daggers or thrusting spears. The emerging hominins did not *make* weapons, they found them in their surroundings, picked them up, and used them (Chapter 4). Repetition of this innovative behavior of using hand-held weapons to inflict harm during conflicts with other males would have hastened the rise to higher rank of those who employed it.

The ascent in dominance facilitated by this new mode of combat yielded an increase in reproductive benefits, including more breeding opportunities and preferred access to food. Mimicked by other males, the behavior spread within the community, leading to a band of apes who settled disputes by throwing and striking (or threatening to do so). When conflicts arose with neighboring communities who did not use weapons, the new behavior provided more advantages in proliferating their kind: more food, more females and improved, expanded territorial resources. Females in the community also had enhanced reproductive prospects due to better nutrition that promoted the reproduction of their genes and the development of their progeny.

This first community of hominins, distinguished by a new behavior, had embarked on a multimillion year process of natural selection for improved bipedal use of weapons which endured as long as this behavior yielded reproductive advantages. In each succeeding generation, those who were more adept at bipedal combat left more offspring, their genes increased in the gene pool, and their bodies were increasingly adapted for that behavior.

Adaptation to bipedal use of weapons would have led to improved dynamic vertical balance and more stable upright posture, controlled by altered pelvic muscles which regulated lateral tilt of the torso, increased ability to rotate the hips, enhanced capacity to extend the femur at the hip and knee joints, with larger leg muscles for more powerful thrust and a foot transmuted from a grasping appendage to a resilient, stable, platform. Changes in the pelvis, lower limbs, feet and hands reduced the ability to locomote in trees and made quadrupedal locomotion on the ground more difficult, but bipedal locomotion was enhanced in numerous ways, leading to its more frequent use. In the transition to bipedal walking which followed prior selection for bipedal weapons use, hominins moved through a series of compromise forms in which each activity was effective, each constrained the other, and both were improving.

Upright use of hand-held weapons leads to upright walking *indirectly* in this proposal. Natural selection for bipedal stance (posture) is asserted to be the forerunner of bipedal locomotion because natural selection that acted to improve one form of bipedal behavior promoted the transition to another. As bipedal locomotion became a more efficient and common activity, it began to yield non-locomotor advantages, such as the ability to carry weapons, food and babies while moving across the landscape. Eventually, perhaps early in the hominin lineage, bipedal locomotion became sufficiently crucial to reproductive success that it also began to affect natural selection, providing benefits not associated with armed conflict but rather with efficient travel. From then on, the two functionally different behaviors evolved simultaneously.

What if bipedal locomotion came first? Two notable factors seem to argue against this alternative.

First, no compelling reproductive advantages have yet been proposed that would immediately provoke natural selection to improve the initial, rudimentary stage of bipedal locomotion and would continue to act for millions of years thereafter. These are the barriers that impeded acceptance of previous hypotheses.

Second, one or more additional explanations would be required to account for the ontogeny and prowess of modern humans in throwing and striking, our distinctive hand structure and many other present and previous hominin features that are all beyond the scope of existing hypotheses of the origins of upright walking. The concept that bipedal use of weapons came first accounts for bipedal walking *and* numerous additional features of human evolution.

Some might complain that this is just another untestable hypothesis, akin to earlier hypotheses listed above. However, it is not a hypothesis and it has been tested. It is not a tentative construal of a single item of experience. Rather, this explanation of bipedal locomotion emerges seamlessly from a broad explanatory structure based on the general principle that hominins were adapted by natural selection for bipedal throwing and club-swinging because of the reproductive benefits it provided (Chapter 16). Secondly, this principle has been tested. It has been shown in previous chapters to offer a consistent account of numerous features of hominin origins and evolution, as well as modern human anatomy and motor behavior.

The next chapter will show how it can illuminate the exquisite control of the arm and hand that is an inherited characteristic of the human central nervous system.

CHAPTER 15

The Central Nervous System and Bipedal Use of Weapons

"The soundest argument by far for the throwing hypothesis is the enormous [neural] demands of the human high-performance throw...The brain entered entirely new realms of achievement" [Kirschmann, 1999, 1.2].

"Clearly, a perfectly accurate, real-time internal model of arm mechanics would need enormous computational power..." [Hore, et al., 1999b, p. 1187].

Introduction. The central nervous system is the major regulator of behavior. If natural selection acted to improve throwing, then heritable variations in the central nervous system that facilitated throwing would have been naturally selected due to the enhanced reproductive success of individuals who possessed these genes. If so, distinctive traces of this adaptive process should be demonstrable in the central nervous system in modern humans. The following discussion shows unequivocally that convincing evidence in support of this prediction is available.

Kirschmann [1999] was the first to propose and elaborate the idea that natural selection for aimed throwing would have had important consequences for the evolution of the brain. Only humans throw well, he wrote, and the human act of throwing is exceedingly complex, involving numerous rotations of different parts of the body. Furthermore, it happens very rapidly. A high-performance, well-aimed throw makes enormous demands on the central nervous system. Possibly no greater feat of coordination than the human throwing motion has ever evolved. This must have involved highly significant modifications of the brain [Kirschmann, 1999, 3.3.1].

Among the changes Kirschmann anticipated was improvement in calculating target distance, which is crucial to an accurate throw [1999,1.4]. In his view, this was a major factor in remodeling brain structure and causing brain expansion. Target distance is partly determined from a comparison of remembered size of objects with the size of the object's image on the retina [3.3.4]. This means it is dependent upon *memory*, which contributes importantly to human cognition. Because cognitive ability that evolved to solve one task can be applied to others, the human ability to devise complex scenarios may have arisen from this mental faculty. This and the evolved thrower's abilities to process sequential data may play a role in the human ability for advanced planning and the development of language [3.3.4; 6.8; 7]

Kirschmann thought the adaptive traits for throwing were substantially more demanding of the brain than those in the rest of the body [1999, 3.3.4]. The throwing adaptation involved decisive changes in the hominin body plan, but the required parts were already present. In contrast, neural pathways in the brain needed substantial revisions to meet the enormous demands of a high-performance, aimed throw. Because it was a novel behavior, absent in hominin ancestors, unprecedented neural interactions had to be created. This adaptation within the brain was very significant in the evolution of human mental capacity [Kirschmann, 1999, 1.2; 3.3.1].

When Kirschmann's book was published, neurophysiological analysis of the control of throwing had just been initiated by Jonathan Hore and his colleagues at the University of Western Ontario. They focused their research program on central nervous system control of the arm and hand during aimed throwing, using high-speed analysis of the throwing motion based on magnetic-field search-coil technique. Subjects equipped with search coils taped to the back of the third distal manual phalanx, the back of the hand and forearm, the lateralaspect of the upper arm, the scapula and sternum threw balls while seated in three orthogonal alternating magnetic fields. Coil voltages were used to calculate the angular positions of each segment in 3D space. Timing of ball release was measured with a microswitch attached to the distal phalanx of the third finger; ball speed was recorded by a radar gun. EMG activity was also analyzed [Hore, et al., 1999b; Debicki, et al., 2011].

Control of hand and arm movement during throwing. A feature of skilled throwing is fast speeds, which approach 160 km/h in the most rapid throws. Wrist flexion is important in achieving a high ball velocity [Gray, et al., 2006]. As the forearm accelerates forward, the hand lags behind, moving into full extension at the wrist joint (~ 30°). This represents an *interaction torque*, which occurs when a muscular torque at one joint elicits a torque at an adjacent joint. Interaction torques are maximized when the adjacent limb rapidly changes its angular velocity or reverses it [Sarlegna and Sainburg, 2009]. Many interaction torques occur throughout the throwing motion.

When the forearm accelerates and the wrist lags behind, wrist flexor muscle activity begins, contributing an active (muscular) flexion torque at the wrist joint [Debicki, et al., 2004; Hirashima, et al., 2003]. This adds to a passive, interaction flexion torque that results when forearm deceleration occurs (a whip-like effect) [Debicki, et al., 2004; Gray, et al., 2006; Hirashima, et al., 2007]. Skilled throwers produce a large elbow extension deceleration before ball release [Hore, et al., 2005a, b; Gray, et al., 2006; Hirashima, et al., 2007, 2008]. This deceleration may also counteract wrist extensor interaction and viscoelastic torques [Debicki, et al., 2011]. (Deceleration of elbow extension after ball release by a passive interaction torque initiated by the CNS at the shoulder joint acts to prevent injury [Hore, et al., 2011]). At the release point, wrist flexion velocity reaches its peak and the hand is aligned with the forearm [Gray, et al., 2006; Debicki, et al., 2004]. The forearm is aligned with the upper arm and the plane of the shoulders for a leverage effect that maximizes the speed transferred to the arm and hand from rotation of the trunk

[Atwater 1979] (Chapter 13 Fig. 12).

In the upper body, the velocity of ball release is mainly produced by forward rotation of the trunk, internal rotation of the humerus, elbow extension and wrist flexion. Skilled throwers configure the limb to maximize all four of these angular velocities [Hirashima and Ohtsuki, 2008] and all of these joint-rotation angular velocities are the net outcome of the effects of interaction torques, muscle torques, gravity torques and velocity-dependent torques [Hirashima and Ohtsuki, 2008]. Hirashima, et al. [2007, 2008] use the term "velocity-dependent torques" to emphasize that some interaction torques may be traced back to origins in body segments some distance from where their ultimate effects can be measured.

The velocity-dependent torque is the result of the history of the muscle torques acting at the joint under consideration."The effect of the velocity-dependent torque on the joint angular acceleration is related to the phenomena called 'whip-like effect,' 'proximal-to-distal sequence' and 'kinetic chain' " [Hirashima, et al., 2008, p.2875] (Chapter 13). Some velocity dependent torques affecting the arm may prove to have a history that included muscular torques generated in the legs and hips. However, the participation of several parts of the body during throwing has yet to be examined from this perspective, so that "...future studies need to examine the over arm throwing with a more accurate model that includes the non-throwing arm, finger joints, or the lower extremity with foot ground contacts" [Hirashima, et al., 2008, p. 2882].

Fast throws are associated with large back forces from the ball on the fingers that must be controlled if accuracy is to be achieved and injury prevented. These forces occur because as the fingers exert force on the ball to accelerate it, the ball exerts an equal force on the fingers [Hore, et al., 1999b]. Back forces begin with the start of hand acceleration and increase progressively throughout it. In the process of ball release the back force on the fingers' distal phalanges increases during extension and peaks at final release from the fingertips as kinetic energy is transferred to the missile through the extended index and middle fingers [Hore, et al., 2001]. This back force may contribute to extending the fingers, but the fingers do not extend fully because the CNS anticipates the size of the back forces and generates appropriate finger flexor torque to oppose them [Hore, et al., 1999b, 2001]. Computation by the cerebellum of finger stiffness based on hand acceleration may also be involved [Hore and Watts, 2011].

In addition to release *velocity*, a second feature of skilled throwing is missile *accuracy*. The major factor determining accuracy is the timing of onset of finger opening [Hore, et al., 1996a, 1996b] because timing errors affect the missile release point on the high-velocity flattened-arc hand trajectory, and there is only one point on this trajectory where release will result in a missile path towards the target [Hore, et al., 1995; Watts, et al., 2004]. In recreational baseball players ball release has a timing variability of up to 10 ms [Hore, et al, 1995; Timmann, et al., 2001], which is reduced to 5 ms or less in very skilled subjects [Jegede, et al., 2005]. To throw consistently within the baseball strike zone requires a timing precision of 1-2 ms [Hore, et al., 2002; Hore and Watts, 2005]. In the release sequence, the thumb abducts and the fingers extend. This causes the missile to uncouple from the hand grip and roll distally along the flexor surface of the fingers. During ball rolling, force continues to be applied until final release occurs from the fingertips. The rolling sequence takes about 15-20 msec [Hore, et al., 1996a,b; 1999b].

The central nervous system controls this complex biomechanical apparatus. The CNS determines the motor command for each of the many muscles involved in bipedal throwing. In addition to muscle torques, gravitational torques, interaction torques and velocity-dependent torques, there are also inertial effects which vary as the interacting body segments of the kinematic chain change their spatial orientations throughout the throw [Hirashima, et al., 2007]. All of these variables must be predicted and controlled by the CNS because all affect the end results of both missile velocity and the accuracy of its path toward the target.

Control of the arm by the CNS during throwing represents an extreme manifestation of the regulation of arm movements used in daily activities. In a simple movement such as reaching for an object, a motor plan is prepared in advance [Sarlegna and Sainburg, 2009]. Vision and proprioception contribute to different parts of this plan. Vision is used to identify the spatial coordinates of the target, an essential component in assembly of the motor plan governing the trajectory and motion of the arm. Proprioception provides information about the configuration and motion of the arm and may participate in transforming the spatial plan into a coordinated output of motor commands that produce the torques that propel the hand toward the target [Sarlegna and Sainburg, 2009].

The cerebellum is one CNS component that has been implicated in the control of torques during throwing. A skilled, fast throw requires a complex pattern of interjoint coordination, and that involves regulating interaction torques [Hore, et al., 2005b]. Subjects with cerebellar deficits have slower arm movements which have been attributed to an inability to exploit such torques [Timmann, et al., 2001, 2008]. There is broad agreement that the cerebellum contributes to precision in timing of hand movements, but there were no specific proposals about how this would occur for finger opening in throwing until Hore and Watts [2005] proposed a hypothesis consistent with a current model for reaching and grasping.

The reach-and-grasp model includes a transport phase in which the hand travels on a predetermined path and a grasp phase that involves finger motion. The reach component uses sensory information about the location of the target to generate the handpath movements, whereas the grasp component uses information about the state of the arm while the movement is in progress. Cerebellar patients seem to lack the ability to combine reach and grasp movements into a single motor program [Zackowski, et al., 2002]. Similarly, studies of ball release during throwing support the idea that precisely timed finger opening is controlled centrally by a process in which a spatial controller (possibly the cerebellum) matches angular positions of finger opening to the intended handpath [Hore and Watts, 2005]. The effect of each muscle activation on joint angular acceleration depends on the joint angle at which the muscle acts and others at which the muscle does not act. The CNS must consider joint angles throughout the throwing apparatus to determine the motor command for each muscle [Hirashima, et al., 2007] because the angular acceleration by torques depends on the system mass matrix as well as the magnitude of muscle and velocity-dependent torques [Hiroshima and Ohtsuki, 2008]. Granted that the musculoskeletal movement of the throwing motion is exceedingly complex, and the CNS controls it, what can be said about the organization of this control?

The brain contains an internal model of the throwing motion. A current view is that the nervous system acts through *predictive* mechanisms that involve *an internal model* of limb and movement dynamics [Hore, et al., 1999b; Debicki, et al., 2010]. According to this concept, the brain analyzes incoming information from several sensory modalities (especially vision and proprioception), then sends out signals to muscle units based on a prediction it has made from its analysis. During the relatively slow movements of normal activities, incoming sensory information can be used by the CNS to update the internal model of limb mechanics to make adjustments to the initial plan [Sarlegna and Sainburg, 2009].

In contrast, once the throwing motion is underway, by the time it reaches the upper body, it is too rapid for proprioceptive feedback from wrist, elbow, or shoulder joint rotations to control finger opening [Hore and Watts, 2005]. It is also too rapid for corrections based on visual input. The interval between front foot contact and missile release is 145 msec, less than the time required for a visual-motor reaction (estimated at ~ 150-170 ms: [Saunders and Knill, 2004]). The acceleration phase lasts only 30-50 ms (Chapter 13). Feedback adjustments are physiologically impossible. The brain has to *predict* when finger-opening should occur for a fast, aimed throw and integrate all of the muscle, gravity, interactive, and velocity-dependent torques and changing inertial factors into its prediction, then issue the neuromotor commands to each and every muscle group that are appropriate to this outcome. It is an awe-inspiring feat, unique to the brain of our species. It shows unequivocally that natural selection for improved throwing during hominin evolution had profound effects on the brain.

Neural control of handedness. Another singular aspect of human behavior which appears to be linked to the throwing adaptation is handedness (Chapter 10). Distinct CNS control mechanisms may be employed for dominant and nondominant arm movements. Hore and coworkers [1996b] found that when subjects threw balls with either arm, hand trajectories and rotations at all joints were more variable in the nondominant arm and accuracy was decreased. The major cause of throwing inaccuracy was more variable timing of ball release due to diminished control of the timing of onset, amplitude and velocity of finger extension. Because the fingers of each arm are activated primarily by the contralateral hemisphere, the nondominant arm's lack of accuracy may be due to a lack of precision in commands from the contralateral hemisphere [Hore, et al., 1996b]. Hore and colleagues subsequently reported that when skilled throwers threw with the nondominant arm, the result was slower arm movements and an inability to exploit interaction torques. In the dominant arm a more complex pattern of interjoint coordination exploited such torques [Hore, et al., 2005a, b; Timmann, et al., 2008; Gray, et al., 2006]. "Given that fast and accurate throws are made with the dominant arm...the mechanisms that control the dominant arm are specialized for exploiting limb dynamics in feedforward situations" [Hore and Watts, 2011, p. 2032].

Studies of other types of movements by the dominant and nondominant arm have yielded similar results. Sainburg and colleagues have demonstrated that the dominant arm is better able to coordinate intersegmental dynamics for specifying trajectory speed and direction. The nondominant arm limitation is not due to a torque production deficit. Rather, it is the result of less efficiency in producing such torques [Bagesteiro and Sainburg, 2002]. During rapid target-reaching movements, dominant arm advantages seem to involve more accurate predictions of the effects of arm intersegmental dynamics by the CNS during movement planning [Sainburg and Kalakanis, 2000]. Although differences in coordination might also reflect more practice in use of the dominant limb, these authors emphasize the genetic aspects of handedness, proposing that it emerges from unique neural circuits in each hemisphere that are specialized for controlling different aspects of limb movements. Handedness may confer direct advantages for the control of movements [Sainburg, 2005]. According to this scheme, handedness developed in response to a need for more precise coordination.

Conclusion. Apart from recent studies dealing with the arm and hand, summarized above, the promising research field that links the central nervous system with upright throwing has yet to be actively exploited. Neural control of the lower body during the throwing motion has not been investigated. There are no studies of the neurophysiology of the club-swinging motion or of the control of upright dynamic balance during throwing or clubbing. Simple considerations, however, suggest that the control of these behaviors will be found to be widely distributed in the CNS in sensory, motor and executive components, conscious and unconscious alike.

For an early hominin to throw or swing a club effectively he must detect the target animal, transfer attention to it, process the visual input to extract information about recognition, distance, size, orientation, motion, and other variables, then decide what action to take based on information from visual analysis, other senses, memories and emotions. If the decision is made to attack, he must decide when and how close to advance, which weapon to use and where to aim it. If he chooses to throw, he must decide when to do it.

When the "throw now" decision is made, unconscious mechanisms produce the throwing motion, which requires an aiming system. The foveal image on the retina, where visual acuity is greatest, presumably defines the target for the neural computations that control the trajectory of the hand and weapon. Because the throwing motion is too complicated and too rapid to be newly created in an instant or regulated by sensory feedback as it occurs, it must be run off automatically from a stored motor program. This program is adjusted before the motion begins, taking into account target direction, distance and movement, as well as muscle, gravity, interaction and velocity-dependent torques, inertial effects, the size and mass of the missile and the back force it will exert on the fingers. Asymmetric movements of both sides of the body, from foot to fingertip, must be predicted, programmed and coordinated in proper sequence to convey the most energy to the weapon and launch it towards the target.

Timing of missile release appears to be controlled by a mechanism arising from CNS calculations based on positional variables related to the path of the hand. The throw consists of a series of muscular events driven by coordinated firing of motor neurons in different brain sectors with precise patterns, durations and frequencies. Numerous components of the CNS act simultaneously to maintain dynamic vertical balance: sensory input from the visual system, vestibular apparatus, prop-rioceptive sensors, golgi organs, plantar pressure sensors, cutaneous receptors in the feet, and reafferent motor signals are sent to CNS equilibrium control centers, where calculations determine which muscles need to be activated or deactivated to maintain upright bipedal balance as rapid displacements of the center of gravity occur during the explosive motion.

It is difficult to comprehend the magnitude of the computations that must occur during the fleeting interval after the decision to initiate a throw has been made until it is completed. The forward movement of the arm and hand during the acceleration phase requires enormous computational power [Hore, et al., 1999a, b], and these segments are only part of the human throwing motion. Although the modern human brain clearly exhibits this computational capacity, it was not present at the outset of hominin evolution. In agreement with Kirschmann [1999], it seems that these and other mental properties, developed during several million years of hominin adaptation to bipedal use of weapons, provide the foundation of the modern human mind.

CHAPTER 16

The Structure of Evolutionary Explanations

"It is the desire for explanations which are both systematic and controllable by factual evidence that generates science; and it is the organization and classification of knowledge on the basis of explanatory principles that is the distinctive goal of the sciences" [Nagel, 1979, p.4].

"If any problem in the philosophy of science justifiably can be claimed the most central or important, it is that of the nature and structure of scientific theories, including the diverse roles theories play in the scientific enterprise" [Suppe, 1977, p. 3]

Introduction. This book addresses both the evidence and the explanation of human origins and evolution. Evidence forms the basis of science, but the goal of science goes beyond collecting facts to exploring the significance of the evidence. What do the hard-won hominin fossils mean? What can they tell us about our ancient ancestors' behavior? Can they support predictions about future evolution? How can the distinctive changes in hominin size and shape through the millenia be explained? *What does it mean to explain something?*

There is no precise definition of scientific explanation or what structure it should have [Hempel, 1965b]. Explanation can take different forms, depending on what is being explained and because there is always more than one way to explain it. However, it is clear that the *goal* of explanation is to increase comprehension by reducing a situation to elements that are so logical or so familiar that we accept them as understood. The evidence is not enough; we want to understand its significance.

Explanation may be accomplished in various ways, but there are some general rules. Scientific explanations share a basic two-part form: There is (1) something that needs to be explained (called the *explanandum*, plural *explananda*), and (2) there are statements that provide the explanation (the *explanans*). The explanation should reduce the number of unexplained items by at least one, otherwise no progress has been made. Some regularities can be explained by showing they are manifestations of another, more comprehensive regularity. That is, a deeper understanding of the matter is achieved if it can be shown to be an outcome of more fundamental processes. Explanations are not true or false; rather they are accepted or rejected according to whether they are more or less adequate than other possible explanations [Suppe, 1977].

Biological explanations require elements that are not necessary in explanations of the inanimate world. Living organisms are enormously more complex than non-living entities such as atoms, molecules and rocks. Life involves phenomena not found in physics and chemistry so we cannot expect to explain them with purely physical laws. They are not just *ordered*, like the periodic table of the elements or atoms in a crystal, they are *organized*: their multitudinous parts have *functions* which act to sustain the organized structure and to reproduce it. Living organisms obey the laws of nature, but they also are subject to biological laws or concepts, such as genetic inheritance, reproduction, the succession of generations, natural selection, sexual selection, competition, adaptation, and reproductive success.

It is a commonsense notion that to explain something is to identify its cause [Salmon, 1990]. This also applies to scientific explanation. One seeks to identify causes: factors that by their own direct effects produce change. Scientific evolutionary biology is founded on the Darwinian causal process—natural selection of inherited variations. This accounts for the origin of species and the emergence of evolutionary adaptations [Beckner, 1959; Goudge, 1961; Sober, 1984; Rosenberg and McShea, 2007]. Natural selection of inherited variations that enhance reproductive success is the causal concept that underlies evolutionary theory. Therefore, in specific cases of evolutionary change (such as the transition to bipedal locomotion, canine diminution, throwing prowess and the like), we seek to identify the factors that yielded more food or mating opportunities to those whose genes were preferentially copied. We ask, in other words, "where was the reproductive advantage?" This represents the search for *ultimate* causes in biology [Mayr, 1961].

Hypotheses and theories. Hypotheses and theories are both forms of explanation, but despite the frequent tendency to use these terms interchangeably, in strict scientific practice they are very different.

The *hypothesis* stands as the lowest tier in the hierarchy of scientific explanation. Simple hypotheses are *ad hoc* explanations of individual events without any wider application. Their scope is a single object of interest. They are tentative explanations of a singular item of experience, based upon causes that seem plausible or at least possible. *Theory-based* hypotheses are another matter. They are predictions of the form "*if* the theory is true, *then* the following outcome should occur," but a hypothesis unattached to a theory is only a speculative guess.

The *theory* resides on the highest level of scientific explanation. In biology, the overriding theory is the modern Darwinian theory of evolution. Subsidiary Darwinian theories pertain to particular aspects of biology; they are all special cases of the general theory of evolution. I shall now describe the classical structure of a theory and present the bipedal weapons proposal in that format later in this chapter. Science welcomes other explanatory schemes with different structures [Kuhn, 1962]. However, some are more convincing and useful than others.

Theories come in different sizes, bounded by the range and detail of evidence which they can explain. Depth, scope, specificity of application and the parsimony of explanatory principles are the measures of a theory's success. Darwin's theory clarified all of biology, explaining it by the principle of natural selection through differential reproduction. It is the quintessential theory.

Scenarios. Somewhere between an unsupported hypothesis and a scientific theory lies an informal explanatory structure referred to as a "scenario". Some writers employ the term in the cinematic sense to mean the summary of a story. Others apply it to evolutionary accounts that lack the structure or scope of a theory but are more elaborate than hypotheses. One variety is the "historical narrative" which Mayr [2004], believed could provide a "quite plausible" story of human evolution. It had the form of a scenario in which past events are accounted for by their consequences: one stage might have led to the next stage, which could have been followed by another, and so on. This format was used by Kingdon [2003] to explain hominin bipedalism. Here are some examples of scenarios which seek to account for more than one hominin trait by describing a sequence of events that might have led to their emergence.

Man the Hunter. Dart [1925, 1949a, b, 1953, Dart and Craig, 1959] created a scenario involving the use of weapons and used it to explain upright stance, diminution of the canine teeth, hunting and fighting prowess of later hominins, and modern human skill at aimed throwing. It was part of Darwin's vision, but devoid of Darwin's emphasis on intellectual powers and social qualities [1871, p. 157]. Male hunting prowess subsequently was said to explain a wide variety of human social attributes [Washburn and Lancaster, 1968; Laughlin, 1968]. Our intellect, interest, emotions, division of labor between genders and basic social life were all ascribed to an adaptation to hunting. Males hunted, females gathered vegetable food and the results were shared with each other, forming the basis for the human family.

Woman the gatherer. This scenario shifted emphasis to the female gender. Gathering and sharing by females was the foundation of human society. Food sharing and the family developed from the bonds between a mother and her children. Females nourished themselves and their young; they were the providers and the inventors. Gathering, carrying and sharing food spawned bipedal locomotion. Females chose to copulate with more sociable, less disruptive, food-sharing males—in other words, with males more like themselves. This resulted in the diminution of male canine teeth. Females may have chosen to mate with males having smaller canine teeth because they appeared to be more sociable [Linton, 1971; Tanner and Zihlman, 1976; G. Isaac, 1978; J. Lancaster, 1978; Zihlman, 1978, 1981; Fedigan, 1986].

Man the provisioner. Lovejoy reversed genders again. Males were the providers. They brought plant food to females as part of a food-for-sex exchange based on pair-bonded monogamy with dedicated sexual privileges. Food provisioning led to a gradual shift from a female-centric group to a "bifocal" one. Continually receptive females with no visible signs of ovulation ("ovulatory crypsis") enter this scenario. These innovations would require "copulatory vigilance" in both sexes to ensure fertilization. Monogamous copulation would increase pair-bonding and serve as a social display asserting that bond. Canine diminution resulted from this cooperative behavior. Bipedal locomotion emerged from males carrying provisions to their mates [Lovejoy, 1981, 1993, Parker, 1987].

The cooking scenario. Wrangham [2001, 2009] proposed that the control of fire and its subsequent use in cooking were defining events in human evolution, forming the basis of the transition from *Australopithecus* to *Homo* (Chapter 7). The benefits of cooked food resulted in smaller digestive tracts, and large body size. Shoulder, arm and trunk adaptations that enabled hominins to climb well disappeared, leading to dedicated bipedal locomotion when hominins controlled fire and slept on the ground.

Man the thrower. A scenario with expanded scope that avoids social assumptions, is simple in its explanation and more overtly linvolves natural selection was presented by Kirschmann [1999], who refers to it as a "model." Throwing rocks from a bipedal stance as a defensive measure by early hominin males plays an important role in his proposal, which extends its explanatory range to include the central nervous system (Chapter 15). Several features of hominin fossils and modern humans (including bipedal locomotion) are attributed to an adaptation to throwing. Hominins collected rocks for throwing and carried them when they crossed open country. This required upright walking, a behavior that was facilitated by physical adaptations to the improved use of projectiles, such as upright orientation of the body. The longer, stronger legs of *H. erectus* not only improved bipedal gait, they also served as a counterweight that permitted the use of a windup in throwing. The transition to bipedality rests upon the multiple advantages of walking upright for the use of throwing weapons against predators [Kirschmann, 1999; 1.4, 3.6] (Chapter 14). Diminution in size of hominin canine teeth resulted from the loss of their function in fighting which can be traced to the intensive use of rocks as weapons. The weapon characteristic had been transferred from the teeth to the hands [Kirschmann, 1999] (Chapter 11).

The Adaptive Suite. An innovative type of explanatory structure has recently been applied to human evolution in Lovejoy's latest version of his Man the Provisioner scenario [2009a, b], which he calls an "adaptive suite." This format can be traced back to an ecological account of weasel metabolism [Brown and Lasiewski, 1972]. Lovejoy defines adaptive suites as "semiformal, largely inductive algorithms that causally interrelate fundamental characters that may have contributed to an organism's total adaptive pattern" [2009b, p. 74e1]. He asserts that "the effective power of adaptive suites is demonstrable by their explanatory success" [Lovejoy, 2009b, p. 74e2], citing Pianka's [1988] example of the horned lizard.

The last common ancestor of humans and African apes probably exhibited a multimale, multifemale mating structure with moderate canine dimorphism and minimal male-to-male agonism [Lovejoy, 2009b]. *Ar. ramidus* differs by the absence of large canine teeth in males and its ability to walk upright [Lovejoy, 2009a]. These are the two traits he endeavors to explain: Why did early hominin canine teeth become smaller and why did hominins adopt bipedal locomotion?

Ar. ramidus had small canines. Because large canines are correlated with male-male conflict, *Ardipithecus* males must have stopped competing aggressively with each other due to a major shift in social structure, including reduction of fighting, regular provisioning of females, pair-bonding and male parental

CHAPTER 16: THE STRUCTURE OF EVOLUTIONARY EXPLANATIONS

investment [Lovejoy, 2009a]. Temporary pair bonds based on sex-for-food exchanges encouraged copulation with males who provisioned. Females "may have become progressively more solicitous of smaller-canined (and thereby less agonistically equipped) males, particularly if they could encourage such males to habitually target them [with provisions] in preference to other females" [Lovejoy, 2009b, p. 74e5]. These multiple developments explain canine diminution.

Regarding bipedal locomotion, Lovejoy rejects any role for weapons because "habitual bipedality evolved millions of years before any evidence of stone tools" [Lovejoy, 2009b, p. 74e5]. This is the familiar no hand-made tools = no weapons argument (Chapter 4), but Lovejoy adds a new one: human scrota are more pendulous than those of chimpanzees, making them "extraordinarily vulnerable during upright combat. [Thus]...it seems illogical to attribute habitual uprightness to weapons" [Lovejoy, 2009b, p. 74e3].

"Because upright walking provided no energy advantage for *Ar. ramidus*...reproductive success must have been central to its evolution in early hominids" [Lovejoy, 2009a, p.74]. Some *other* behavior must have provided the benefits that promoted bipedal gait. It was *carrying things*. Bipedality "permits food transport over long distances, a behavior not generally feasible for an arboreal or quadrupedal hominoid. Bipedality also facilitates the regular use of rudimentary tools" [Lovejoy, 2009b, p. 74e5]. However, since carrying objects does not by itself yield reproductive benefits, the argument requires another element: "vested provisioning".

In a flow chart, two boxes contain the explananda (bipedal locomotion and loss of honing canines) [Lovejoy, 2009a]. Three other boxes read "vested provisioning, ovulatory crypsis" and "reduced intra-sexual agonism and increased social adhesion". Vested provisioning sends arrows to the other four boxes. Canines and bipedality receive arrows only from vested provisioning. Bipedality sends one back. The caption states that "...females preferred nonaggressive males who gained reproductive success by obtaining copulation in exchange for valuable foods (vested provisioning). Success would depend on copulatory frequency with females whose ovulatory cycles were cryptic. The result would be reduced agonism in unrelated females, and cooperative expansion of day ranges among equally cooperative males, eventually leading to exploitation of new habitats" [Lovejoy, 2009a, p. 74]. Additional details were displayed in a more elaborate flow chart [Lovejoy, 2009b, Fig. 2, p. 74e4] which contains 28 text-containing boxes associated with 56 arrows. Ovulatory crypsis is introduced because while males were in outlying areas foraging for provisions, their mates might be inseminated by males who stayed behind. "That is, males could only succeed by provisioning mates with self-crypsis; they would otherwise be unprotected from female copulation with more dominant/aggressive males while ovulating" [Lovejoy, 2009b, p. 74e6]. Evidently, male provisioning had to await the evolution of ovulatory crypsis.

A critique of the adaptive suite. This explanation omits the role of natural selection. However, the focus here is on the *structure* of scientific explanations. As noted above, an explanation should reduce the number of unexplained items by at least one, otherwise no progress has been made. There are two items in the explananda (bipedal locomotion and canine diminution). How many unexplained factors are introduced to explain these two traits? There are many. First, there is ovulatory crypsis. Without it males would not have wanted to provision a female and her children, yet ovulatory crypsis is unexplained. If it is accounted for by "fear of cuckoldry", how is that explained? Did males understand paternity? If so, did they care about it? This needs explaining. Males foraged in outlying areas to avoid competing with females and children. What accounts for such behavior? The reason why males lost their ancestral aggressiveness requires explanation, as does permanent pair-bonding between mates, females who were continuously receptive but only to one male, and male investment in parenting behavior.

Males not only lost their aggressiveness and their dental armament, they had to forage in more dangerous, far-flung regions without weapons and had to expend further time and energy acquiring more food than they could eat and carrying it back to a female who gave them additional dependents every three or four years. Why did hominin males behave in such a non-apelike way? Why would males who were skilled at fighting

CHAPTER 16: THE STRUCTURE OF EVOLUTIONARY EXPLANATIONS

reap fewer reproductive benefits than males of gentle demeanor who lacked threatening canines?

This adaptive suite does not meet the requirement that a scientific explanation should reduce the number of unexplained items. Also absent is a scheme of causality. Evolution results from a change in gene frequency due to natural selection, but instead of causality based on differential reproduction, the adaptive suite provides a complicated network of elements interacting with each other in unspecified ways. It resembles a mode of explanation once likened to a crossword puzzle, whose numerous components are not causally related, but provide evidential support for their neighbors [Beckner, 1959; Goudge, 1961]. Furthermore, the adaptive suite provides no basis for predictions.

Explanation by theory. Theories are the most inclusive of scientific explanatory structures and stand among the highest goals of the scientific endeavor. A theoretical explanation broadens and deepens our understanding. It increases the breadth of comprehension when it covers a wide range of scientific evidence and offers a systematically unified account of diverse phenomena [Hempel, 1965b, 1970]. It deepens understanding because it shows that the empirical observations within its scope are the result of underlying causal processes [Hempel, 1966; Suppe, 1977]. The task of a theory is to present a generalized explanation of the phenomena within its scope which will provide answers to a variety of questions about them and their underlying mechanisms, including requests for predictions [Suppe, 1977]

An *evolutionary* theory must be consistent with modern Darwinism. Theories concerning behavior will be more convincing when they show how natural selection brought about an evolutionary adaptation due to reproductive benefits that occurred at the outset of the behavior and continued during the subsequent period of adaptation.

Theoretical principles. The number of principles in the explanans (the basis of the explanation) should be as few as possible. *Parsimony* is the goal. Einstein believed that as science progressed, parsimony increased. In 1936 he stated that "The noblest aim of all theory.[is] to make these irreducible elements as simple and as few in number as possible, without having to renounce the adequate representation of any empirical content" [Holton, 1981, p. 13]. In addition to parsimony, the theory gains acceptability if the explanation itself is simpler than more complicated alternatives that may account for the same phenomena [Hempel, 1966]. It should be both parsimonious and perspicuous.

Theoretical scope. The scope of a theory is the range of subject matter through which the explanation appears to have correct application [Suppe, 1977]. A wide scope means the theoretical principles touch on something of basic importance. A new theory will gain support if its scope includes that of earlier theories and expands the range of explanation into new areas and qualitatively disparate subject matters and disciplines [Goudge, 1961;Hempel, 1965a, 1966; Nagel, 1979; Harris, 1981]. When a theory is first proposed, its scope can only be estimated. Subsequent attempts to use it for explanation or prediction will sculpt its edges, expanding or contracting its applicable range. The intended scope of the bipedal use of weapons theory includes the topics discussed in this book and the predictions described in the appendix. Family structure, provisioning, language, the rapid encephalization that occurred in *H. heidelburgensis*, and the cultural florescence in *H. sapiens* seem to lie outside its range. The true explanatory scope of any theory must be "discovered". If the scope proves to include a large variety of phenomena gathered using a variety of different scientific methods, and it deals with important issues, then the theory can be considered fruitful [Suppe, 1977]. This concept of theoretical structure continues to be viable today: A recent compendium of scientific explanations and theories shows that the common thread involves a simple (and nonobvious) idea that is proposed as an explanation for a diverse and complicated set of phenomena [Brockman, 2013].

Prediction. A prediction in science is a calculated guess about the future based upon information available in the present. Theories can be tested by observing whether their predictions are realized or not. A successful theory serves as a

CHAPTER 16: THE STRUCTURE OF EVOLUTIONARY EXPLANATIONS

basis for such forecasts and gains support when test findings are favorable. Predictive success will strengthen confidence in a theory that already has supplied a systematically unified explanation [Hempel, 1966; NAS, 2008]. Darwin's evolutionary theory is often cited for its prodigious explanatory strength which contrasts with its limited predictive powers. Explaining is always easier because one may adjust the emphasis on one or the other factor and erect a plausible explanation, whereas in predicting one has to know in advance the relative contributions of many variables. Evolutionary forecasting is conditioned by a large number of factors, and a full inventory of them cannot be provided [Goudge, 1961]. Mayr [1961] suggested that nothing in biology is less predictable than the future course of evolution.

There are ways to escape these limitations. Predictions can be made about the outcome of future *research* that tests existing theories. Such tests can be applied at different levels of complexity with a large array of scientific methods. This approach, first described in relation to human evolution studies by Kirschmann [1999], will be elaborated in the Appendix.

Darwin's theory. There is a distinction between Darwin's argument for the existence of evolution and a complete theory [Hull, 1974]. Darwin had two different goals in 1859: first, to establish the fact of evolution and second, to propose natural selection as its primary mechanism [Gould, 1982]. Nowadays, Darwin's work can be stated in theoretical form. Darwin conceived the idea that evolution could proceed if inherited traits were evaluated by a process in which breeding populations gradually acquired the characteristics of those who left more progeny. This reproduction-related rule, which he called natural selection, is *the main explanatory principle in his theory*. As such it is the apotheosis of parsimony and simplicity, connecting all of biology in a unified web—past, present and future—the epitome of scientific theorizing.

Formalization. The logical features of a theory can be understood by examination of its structure to see how it is organized. Formalizing the structure exposes the theory's explanatory framework, scope, causal aspects, and capacity to generate testable predictions. It brings out meanings in an explicit fashion, reveals their essential features and makes clear what is being assumed [Goudge, 1961; Suppe, 1977]. A scientific theory necessarily involves assumptions. It is assumed, for example, that the known laws of nature are true and applicable. In biology, some levels of organization are simply assumed to be in force in their normal mode. Darwin assumed that part of the variability observed in nature was inherited. Limitation of special assumptions is one of the goals of scientific explanation.

The bipedal use of weapons proposal, which I shall now formulate as a theory, is a subset of Darwin's theory. What better foundation could we choose in a quest to understand our own evolution? We can *assume* that we are the result of natural selection of inherited variations that enabled our ancestors to reproduce more copies of their genes.

The bipedal use of weapons theory. The theory is constrained and supported by the laws of physics, chemistry and biology. It is assumed that the hominin ancestor was an arboreal ape that foraged for food in trees and on the ground, mainly plant food; that it lived in a society of both genders with a male-dominated hierarchy established by aggressive behavior; that males who gained higher rank had greater access to ovulating females but otherwise mating was promiscuous. Gender dimorphism in body size, present in chimpanzees and modern humans is attributed to their common ancestry and assumed to reflect the greater genetic dividends that accrued to males who fought and females who did not. Male coalition aggression (loyalty to one's own male group and hostility to outgroup males) is also assumed to be an ancestral trait.

Theoretical Principles. There is only one: *The hominin lineage began with an innovative behavior, the bipedal use of hand-held weapons, to which it subsequently became adapted by natural selection.* The first hominins used hand-held sticks and rocks as weapons from an upright stance during conflicts with adversaries. This behavior yielded reproductive advantages immediately and for millions of years thereafter. From this process, there gradually evolved a bipedal creature resembling ourselves.

CHAPTER 16: THE STRUCTURE OF EVOLUTIONARY EXPLANATIONS

Theoretical Scope. The theory is applicable to *Homo sapiens* and our predecessors since the divergence from the last common ancestor with chimpanzees. The chronological scope extends from hominin origins to *Homo erectus*—about 7 Mya to 1.5 Mya—a large segment of our lineage. Most of the major changes in hominin anatomical structure documented in the fossil record occurred during that period.

Causal effects. A causal factor is an agent that by its own direct effects produces changes. Many of our evolved traits appear to have been directly caused by natural selection acting to improve adaptation to bipedal use of weapons. Others seem to have been indirectly the result of this process. In these instances, natural selection for bipedal use of weapons use was a contributing factor which modified the effects of natural selection for other traits. Numerous relics of this long-acting selective process are evident in modern human development, behavior, anatomy and physiology.

Hominin Origins. The theory accounts for the onset of the lineage that led to *Homo* by ascribing it to the reproductive benefits that were derived by the earliest practitioners of the bipedal use of weapons.

Direct Causal Effects. These are due to natural selection that acted to improve the bipedal use of hand-held weapons.

1. *Evolution of the hand.* The theory explains hominin hand evolution and the modern human hand in impressive detail. It reveals the evolutionary origins of the two unique human grips, which were adapted for grasping a spheroidal missile or a cylindroidal club-handle. One is the throwing grip; the other is the clubbing grip [Young, 2003, 2010,] (Chapter 9).

2. *Handedness.* We are all handed, left or right. This novel human characteristic is a direct effect of natural selection for bipedal weapons use. It results from the asymmetrical motion in which a dominant hand was selected to dispatch a hand-held weapon towards a target [Young, 2010] (Chapter 10).

3. *Ontogeny of throwing and clubbing.* The theory accounts for the evidence that modern human children begin to

throw before they can stand, subsequently throw bipedally with a motion that increasingly resembles the one used by elite adult throwers, that teaching has little effect on the development of this behavior, and that boys outperform girls (Chapter 13).

4. *Modern human throwing and clubbing prowess.* Adult humans in all ethnic and cultural groups throw with the same motion, which is unique to our species. This is explained by the theory, as is the prowess of elite throwers and the exceeding complexity and rapidity of the motion. These are direct results of natural selection for bipedal use of weapons (Chapter 13).

5. *The transition to Homo.* This transformation also lies within the theory's explanatory scope. After 5 million years of adaptation to the bipedal use of weapons, hominins began to use this behavior to exploit a source of food that was abundant on the grasslands. This resulted in a profound remodeling of the body in which arboreal traits were lost and numerous features that enhanced the bipedal use of weapons were selected (Chapters 6, 7).

6. *The acquisition of animal food.* Driving predators from carcasses and hunting prey to acquire animal tissues was made possible by the adaptation to bipedal use of weapons. Access to high-quality food promoted successful reproduction. Coalition aggression among males was likely used for obtaining meat, and for fighting with outgroup males and predators (Chapters 3, 8). Adding nutritious food to the diet yielded reproductive benefits that enhanced natural selection for weapons prowess.

7. *Musculoskeletal robusticity of the hominin lineage.* Robusticity of bone and muscle characterizes all hominins from the most ancient specimens until late *Homo sapiens*. The theory explains this physical attribute as a direct effect of selection for throwing and clubbing prowess (Chapter 12).

8. *CNS control of throwing and clubbing.* The central nervous system controls the acts of throwing and club-swinging. These behaviors are embedded in human neural systems due to natural selection because they yielded reproductive advantages to those who performed these acts most proficiently, as explained directly by the theory (Chapter 15).

Indirect Causal Effects. These are evolutionary alterations that were importantly *influenced* by natural selection for bipedal use of hand-held weapons.

    1. *Bipedal locomotion*. Adaptation to use of weapons from a bipedal stance, driven by the reproductive advantages it provided, *promoted* the subsequent transition to bipedal walking. Body systems selected for bipedality during use of weapons ultimately led to bipedal locomotion becoming the most efficient gait (Chapter 14).

    2. *Canine diminution*. Natural selection augmenting the use of hand-held weapons resulted in that behavior becoming more effective for threat and combat than long, pointed canine teeth. Consequently, large canine teeth were no longer maintained by natural selection and were modified in size and shape by dietary selective factors (Chapter 11).

    3. *Diminution in the chewing apparatus*. In the transition to *Homo*, the massive australopithecine chewing apparatus was reduced, indicating a change in diet. The new dietary source which required less mastication seems to have been animal tissues acquired by the bipedal use of weapons. Reduction of the chewing apparatus was an indirect effect of this behavior (Chapters 7, 8).

    4. *Gender disparity in throwing and striking* (Chapter 13). Natural selection improved throwing and clubbing ability more in males than in females. Modern males outperform females from early childhood onward. This is explained as the result of weapons producing benefits primarily for males, whereas selection for gestating and nurturing children yielded greater reproductive benefits in females. Other gender disparities linked to male fighting with weapons are size dimorphism, strength, and differences in body form and composition (Chapters 2, 5, 12, 13).

Conclusion. The bipedal use of weapons proposal meets the requirements of a scientific explanation and a scientific theory. It rests on a single explanatory principle and has an extensive explanatory scope which encompasses a wide variety of subject areas and numerous scientific disciplines. It is explicitly congruent with modern Darwinian evolutionary theory. In this way, the new theory increases understanding ofhuman origins and evolution, and provides a fertile source of predictions.

CHAPTER 17

Summary and Conclusions

"There is grandeur in this view of life, with its several powers, having been originally breathed into a few forms or into one; and that, whilst this planet has gone cycling on according to the fixed law of gravity, from so simple a beginning endless forms most beautiful and most wonderful have been, and are being, created"
[Darwin, 1859, pp. 459-460].

Introduction. Where did we come from? How did we become the way we are today? Why did we diverge from our ancestors? What happened next? These are questions I have grappled with in the foregoing pages, guided by Darwin's theory, aided by fifteen decades of scientific evidence that has accumulated since his day, building on analyses and explanations created by other scientists, and propelled by my own attempt to explore a new way of approaching these provocative queries.

There are two elements to the method I used that contributed importantly to the explanatory structure which emerged. The first is a strict application of the central rule of modern Darwinism: Every evolutionary change resulted from the reproductive benefits it provided. Second, I examined the origin and evolution of our lineage from a novel perspective—the idea that the bipedal use of hand-held weapons marked the onset of our lineage and that natural selection acted to refine and enhance this behavior because it promoted reproductive success. During millions of years, hominins underwent an evolutionary adaptation to throwing and striking because those with greater prowess in this behavior replicated more copies of their genes. (All ages of both genders derived benefits from this behavior, although it was mainly used by adult males). This simple proposition has made it possible to provide a coherent, Darwinian explanation for some of the most important issues in the study of early human evolution at the present time.

The structure of this explanation has been formulated as a theory. Its major strengths are first, it fills a void in the scientific study of our species, which lacks a theory of human origins and evolution [Tooby and DeVore, 1986; Bowler, 1986; Cartmill, 1990]. Second, it is parsimonious in the simplicity of its single explanatory principle (the adaptation to bipedal use of weapons). Third, it has a wide scope of application. Fourth, it is explicitly based on the modern concept of natural selection. Fifth, it can serve as a fertile source of predictions.

An explanation of a new evolutionary adaptation requires identification of the reproductive advantages associated with its onset. The adaptive feature should provide *immediate* benefits to get the process started. Then, the explanation must demonstrate that these advantages would continue through the succeeding generations during which the adaptation was established. In the case of bipedal throwing and clubbing, the duration appears to be about *five million years*—from hominin origins until some time after the emergence of *Homo erectus*. With a 15-year generation time, this would represent 333,000 reiterations of natural selection.

The behavior of using hand-held weapons from a bipedal stance fulfills both these stringent requirements: the immediate benefits and the enduring utility. It would have provided reproductive advantages to those who first employed it, and would have continued to do so for an indefinite period, because it was an effective way of settling disputes over scarce resources necessary for reproduction—namely, food, mates and safe habitats.

The assumption that it was primarily males who engaged in fighting with weapons fits all the evidence. The bipedal use of hand-held weapons, such as clubs and throwing-stones, would have given hominin males an advantage in conflicts with conspecific adversaries who did not use weapons or used them ineffectively, and in defense against predators, who could be driven off or killed with such weapons, particularly when they were employed in concert by groups of males. This strategy was effective in any habitat. Those who were most proficient in such behavior would have proliferated because they were more likely to survive and reproduce, a process that gradually yielded a unique animal, skilled in the bipedal use of weapons. While no longer adaptive in most modern environments, it remains an option against adversaries that lack firearms or elect not to use them. Mobs of protestors facing off against security forces are likely to include males throwing rocks or other projectiles and swinging clubs of some kind, often opposed by police also swinging clubs. This behavior is a relic of our evolutionary history.

The bipedal use of hand-held weapons theory. A distinctive goal of science is to increase understanding by explaining the regularities exposed by analysis of empirical evidence (Chapters 1, 16). The highest level of scientific explanation is the theory. Theories, like explanations in general, have two parts: the part that explains (the principles) and the part that is explained (the scope). The goal of theorizing is an explanation with a minimum of simple principles that explain a vast body of evidence by showing it to be the result of a few underlying causal factors. The epitome of such a theory in biology is Darwin's theory of evolution, of which the present theory is a derivative focused on our own lineage.

The single explanatory principle in the bipedal use of weapons theory is this: *The hominin lineage began with an innovative behavior (the bipedal use of hand-held weapons) to which it subsequently became adapted by natural selection.* A number of features of the human odyssey can be explained in whole or in part by this principle. That is, they fall within its scope.

Human origins. What caused the first hominins to diverge from the ancestral ape lineage? Currently there is no accepted explanation, although many believe that "bipedalism" was involved. The fossil record supports this conviction, but no

CHAPTER 17: SUMMARY AND CONCLUSIONS

Darwinian account based on *bipedal locomotion* has yet been successful. Many hypotheses have been proposed, but none has proved compelling (Chapter 14). The theory presented here states that the behavior associated with the origins of the hominin lineage was another form of bipedalism: *the bipedal use of weapons*. This eliminates the obstacle that has stymied previous efforts to show how the first hominins could have gained reproductive advantages from bipedal walking. In contrast, there are numerous *immediate and continuing* benefits that would have ensued from the addition of weapons to existing methods of fighting—benefits that would have enhanced reproductive success, primarily by increasing access to food, mating opportunities and safe habitats (Chapter 3).

For males, acquiring sufficient nutritious food and fertile mates are keys to reproductive success. For females, food is the crucial biological necessity for personal survival and the process of reproduction itself. For children, food is essential for normal development. Using weapons from a bipedal stance can address these issues; bipedal walking cannot. This line of reasoning has led me to suggest [Young, 2003, 2009, 2010] (Chapter 3) that human origins can be accounted for by the onset of bipedal use of weapons. In subsequent generations such behavior would have escalated due to the advantages it yielded in the competition for scarce resources. Natural selection must have become involved at an early stage among the most capable armed apes, who were better able to nourish themselves and reproduce their kind, resulting in differential preservation of the genes of individuals whose heritable traits facilitated throwing and club-swinging prowess.

Bipedal locomotion. The idea that the first hominin specialty was bipedal locomotion implied that if a reproductive benefit could be discovered for such behavior the problem of human origins would be solved, but no solution has been found. This obstacle to understanding is eliminated by the concept that bipedal use of weapons *preceded* bipedal locomotion. Natural selection that improved ability to transmit an explosive burst of energy to a hand-held weapon while balanced in an upright stance on sturdy legs set the stage for a subsequent transition to upright walking. Bipedal locomotion emerged later, facilitated

CHAPTER 17: SUMMARY AND CONCLUSIONS

by adaptive changes in anatomy and physiology that were the result of selection for improved bipedal throwing and striking.

Although bipedal walking and bipedal use of weapons share the fundamental element of upright behavior balanced on two legs, there are significant differences between them (Chapter 14). Throwing and clubbing motions involve rapid sequential patterns of muscular contractions adapted to maximize the amount of energy transferred to a hand-held implement in a single explosive act while the body remains in one place. In contrast, bipedal locomotion consists of a comparatively slow, rhythmic, repetitive pattern of muscular contractions adapted to minimize energy expenditure while moving the body to another place.

However, it is their *similarities* that are important in this context. Both are performed with the body oriented vertically on strong, robust legs and stable, springy feet which supply propulsive force. Each involves rotation of the femur, pelvis and torso, and coordinated arm and leg movements. Both require maintenance of vertical balance during movement. Natural selection preserved inherited variations that yielded these attributes because they improved upright throwing and club-swinging. This provided the foundation for their subsequent use in upright locomotion.

Adaptation to bipedal use of weapons led to improved dynamic vertical balance, more stable upright stance, altered pelvic muscles which controlled lateral tilt of the torso, increased ability to rotate the hips and torso independently, enhanced capacity to extend the femur at the hip and knee joints, augmented leg musculature for more powerful thrust. and transformed the foot from a grasping appendage to a resilient, stable platform. Changes in the waist, pelvis, lower limbs, feet and hands reduced the ability to locomote in trees and hindered quadrupedal locomotion on the ground, while bipedal locomotion was enhanced in many ways, leading to its more frequent use.

As bipedal locomotion became a more efficient and common activity, it began to yield non-locomotor advantages, such as an improved capacity to carry weapons, food, babies and provisions. Eventually, the ability to walk bipedally began to have an effect on reproductive success. It then began to experience the direct effects of natural selection. Subsequently, the two distinctive bipedal behaviors evolved concurrently (Chapter 14).

The human hand. Natural selection for bipedal use of weapons was an *indirect* cause of the change in gait. The evolution of the human hand, to the contrary, is an example of the *direct* causal effects of natural selection for enhanced prowess in the bipedal use of hand-held weapons. The theory explains the remodeling of the hand as an adaptation to gripping spheroidal missiles for throwing and cylindroidal clubhandles for striking. It accounts for the transformation of an ape-like hand into a human hand, including the iconic opposable thumb and the exotic styloid process on the dorsal aspect of the base of the third metacarpal (Chapter 9). It explains the shortened palm, reduced finger length, reduced finger curvature, large apical tufts on the distal phalanges with their broad palmar fibrofatty pads, full opposability of the distal pad of the thumb to those of the fingers, the lengthened, more mobile and muscular thumb, the three new muscles that control its motion, modification of the palm which produces an obliquity of the hand when it is flexed, increased robusticity of the metacarpals and bases of the second and third fingers compared to the fourth and fifth, rotation of the fingers around the central axis when they are flexed, and the styloid process on the third metacarpal.

It was Napier [1956] who discovered that the modern human hand displayed two new basic grips, which he called the precision grip and the power grip. He illustrated these grips by depicting a hand grasping a sphere and a cylinder. Young [2003, 2010] showed that the precision grip is a *throwing* grip and the power grip is a *clubbing* grip, thereby providing an evolutionary explanation that is perfectly congruent with the fossil evidence. The hands of the earliest hominins had curved fingers (*Orrorin tugenensis, Ar. kadabba* ~5-6 Mya) and diminutive thumbs (*Ar. ramidus*, 4.4 Mya). In *A. afarensis* (3.5-3.2 Mya) an improved capacity to form the hominin throwing and clubbing grips is unmistakable and in *A. africanus* (~3.5-2.5) the classic grips are well developed. The earliest stone tool-makers possessed hands that already closely approached their current form.

CHAPTER 17: SUMMARY AND CONCLUSIONS

The transition from *Australopithecus* to *Homo*. The emergence of *Homo* in Africa was the culmination of the most profound change in body form in the entire lineage. It began about 2 Mya, some 5 My after hominin origins, and the transition appears to have been rapid (Chapter 6). When completed, hominins had abandoned their ancestral arboreal adaptation and were dedicated to life on the ground with a body structure similar to ours. All of the major features of this metamorphosis, including alterations in body size and in the structure of the shoulder, arms, legs, thorax and waist can be explained as the direct result of a period of intense natural selection for bipedal use of weapons made possible by the end of selection for life in the trees.

Australopithecus species and their forbears had maintained a body type characterized by small size, long arms, short legs, stiff waist, and curved fingers and toes until the transition to *Homo*. This body plan is a recognizable adaptation to arboreal life. Why did they abandon the trees? The new lifestyle must have provided more reproductive benefits than the old. It appears that hominins gained access to a nutritious food source that was more abundant on the grasslands than in the ancestral woodlands. The consensus opinion is that the new dietary staple was *animal tissue* ("meat"), plentiful in the herds of animals that were proliferating on the expanding savannas. How did they *acquire* this food so they could eat it? This question will be addressed below, after examination of the fossil evidence of the body remodeling that occurred.

Hominin body size increased. Life on the ground removed the limitation imposed by large body size in trees, but larger size could only enhance reproductive success if it was part of a package that compensated for loss of access to arboreal foods while augmenting food supplies to support a larger body. (This is why access to animal food is part of the argument). The explanation for increased body size in *Homo* has two, gender-related parts.

Larger, more massive males would have gained an advantage in fighting with smaller males for access to food and females, promoting natural selection for size increase. This traditional explanation for gender size dimorphism is applicable here, with the added element that males were fighting with weapons. Combat with predators and prey also would have favored larger body size and strength, which enabled hominin males to apply greater force to their missiles and clubs, accelerating them to greater velocity. This enhanced their fighting prowess which increased their likelihood of eating well and mating often, enabling them to reproduce more copies of their genes.

Females also became larger than their australopithecine predecessors in the transition to *Homo,* but relatively more so than males (size dimorphism declined), indicating the presence of a selective factor specific to females—possibly the reproductive process itself. In many living species large females have more success in gestating, birthing, lactating and rearing larger, healthier offspring. They have larger birth canals and more nutritional resources that can be transferred to their children through the placenta or in milk and typically have more surviving offspring. The benefit to females of larger size seems linked to successful reproduction.

The thorax changed shape and shoulder mobility increased. Apes and early hominins had "funnel-shaped" rib cages, narrow at the top and broad at the bottom. In *Homo* the thorax became "barrel-shaped". Expansion of the upper thorax increased the separation of the shoulder joints, facilitating an increased range of motion of the upper limb. It may seem remarkable that when hominins abandoned the trees natural selection increased shoulder mobility. Modern humans have the most mobile shoulder among all primates. This appears unrelated to bipedal walking, in which the shoulder plays no significant role. What was the reproductive advantage?

The concept that natural selection acted to adapt hominins for improved use of hand-held weapons highlights the benefit of mobile shoulder joints. In this behavior the shoulder is a crucial element (Chapters 7, 13 and below). The changed shape of the thorax in *Homo* has been attributed to natural selection for throwing: greater width of the upper thorax contributes to broadening the shoulders, which increases the leverage of upper body rotation around its long axis [Kirschmann [1999]. The exceptionally rapid internal rotation of the humerus in the modern

Homo shoulder joint is a salient feature of the throwing and overhead clubbing motions. Following leg thrusts, pelvis and thorax rotations, kinetic energy is transferred to the shoulder region, where the weapon-bearing arm is extended backward. Next, powerful muscular contraction produces a rapid forward translation and internal rotation of the humerus. Nothing like this has been reported in arboreal primates. It is part of the adaptation to bipedal use of weapons.

The axis of the shoulder joint was redirected laterally. In arboreal primates and hominins prior to *Homo*, the shoulder joint (glenoid fossa) faced upward, indicating an adaptation to tree-climbing and below-branch suspension. In *Homo* the shoulder joint changed to a lateral orientation. What was the adaptive advantage of reorienting the socket for the humerus at right angles to the vertical axis of the upper body? This is exactly its position in the throwing motion when the pulse of energy generated in the lower body is transmitted through the shoulder joint to the upper arm (Chapter 13). During the acceleration phase of the throw until missile release from the hand, the humerus is held in a position of 90° abduction, in line with the plane of the shoulders. This enhances lever-arm action and the transmission of maximum energy to the hand-held weapon. Hominins whose shoulders were broader, whose shoulder joints were more mobile, and whose glenoid fossae were oriented more laterally were able to convey more energy to their weapons when throwing or club-swinging. As a result, they were able to transmit more copies of their genes to succeeding generations.

The waist was modified. In *Australopithecus* as in arboreal apes, the region between the lower ribs and the ilium was stiff, due to a short lumbar spine, a firm binding between thorax and pelvis, and the arrangement of muscular attachments in this region. This limited mobility is believed to enhance arboreal locomotion. In early *Homo*, lengthening of the lumbar spine separated the lower rib cage from the pelvis, a change accompanied by altered muscle attachments on the ribs and pelvis, resulting in a distinctive waist region and an unprecedented ability to rotate the pelvis and thorax independently. This too can be explained by natural selection that acted to improve throwing and club-swinging. These complex, full-body motions involve independent, successive rotations of the pelvis and thorax as part of the chain of kinetic energy transmission from the lower to the upper limbs and ultimately to the hand-held weapon.

The relative size and strength of the arms and legs changed. In the transition to *Homo*, mass, muscularity and strength of the upper limbs was reduced whereas the legs became longer and their mass, muscularity and strength was increased. Early hominins had long, powerful arms coupled with legs that were relatively diminutive, a typically arboreal complex of traits. Small legs reduce the body mass that is supported by the arm during suspension, locomotion, and feeding. *Homo* became adapted for a behavior with other requirements. Longer, more powerful legs in humans have been ascribed to selection for improved bipedal locomotion. The presumption is that longer legs would aid walking by increasing stride length if the mass of the legs remained the same, but it did not.

In *Homo* the legs became more massive, which would hinder leg-swinging due to increased inertial forces. As noted below, premodern hominins were characterized by extraordinary muscularity and robust bones. When the inertial disadvantage is compounded with the elevated body mass in *Homo*, it suggests that some other bipedal behavior, not walking, was primarily driving natural selection to produce this body type.

The alternative explanation offered here is that longer, more muscular and robust legs were naturally selected because they improved throwing and club-swinging prowess, which yielded reproductive benefits. Long, powerful legs are advantageous during the windup, the rear leg thrust, the stride forward, the front leg thrust, and in the deceleration and follow-through (Chapter 13). Modern males who are elite throwers and clubbers have longer legs and are more muscular than the average male in their age group (Chapter 12).

The same explanation can be applied to the reduction in mass and musculature of the arm compared to the leg in *Homo*. In the modern throwing and club-swinging motions, kineticenergy is generated from muscle contraction in the lower limbs, then augmented as it is transferred successively through body segments of smaller size and reduced mass that move progressively more rapidly until it is conveyed to the hand-held weapon. Natural selection for throwing and club-swinging prowess accounts for the heavier legs and less massive arms in *Homo* compared to apes and earlier hominin ancestors. This process preserved heritable variations that enabled hominins to impart higher velocity to their hand-held weapons.

The forearm became shorter in relation to the upper arm. This explanation also can be applied to diminution of the ratio of forearm to upper arm length (the brachial index) in *Homo*. Reduced length and mass enabled the forearm to move faster during the throwing and clubbing motions. From the upper arm, the accumulated energy is transmitted through the elbow joint to the forearm, and then through the wrist joint to the hand. This indicates that the hand should have diminished in size also. It did [Young, 2003] (Chapter 9).

How did *Homo* acquire animal tissues? With the advent of *Homo*, there was a major rise in hominin access to animal food. Analysis of butchered bones indicates that hominins procured large carcasses by hunting or confrontational scavenging, both of which are dangerous activities. Long considered something of a mystery, the answer to how they obtained these carcasses is provided by the bipedal use of weapons theory [Young, 2010] (Chapter 8). After 5 million years of adaptation to the use of hand-held weapons, hominin males, acting in groups, had become major predators. They controlled the carcasses of animals so large they could not be pulled apart, but had to be butchered. The best throwers and clubbers acquired the most meat and their reproduction profited. If they shared it, their community members would also have experienced improved nutrition and reproductive benefits.

The chewing apparatus diminished in size. Increased access to meat made possible by hominin skill with weapons may have played a role. From *A. anamensis* onward, the hominin jaw and dentition increased in size, apparently due to natural selection for powerful, prolonged chewing of unidentified foods. In the transition to *Homo*, molars began to be reduced in size and the enamel became thinner.

In Wrangham's view [2009], this can be attributed at least in part to processing food by cooking (Chapter 7). Others believe that the chewing apparatus diminished because *Homo* gained access to a high-quality, easier-to-chew dietary staple—specifically, meat (animal tissues). The weapons theory accounts for the method by which the meat was acquired.

Handedness. At some time during human evolution an inherited tendency to use one hand rather than the other for complex tasks began to undergo strong natural selection, accentuating a trend of right-handedness that may have been inherited from the *Pan/Hominin* ancestor. During the Lower Paleolithic, *Homo* was a predominantly right-handed knapper of stone tools. In modern humans, approximately 90% of individuals are right-handed and 10% are left-handed. Population-level handedness of this magnitude is a unique feature of modern humans that is particularly evident in throwing and club-swinging and may be derived from selection of those behaviors [Young, 2010] (Chapter 10).

Handedness arises from asymmetrical neural organization of motor systems. So does the human throwing and striking motion, in which left and right sides of the body move differently. This complex task coupled with an effect on reproductive success suggests that natural selection for right-handed throwing and striking may have led to nearly complete population-level right handedness in hominins.

Why did left-handedness persist in 10% of the hominin lineage? Analysis of sports where the prevalence of left-handers exceeds 10% shows that all are interactive competitions where two adversaries compete face-to-face. Left-handed combatants, accustomed to right-handed opponents, have the advantage of surprise. Their throws, punches, or thrusts come from directions that differ from those of the more prevalent right-handers (Chapter 10).

Hominin musculoskeletal robusticity. For millions of years the hominin lineage was characterized by robust bones and powerful muscles. This trait was present in *Orrorin tugenensis* (6 Mya), was maintained in the australopithecines, *Homo erectus,* and subsequent *Homo* species until late *Homo sapiens.* Robusticity also characterized the skeletons of hominin children and infants. This indicates it was due primarily to natural selection, rather than stress from activities such as prolonged running over irregular terrain. (In modern humans elite long-distance runners are characterized by *reduced* musculoskeletal robusticity).

The bipedal weapons theory proposes that hominin robusticity was naturally selected because it improved throwing and striking prowess which provided reproductive benefits. Elite modern throwers and strikers are characterized by large size and robusticity. Professional baseball pitchers and club-swingers are larger in stature and body mass than the average male in their age group [Young, 2010] (Chapter 12). Bone and muscle are expensive tissues in both energetic (caloric) and nutritional requirements. Selection for improved throwing and striking prowess that involved augmenting muscle and bone mass must have been important for reproductive success to compensate for the additional dietary needs and the detrimental effects of increased body mass on locomotor efficiency. Robusticity recently declined in *Homo sapiens* after 0.03 Mya. This has yet to be explained.

Ontogeny of throwing and club-swinging behavior. The bipedal use of weapons theory accounts for a unique feature of modern human child development: overhand throwing is part of the normal ontogenetic pattern in our species. The inherited motor pattern that emerges shortly after birth, and yields a single type of bipedal throwing motion, features a gender disparity favoring boys which increases with age. It involves remarkable complexity and rapidity, and reaches its highest level in a relatively small number of individuals, mostly males. As it develops, it increasingly resembles the throwing motion used by mature athletes.

The emergence of a standardized overarm throwing motion and the gender difference in its development has been recorded in every ethnic group and region where it has been studied, as well as in undernourished children. The gender disparity exceeds that of any other innate motor skill and appears to be resistant to instruction in young children. Girls follow the same sequence of development but lag behind boys in its progress. Societal influences and practice may play a role in fine-tuning the skill, but no child needs to be taught how to throw. Limited evidence of the ontogeny of striking is similar to that of throwing [Young, 2009, 2010] (Chapter 13).

The throwing and club-swinging motion in adults. After puberty, when males increase in stature, mass and muscularity more than females, the gender difference in throwing performance also rises, although all skilled throwers use the same motion. It begins with a weight-shift to the rear leg, which then thrusts into the substrate, propelling the raised front leg into a stride toward the target, shifting the weight forward. When the front foot lands, it thrusts rearward, producing rotation of the pelvis. Pelvic rotation is then blocked, leading to forward rotation of the torso. As the shoulders rotate towards the target, the arm and the hand gripping the missile lag behind. Then shoulder muscles produce rapid humeral rotation coinciding with forward translation of the arm and sudden extension at the elbow joint, transmitting the accumulated energy to the hand and missile.

At the instant precalculated in the brain, the hand unfurls its throwing grip, releasing the projectile on a trajectory towards the target. Injury-prevention mechanisms are automatically activated to resist dislocation and fracture as the pulse of energy passes through the upper extremity. From onset of forward acceleration to missile release in elite throwers the time elapsed is less than 60 milliseconds. The overhand striking motion is fundamentally similar, except that the club is gripped firmly in the clubbing grip and is not released when the weapon strikes the target. These innate human motor patterns can be explained as the result of an evolutionary adaptation that enhanced throwing and clubbing skills because of reproductive benefits they provided to our ancestors (Chapter 13).

The nervous system contains evolved networks that control throwing and club-swinging. Natural selection acting to improve this behavior had a profound effect on the nervous system, which controls this motor skill through central, peripheral, sensory, motor, conscious and unconcious pathways (Chapter 15). The entire throwing motion, in all of its intricate complexity, is stored in the brain in some manner from which it can suddenly be called into action and expressed in a coordinated, exceptionally rapid, full-body motion. Furthermore, the brain must precalculate variations in this motor program to match the variables of each situation.

Kirschmann [1999] was the first to emphasize the role of the brain in throwing. Only humans throw well, he wrote, and because throwing is exceedingly complex and very rapid, it puts enormous demands on the central nervous system (CNS), such as computing target distance and the trajectory of the projectile towards the target. It must adjust the motion after analysis of sensory input, then coordinate the contractions of numerous muscle groups to generate maximum kinetic energy and release the missile precisely. Due to the rapidity of the motion, the brain must use sensory input to calculate the acceleration course in advance. The throwing adaptation should be accorded great significance in the evolution of hominin cognition [Kirschmann, 1999].

Recent neurophysical research on arm and hand movements during a throw (Chapter 15) provides support for these ideas and insight into the role of the brain. The CNS issues the motor commands to coordinate the generation of muscular contraction that provides the force conveyed to the weapon. Because the resulting movements are too rapid for proprioceptive feedback from wrist, elbow, or shoulder to control finger opening, many calculations must be made in advance, as Kirschmann anticipated.

To achieve a fast and accurate throw the CNS calculates and exploits several types of interaction torques in order to generate motor commands that maximize the velocity and trajectory of the missile. Fast throws are associated with large back forces from the missile on the fingers that must be precalculated and controlled if accuracy is to be achieved and injury prevented. The CNS anticipates their size and generates appropriate finger flexor torque to oppose them. A major factor in throwing accuracy is control of the onset of missile release, when the thumb abducts and the fingers extend, uncoupling the projectile from the grip. As the missile rolls distally along the flexor surface of the fingers, force continues to be applied until final release from the fingertips. To throw within the baseball strike zone requires a timing precision of 1-2 ms. It seems likely that mental properties developed during several million years of hominin adaptation to bipedal use of weapons provided a substantial part of the foundation of the modern human mind.

Canine diminution. Reduction in size of human canine teeth compared to those of apes and monkeys was first noted by Darwin, who speculated that early humans probably had long, pointed canine teeth, but as they increasingly used stones, clubs, or other weapons for threat and fighting, the teeth became reduced in size. This conclusion, supported by the vast majority of commentators since Darwin, is consistent with the bipedal use of weapons theory. Reduction in size of hominin canine teeth commenced so early in the lineage that relatively small canine teeth have been used as a marker of hominin status in ancient fossil specimens (Chapter 11). The subsequent diminution in size of the canine teeth in both genders, but particularly in males, is well documented. Modern dimensions appear only in late *Homo erectus*. When dagger-like teeth lost their fighting function, selection for long, pointed canines ceased and was redirected to preserve inherited variations which promoted smaller, more incisor-like canines. These contributed to reproductive success because they improving food-processing.

A simple explanation with broad scope. The foregoing summary shows that the bipedal use of weapons theory has a broad explanatory scope derived from the simple principle that hominins diverged from their ancestors when they began to use weapons from a bipedal stance—a behavior to which they subsequently became adapted by natural selection because of the reproductive benefits it provided.

A theory gains support if its explanatory principles are few and simple, and if its scope includes that of earlier explanations and expands the range into new areas, disparate subject matters and disciplines (Chapter 16). The present theory fulfills these goals. The scope of its single explanatory principle subsumes topics that have been the subject of earlier

CHAPTER 17: SUMMARY AND CONCLUSIONS

hypothesesand includes issues that have not been explained before. It brings together and connects information that formerly had not been shown to be interrelated. It blends evidence from genetics, ontogenetics, paleontology, biomechanics, neurophysiology, comparative anatomy, comparative behavior and other scientific disciplines. The theory is also distinguished by the plentiful predictions that can serve as a means of testing its scope and validity (see Appendix).

I now turn to some brief closing remarks on a few related issues.

Did females fight with weapons? All indications are that hominin males did most of the fighting. Females may have occasionally resorted to use of missiles and clubs, but those who avoided armed conflict presumably were able to raise more children, reproducing their genes more successfully. They inherited the genetic basis for use of weapons, although their skill was mitigated by sex-related genetic modifiers (Chapter 5). Nevertheless, modern females are more skilled at throwing and club-swinging than either gender in any other species.

Is the theory sexist? Some critics of earlier scenarios objected that females were sidelined, bearing and raising children and waiting for males to bring them meat (Chapter 16). Even if numerous human traits are due to natural selection that acted to improve a predominantly male behavior, this does not mean that women played a subsidiary role. The genders are inherently specialized. Females grow the babies, provide their nutrition during gestation and lactation, then nurture them until they are able to survive on their own; these functions are essential to preservation of the genetic lineage. Males do not and for the most part cannot perform these functions. Apart from donating sperm and fighting, the male role appears to be subsidiary in the typical mammalian way. What marks the human odyssey as unusual is not that males were fighting but the way they were doing it—with hand-held weapons. This provided a new manner of obtaining reproductive benefits which endured for a very long time; *then it became obsolete.*

With the ascendancy of firearms a few centuries ago the throwing and clubbing era came to an end. It was no longer a reasonable alternative in agonistic conflicts after guns and other explosive devices became available, just as fang-like canine teeth had proved no match for rocks and clubs. Nowadays the classic male hominin behavior—although it may still be observed in sports, exercise, and urban street demonstrations—has little or no effect on reproductive success.

Was the use of weapons to settle disputes immoral? Linking our origins to the use of weapons may seem distasteful to our modern mentality. We tend to take "personally" the behavior of our ancient ancestors [Symons, 1979; Lewin, 1987], which is why human evolution is the most "emotionally loaded" part of the study of evolution: it affects the way we think about ourselves [Cartmill, 1983; Bowler, 1986].

No doubt an image of our earliest progenitors living in permanent, pair-bonded mating relationships in which males brought provisions to their spouse and children in a peaceful, cooperative society has a pleasant "feel" to it, but the concept that early hominin males were gentle pacifists, making love not war, seems to have little power to explain ancient fossil evidence or modern human behavior. Furthermore, the thought that our forefathers behaved unethically seems misguided because, after all, they were arboreal apes! Morality was not part of their mode of operation. They lived by natural laws, not cultural admonitions. They sought food, shelter and mates by whatever means were available, impelled by their inherited proclivities. Some early glimmerings of morality, such as generosity and kindness, might have originated from sexual selection [Miller, 2001], but modern human moral concepts are predominantly cultural products [T. Huxley, 1894; Ayala, 2010; Bloom, 2010].

Science does not tell us how we should live our lives; it seeks to discover the nature of reality, the facts about the world. Such facts have no direct moral or political implications [Cartmill, et al., 1986; Okasha, 2002]. Just as science endeavors to exclude personal opinions, expectations, preferences, unverifiable forces and entities from its deliberations, it also rules out emotions and value judgments. Ethical considerations represent claims about what *ought* to be, based on opinions about what is right or good. Evolutionary science, based on the concept of differential reproduction, does not even suggest that reproduction is intrinsically good! There is nothing inherently valuable about having descendants. It may make a person feel good to have grandchildren, but this does not mean that maximizing the number of one's descendants is the morally right choice [Rosenberg and McShea, 2007; Overall, 2012].

Are we doomed by our DNA to make war forever? If natural selection acted for millions of years to improve the effectiveness of a way of life involving the use of weapons, that does not mean we are condemned to continue such behavior because of "human nature". Our inherited behavioral systems are mainly unconscious predispositions, shortcuts to survival that adapted us to a world long past [Potts and Hayden, 2008]. Whatever the behavioral propensities of early hominins may have been, they have no moral implications today. We can choose to express these impulses in peaceful, creative ways. We have culture, knowledge, science, and the ability to reflect on and moderate our own behavior [Cartmill, et al., 1986].

Scientific assessment of modern human violence suggests that it has markedly subsided in recent centuries [Pinker, 2011]. It has declined in the family, neighborhood, and among tribes, other armed factions, and major nations and states, as have attitudes that tolerate it. Movements to abolish violence like despotism, slavery, dueling, torture, burning witches, human sacrifice, sadistic punishment and cruelty to children and animals have had an impact on Western life. Resistance to aggression against ethnic minorities, women, children, homosexuals and animals has occurred more recently.

A personal view. My research has led me to believe that our ancient predecessors survived and proliferated in part because males did use violence when anticipated benefits outweighed expected costs. Hominin males who used weapons in competition for scarce resources left more copies of their genes than those who did not. We are here because of their reproductive success. Emotionally, I feel a complex mixture of admiration, respect, pride and gratefulness to these ancestors of long ago. In the challenging natural habitats in which they spent their lives, each day presented dangers that could maim, cripple

CHAPTER 17: SUMMARY AND CONCLUSIONS

or kill, including falls, invading marauders, hungry predators, droughts, fires, starvation, infections, parasites, and diseases—difficulties that from a modern perspective seem horrendous.

Life in a state of nature may sometimes have been poor, nasty, brutish and short, but it was never solitary. This I too admire, that our ancestors lived in social groups of both genders who raised children and gradually forged unprecedented ways of cooperating with each other and aiding those who needed help. I am grateful to those who kept our lineage from the common biological fate of extinction, leaving to us the privilege of serving as vehicles of the precious genome that has been preserved and renovated during millions of years and hundreds of thousands of iterations of natural selection.

If some among our pre-hominin relatives began to use weapons and others did not, which would be most likely to stay alive, find food, acquire mates and raise children to continue the lineage? The answer I have suggested is that those who used weapons from a bipedal stance gained an advantage that increased their chances of survival and reproduction of their genes. That is the central principal of the explanation I have presented for human origins and early evolution. My wish is that this simple proposal will be provocative, thoughtfully examined, transformed in ways that enhance its explanatory value, and will promote understanding of our place in nature.

APPENDIX

PREDICTIONS FROM THE THEORY

Introduction. A major function of theories is to provide a foundation for prediction. The bipedal use of weapons theory readily yields predictions about the results of future research. These can be used to test its scope and explanatory power, and to suggest new approaches to its subject matter. Such predictions are *theory-based hypotheses*.

In 2003 I wrote that if natural selection acted to enhance throwing and clubbing, this "leads to the prediction that the human hand should be adapted for throwing and clubbing" [p. 165], a forecast now fulfilled (Chapter 9). Fossils from *A. afarensis* [Drapeau, et al., 2005)], *Ar. ramidus* [Lovejoy, et al., 2009e, f], and *A. sediba* [Berger, et al., 2010a; Kivell, et al., 2011a, b] described since that prediction are consistent with it. I also predicted that selection for throwing and striking could account for the unprecedented ability of modern humans to throw missiles and swing clubs with power and accuracy" [p. 166]. This too was confirmed, as was the prediction that these motor behaviors have a strong genetic basis [Young, 2009, 2010] (Chapters 10, 13). Selection for upright stance and dynamic upright balance to enhance bipedal use of weapons was predicted to shed new light on the origins of bipedal locomotion. It "would have made upright locomotion more efficient, leading to its increasing use and eventually culminating in habitual bipedalism" [Young, 2003, p. 166]. This prediction has been examined in detail [Young, 2010] (Chapter 14) and shown to be compatible with current evidence. In what follows, additional features of human evolution are considered and twenty predictions about the outcome of future research are presented.

Predictions Concerning Ontogeny

PREDICTION: Evidence of the ontogeny of striking (scanty at present) will reveal that striking develops with a pattern similar to that of throwing, including the gender disparity and biomechanicalfeatures of the motion.

PREDICTION: Research on the development of accuracy in throwing and striking will show that it matures with the same development pattern as these motor behaviors and with a similar gender disparity.

Predictions Concerning the Central Nervous System

PREDICTION: The retinal image centered on the fovea will prove to be the center of the central nervous system's aiming mechanism for throwing and club-swinging.

Comment. Aiming is a crucial part of throwing and striking. An aiming mechanism calculates the target's direction, size, distance, and movement. It is predicted that sensory input from the high-resolution visual center in the retina, the fovea centralis of the macula, defines the target for the CNS aiming system.

PREDICTION: Comparative anatomical studies will show that non-human primates lack specialized neural structures and central control systems comparable to those that regulate the human throwing and club-swinging motions.

Comment. The human brain contains a widely distributed network of neural pathways that control these motor patterns (Chapter 15). Since the human throwing and club-swinging motions are unique, there should be some distinctive aspects in neural organization of the human CNS that are lacking in other primates.

PREDICTION: Neurophysiological studies of the lower limbs and pelvis during the act of throwing will be shown to be part of the same motor behavioral system that activates the upper body, with a similarly wide distribution in the brain.

Comment: This means the lower body will prove to be controlled by neuromotor commands that take into account torques (muscle, gravity, interaction, velocity-dependent) in a manner that generates a controlled pulse of energy that begins in and passes through the lower body before it is transferred to faster-moving, elements of the upper limb (Chapter 13).

PREDICTION: It will be found that there are regions in the unconscious component of the CNS where the complex, inherited movements of the throwing and club-swinging motions are in somesome sense *stored*, and from which they can be activated by the conscious mind.

PREDICTION: **Dynamic** vertical balance (upright balance during body movement) will be shown in humans to involve specific, highly complex, peripheral and central neural system pathways that are lacking in other animals, including apes.

Comment: The human throwing and club-swinging motions require a highly developed dynamic balance system to prevent falling as the center of mass changes position rapidly and significantly during these actions.

Predictions Concerning Movement of Body Components

Introduction. The following predictions will be presented for different body segments in the sequence in which they contribute to the throwing and striking motions, beginning in the lower extremity, continuing through the body to the hand, and followed by the deceleration that brings the motions to an end.

The lower limbs, pelvis and trunk.

PREDICTION: Analysis of lower limb structure in modern humans will show that the increased length and muscularity of the modern human lower limb has been significantly influenced by natural selection for improved throwing and club-swinging.

PREDICTION: The remarkable human ability to rotate the pelvis on fixed femurs and the trunk upon a fixed pelvis will prove to be the result of natural selection for throwing and striking.

Comment. The hindlimb elongated and became more massive in *Homo* (Chapter 6). Because it takes more energy to move a larger mass, this change is unlikely to be the sole result of selection for bipedal locomotion. Rather, it may also be due to natural selection acting to increase leg thrust and stride length to promote a more powerful throw and club-swing (Chapters 7, 13). The leg thrusts that initiate the throwing and clubbing motions transfer their forward momentum into rotational momentum of the pelvis around the vertical axis in a manner that is much more forceful and enormously faster than occurs in walking.

PREDICTION: In hominins, the gluteus maximus muscle will be found to have undergone evolutionary modification to enhance its role in controlling rotation of the pelvis during throwing and club-swinging.

Comment. The gluteus maximus is exceptionally large in *Homo* [Zihlman and Brunker, 1979]. Its action is minimal during walking [Robinson, 1972], but it is recruited during running [Stern, 1972; Lovejoy,

1988; Lieberman, et al, 2009], when rising to an upright position, or going uphill [Zihlman and Brunker, 1979]. Hypertrophy is due primarily to expansion of the cranial portion high on the iliac blade, which makes it a powerful medial rotator of the pelvis when the femur is fixed [Stern, 1972]. Marzke and coworkers [1988)] found that this cranial portion was strongly activated during throwing and club-swinging, counteracting the rapid forward rotation of the contralateral side of the pelvis and maintaining balance by stabilizing the hip at the completion of the motion. Remodeling and hypertrophy of the gluteus maximus may have raised the ability of hominins to increase the velocity of tools [Marzke, et al., 1988]. That view is offered here as a prediction, with the stipulation that the tools were weapons.

The shoulder region.

Introduction. The shoulder is a crucial link in the transfer of energy from the trunk to the upper extremity in the throwing motion. Kirschmann [1999, 3.3.2] suggested that the demands of throwing explain the changes in chest and shoulder anatomy that took place during hominin evolution. The predictions in this section expand and elaborate his idea.

PREDICTION: The anatomical structure of the human shoulder, insofar as it differs from that of arboreal apes in bone, muscle, and neuromuscular control, will prove to be the result of natural selection for throwing and club-swinging prowess.

Comment. Most bone, muscle and ligament specializations of the chimpanzee upper body are adaptations to arm-hanging and vertical climbing [Hunt, 1994]. Modern humans rarely arm-hang or climb vertically. Yet, with the emergence of *Homo,* shoulder mobility became *greater* than that of modern apes (Chapter 6). The reproductive advantages of throwing and striking can explain this, as asserted by Kirschmann [1999]. Ward [2002] suggested that mobility of the human shoulder joint had been exapted for manipulative or throwing abilities. Perhaps throwing had "an important selective influence" on the shoulder region during hominin evolution [Larson, 2007].

PREDICTION: The unprecedented rapidity with which the modern human humerus can be internally rotated from an externally rotated position will be found to be attributable to an adaptation to throwing.

Comment: During a throw, as forward acceleration of the missile proceeds, the humerus spins in the shoulder joint from full external rotation through 120° of internal rotation in less than 60 msec. The rate of rotation can exceed 7000° per second [Escamilla, et al., 1998]. This type of rapid rotation is lacking in the behavior of apes and missing from human walking. It seems to be the result of natural selection for throwing (Chapters 13, 14).

PREDICTION: The muscles which produce the rapid internal rotation of the humerus during the acceleration phase of the human throwing motion will prove to have a unique orientation and structure that is adapted to produce this movement.

Comment. Selection for throwing prowess could have augmented humeral rotation speed by increasing muscle size, changing muscle insertion points on the humerus, increasing the fast-twitch muscle fiber component, or modifying the growth and distribution of the motor nerves. *A. afarensis* had well-developed insertions of arm muscles that promote internal rotation [Coppens, 1991; White, et al., 1993; Johanson, et al., 1994]. Scapular changes in *Homo* affecting the infraspinatus muscle (an external rotator of the humerus) [Green and Alemseged, 2012] might be part of these selective changes. External humeral rotation was important in Neanderthals, according to Trinkaus [1977], who thought it served to counteract the action of well-developed internal rotators.

PREDICTION: The relatively large size of the pectoralis major muscle in modern humans will prove to be part of the hominin adaptation to throwing.

Comment. The pectoralis major muscle is similar in size in chimpanzees and humans [Miller, 1932] or may be larger (and presumably more powerful) in humans [Swindler and Wood, 1982]. This muscle, which contributes to several movements at the shoulder joint, is among the most important internal rotators of the humerus during throwing [Gainor, et al., 1980; Wilk, et al., 2000; Stodden, et al., 2006b] and the overarm clubbing motion [Elliott, et al., 2003].

The elbow joint.

PREDICTION: The structure of the human elbow joint and the muscles that control it will prove to differ from those of apes in ways that support the idea that hominins were naturally selected to improve their throwing prowess.

Comment. In the throwing motion the elbow joint is extended during the windup, then flexes to about 100°, then extends through an arc of 73° in 40 msec [Gainor, et al., 1980]. Angular velocity reaches ~2,400°/sec before ball release [Werner, et al., 2001] (Chapter 13). No such movement sequence or rapid elbow extension has been described in other primates.

The forearm and hand.

PREDICTION: Analysis of the upper limb in modern humans will show that it differs from the upper limb in other primates in ways that facilitate overarm throwing and clubbing.

PREDICTION: In hominins, forearm muscles will prove to be adapted for throwing spheroids and gripping clubs.

Comment. The forearm muscles in apes are relatively more massive than in humans [Tuttle, 1967], except for the pronators and thumb muscles (Chapter 9). Such differences may reflect freedom of the upper limb from use in bipedal gait [Aiello and Dean, 1990]. Powerful forearm muscles in *A. afarensis* suggests selection for human-like elbow and hand joints, improved grip capabilities, and expanded manipulative capabilities [Drapeau, et al., 2005]. The view advocated here is that the hominin upper limb was adapted for the bipedal use of weapons.

PREDICTION: When *H. erectus* hand bones are found, they will show the final stages of transition to modern human hands, including straight phalanges and a larger, more robust and mobile thumb.

Predictions Concerning Protective Mechanisms

Introduction. Throwing with high velocity puts enormous stress on the upper limb due to the violent and rapid rotations that occur. Joint stress accompanies arm-cocking, acceleration to ball release and deceleration after the throw. Innate protective mechanisms are activated in the upper extremity during these movements (Chapters 9, 13). They reduce the chances of injury by intricate patterns of contraction and relaxation in specific muscles at precise times in the throwing motion. Some muscles act intermittently at different stages of the throw. These injury-prevention mechanisms underlie a general prediction regarding the outcome of future investigations:

PREDICTION: The neuromuscular and skeletal features in the upper limb that act during the throw to help prevent injury from the severe mechanical forces generated will prove to be unique in humans and to result from natural selection for throwing.

Comment. In the shoulder: The rapid internal rotation of the humerus during the acceleration phase is potentially injurious. Rotator cuff muscles resist distraction and help control humeral head motion in the glenoid fossa [Fleisig, et al., 1989; Fleisig, et al., 1996; Wilk, et al.,

2000; Werner, et al., 2001]. Muscular compressive force is applied, together with flexion torque, medial force and anterior force. The biceps brachii resists humeral distraction during arm acceleration and again later during deceleration [Fleisig, et al., 1995].

In the elbow: Just before missile acceleration begins, the throwing shoulder is brought forward, then the upper arm, while the forearm and hand are left behind. This exerts an extreme valgus (outward) stress on the medial elbow musculature, the medial collateral ligament, the medial joint capsule and the elbow joint itself [Tullow and King, 1973]. Varus (inward) torque is applied at the elbow throughout the arm cocking and acceleration phases, to protect these structures [Gainor, et al., 1980; Werner, et al., 1993]. Eccentric contractions by elbow flexors protect the integrity of the joint and decelerate elbow extension [Fleisig, et al., 1989, 1995]. Stopping short of full extension avoids injury to the posterior part of the elbow joint [Feltner and Dapena, 1986]. Contraction of several muscles passing over the joint assist elbow ligaments by applying strong compression force to stabilize the elbow and prevent elbow distraction as the forearm and hand are propelled distally [Werner, et al., 1993].

Deceleration after missile release: This the most violent part of the throwing mechanism, when kinetic energy not conveyed to the missile must be dissipated [Fleisig, et al., 1989]. The scapula shifts forward [Fleisig, et al., 1996; Kibler and McMullen, 2003]; the arm adducts horizontally across the trunk, to decelerate. Then the muscles used for horizontal adduction and internal rotation of the humerus cease contraction, assisted by the rotator cuff muscles and posterior shoulder muscles [Fleisig, et al., 1995]. Wrist and finger extensor muscles decelerate wrist and finger flexion. Deceleration of the throwing arm, flexion of the trunk and extension of the lead knee allow momentum to be absorbed by the large muscles of the trunk and lower extremities while reducing stress on the throwing arm [Fleisig, et al., 1996].

A final prediction: Hominin fossils yet to be discovered or described will prove to be consistent with the bipedal use of weapons theory.

The foregoing predictions range from generalized concepts to very specific forecasts. They challenge the scope of the theory's explanatory power and by their variety show that the theory can be tested at many levels of organization throughout the body. Failure of any of these predictions to be confirmed would not be fatal to the theory, but might show that its explanatory limits are more restricted than envisioned here. Theories are less likely to be falsified than they are to be replaced by better ones. That would seem to require a different but equally parsimonious explanatory principle with a greater range of application than the one proposed here.

GLOSSARY

Abbreviations.
  A.L., Hadar, Ethiopia
  BAR, Tugen Hills, Kenya
  CNS, Central Nervous System
  DIK, Dikiki, Ethiopia
  D, Dmanisi, Georgia.
  EMG, Electromyography
  KNM-ER, East Turkana, Kenya
  KNM-WT, West Turkana, Kenya
  Kya, Thousand years ago
  LCA, Last Common Ancestor
  L.H., Laetoli, Tanzania
  Mya, Million years ago
  O.H., Olduvai Gorge, South Africa
  StW, Sterkfontein, South Africa
  STS, Sterkfontein, South Africa
  SK, Swartkrans, South Africa
  U.W., Malapa, South Africa

Acheulean. An archaeological stone tool industry during the Lower Paleolithic era typically associated with *Homo erectus*.

Adaptation. Evolutionary adaptation is the process by which natural selection, acting cumulatively through many generations, can establish a new inherited feature which enhances reproductive success. As a noun, the term refers to the end result of that process. This book describes how hominins became adapted to throwing and striking, and identifies numerous adaptations in body structure that resulted from the adaptive process.

A./H. habilis. A taxon designation which combines two genera names (*Australopithecus* and *Homo*) that have been assigned by different authors to the species *habilis*. Some *habilis* fossils may be remains of hominins who were transitional between *Australopithecus* and *Homo*.

Anagenetic. Direct continuation of an evolutionary lineage; the evolution of an ancestral population into a new one. No branching or divergence is observed.

Anthropoids. The suborder of primates that includes monkeys and apes.

Arborealism. A lifestyle predominantly dependent on activity in trees.

Archeology. The study of artifacts of past human life. The first artifacts were flaked stones.

Bauplan. The general body plan. The blueprint for an animal's morphological structure.

Biomechanics. The mechanical properties of biological systems. The study of structure and function of living systems by the methods of mechanics.

Bipedalism, bipedality. Upright activity performed while supported by the lower limbs. Two examples of biped- ality are bipedal use of hand-held weapons and bipedal locomotion.

Bonobo. A species of modern chimpanzees, *Pan paniscus*, native to central, equatorial Africa.

Brachial index. The ratio between radius length and humerus length (radius length ÷ humerus length × 100).

Bridging. An act of arboreal locomotion in which an ape in the top branches of a tree grasps branches of an adjacent tree and then transfers itself to the latter.

Cause. An action or event that produces another action or event. Natural selection is the primary causal agent that produces evolutionary change. An indirect cause of an evolutionary change is one that influences or promotes a change but does not directly cause it. For example, adaptation to use of weapons from a bipedal stance, driven by the reproductive advantages it provided, promoted the subsequent transition to bipedal walking.

Clade. A group consisting of an organism and all of its descendants.

GLOSSARY

Clambering. Cautious arboreal quadrupedalism with weight support over multiple branches.

Close-Packing. The stable position of a joint when bones are maximally congruent and ligaments are taut. In the clubbing (power) grip, when the thumb exerts pressure towards thepalm, the lateral ligament of the carpo-metacarpal joint is taut, and the joint is in one close-packed position. In the throwing (precision) grip, as the thumb abducts and moves into opposition, the anterior and posterior oblique ligaments are tightened, and the joint is in its second close-packed position.

Clubbing grip. A firm grip of a cylinder lying obliquely against the palm of the hand, squeezed in a clamp formed by the flexed fingers and the palm, with the thumb wrapped over the fingers for reinforcement. It is one of the two unique human handgrips (also called a "power grip").

Conspecific. A member of the same species.

Derived. In paleontology, this term denotes a change from the ancestral condition. It contrasts with primitive (ancestral) traits which appeared earlier in the evolution of the lineage.

Diastema. Plural, diastemata. A space between teeth in a jaw.

Dorsiflexion. Movement of the upper (dorsal) surface of the foot towards the shin at the ankle joint or of the upper (dorsal) surface of the hand towards the forearm at the wrist joint.

Electromyography. A technique used to evaluate and record the electrical activity produced by skeletal muscles when they are activated.

Estrus. The period during which a non-pregnant female is sexually receptive and in which ovulation occurs.

Evolution. Biological evolution refers to the process of change across successive generations in the heritable characteristics of populations.

Exapted. Traits which evolved under natural selection to carry out a particular function that later are used for a different, even unrelated function. For example, hands selected for gripping missiles were subsequently exapted for chipping stones and playing pianos

Explanation. The goal of explanation is to increase comprehension by reducing a situation to elements that are so logical or familiar that we accept them as understood. Scientific explanations share a basic two-part form: One describes what needs to be explained; the other provides the explanation.

Gender size dimorphism. A difference between males and females in body size; also called sexual size dimorphism.

Genes. The term is used in this book as a shorthand for the hereditary elements, including gene alleles (different versions of the same gene). The genes embody the instructions coded in DNA molecules. They provide continuity from one generation to the next.

Genetic lineage. The only part of the organism transmitted to the next generation is the genotype (a person's genes). Individuals are transitory bearers of these genetic elements. The continuity in successive generations during which evolution may occur is in this persisting yet alterable genetic lineage. Currently living modern humans are the temporary repositories of the hominin genetic stock.

Genome. The total of all genetic material in a species.

Genotype. The genetic material in an individual organism.

Gracility, gracilization. The low end of the scale of robusticity. Late modern humans are "gracilized" compared to preceding hominins who were characterized by relatively larger muscles and a more sturdy skeletal structure.

Handedness. One of the two hands is preferably employed, especially for skilled actions. About 90% of modern humans are right-handed and 10% are left-handed.

Hominid. A member of the lineage that includes the great apes and modern humans (the family *Hominidea*). Historically, the term denoted the human lineage (now called hominins) and is still used in this sense by some authors.

Hominin. Modern humans and all of our ancestors, including species which became extinct, after our lineage diverged

GLOSSARY

from from the common ancestor we shared with chimpanzees.

Hominoid. The taxon *Hominoidea* includes fossil and modern apes as well as hominins. It includes two families: *Hominidea* (apes and humans) and *Hylobatidea* (the gibbons). There are some 40 genera of hominoids dating back to the beginning of the Miocene era, 23 Mya.

Homology. Shared, derived similarity due to close common ancestry.

Homoplasy. Shared morphology not seen in the most recent common ancestor of two species. Homoplasy is due to parallel evolution—the acquisition of similar biological traits in different lineages, often due to evolutionary solutions to similar problems.

Humerofemoral index. This index indicates the relative lengths of the upper arms and thighs. It is determined by comparing the lengths of the humerus and femur (humerus length ÷ femur length x 100).

Hypothenar eminence. Situated along the ulnar border of the palm, this slight ridge reflects the underlying presence of three muscles that control the movement of the little finger.

Hypothesis. Stand-alone hypotheses are tentative explanations of a single item of experience, based upon imagined causes that seem at least plausible. They represent speculative guesses. Theory-based hypotheses are predictions: if the theory is true, then the predicted outcome should occur. Theories can be tested in this manner.

Incisiform. Shaped like an incisor tooth.

Interaction torques. A torque (turning force) which occurs when a muscular torque at one joint elicits a torque at an adjacent joint. For example, during a throw, wrist flexor muscle activity contributes an active flexion torque at the wrist joint which adds to a passive interaction flexion torque resulting from forearm deceleration.

Intermembral index. This index indicates the relative lengths of the upper and lower limbs (arm length ÷ leg length x

100).

Kinematics. Analysis of the motions of bodies and body segments independent of aspects of mass or velocity. Children and adults throw with similar kinematics, but adults with more mass and muscle can generate more kinetic energy with that motion.

Kinetics. The study of motions and its causes, including mass, velocity, forces and torques.

Kinetic energy. The energy of motion. Kinetic energy is equal to the product of half the mass times the square of the velocity; that is, $½ mv^2$.

Knapping. Chipping flakes from a hand-held core by striking it with a hammerstone.

Knuckle walking. A form of quadrupedal locomotion in which weight placed on the forelimbs is supported on the knuckles of the hands.

Lateralized. One side or the other is favored, rather than being equal on both sides.

Lineage. Lines of genetic descent during which the gene pool (the genome) is continuous through numerous generations while being subject to modification by natural selection during the reproductive process.

Manuports. Literally, "hand carried". This term refers to stones transported by hominins to sites from regions more distant than a local stream bed or rock pile.

Megadontia. Literally, "large teeth". The term has been applied to the enlarged premolars and molars of the australopithecines from *A. afarensis* until the transition to *Homo*, apparently due to natural selection for processing a diet that required powerful chewing.

Miocene. The period in the geological time scale that ex-tends from 23.0 Mya until 5.3 Mya. It precedes the Pliocene epoch.

Momentum. The product of mass times velocity. If no outside force is applied, the momentum does not change. In the throwing and clubbing motions, the force that increases momentum is derived from muscle contraction. Angular momentum is the product of mass times velocity times the distance from the axis of (circular) rotation.

Monogamy. A family unit consisting of a pair-bonded adult male and female in an exclusive sexual relationship plus their children.

Natural selection. The essential cause of biological evolution. Natural selection acts primarily at the level of individual organisms, selecting heritable phenotypic properties which promote reproductive success. Selection occurs when individuals reproduce. Organisms that produce more viable offspring transmit more copies of their genes to the next generation. Their genes are "naturally selected," compared to the genes of those who produce few or no descendants. In this way a population eventually comes to resemble the individuals who reproduce more frequently.

Neolithic. The New Stone Age, when flaked tools were finished by polishing.

Neurophysiology. The study of nervous system function.

New Stone Age. See Neolithic.

Old Stone Age. See Paleolithic.

Oldowan. The earliest, most primitive hominin stone tool industry, beginning about 2.6 Mya, marking the beginning of the archaeological record. It was associated with australopithecines and was succeeded by the Acheulean industry, a refinement that began in early *Homo erectus*.

Ontogeny. Ontogenesis. The biological development of an individual from an embryo to an adult.

Orthograde. Upright orientation of the body. A stance or mode of locomotion in which the body is oriented verti-cally in relation to the substrate.

Outgroup. Members of a different community from the one under consideration.

Ovulatory crypsis. Having an ovulatory cycle that is devoid of any obvious external signs.

Paleoanthropology. The study of ancient humans (hominins).

Paleodeme. Fossil remains of close geographic distribution and similar age. Examples include the hominins from Dmanisi and *Australopithecus sediba* individuals from South Africa.

Paleolithic. The old stone age. A classification based on stone tools produced by flaking or knapping which began about 2.6 Mya, initiating the Lower (Early) Paleolithic which lasted until roughly 300 Kya. The Middle Paleolithic extends from 300-40 Kya, and the Upper Paleolithic, 40 to 10 Kya. Late species of *Australopithecus* were present and the origins of *Homo* occurred early in the Lower Paleolithic. Neanderthals inhabited Europe and anatomically modern humans arose in Africa during the Middle Paleolithic.

Paleontology. The study of life from past geological periods as known from fossil evidence.

Palmarflexion. Movement of the palmar surface of the hand towards the forearm at the wrist joint.

Pan, Panins. The chimpanzee genus. It includes the common chimpanzee, *Pan troglodytes*, and bonobos, *Pan paniscus*.

Phenotype. Any observable trait of the organism which emerges during growth and development from interaction of the genotype with environmental factors such as nutrition.

Phylogeny. The connections between groups of organisms determined from ancestor-descendant relationships .

Plantarflexion. Movement of the sole of the foot towards the heel at the ankle joint.

Pleistocene. The period in the geological time scale that extends from 2.6 million years ago (Mya) until 11,700 years ago. (In 2009 the start date of 2.6 Mya replaced the former start date of 1.8 Mya). The Pleistocene is roughly coterminous with the Paleolithic.

Pliocene. The period in the geological time scale (the epoch) that extends from 5.3 Mya to 2.6 Mya. It follows the Miocene and precedes the Pleistocene.

Polygyny. Polygyny in the context of early human evolution denotes males mating with more than one partner.

Power grip. A term for the clubbing grip which refers to its

GLOSSARY

modern function rather than its evolutionary origin.

Precision grip. A term for the throwing grip which refers to its modern function rather than its evolutionary origin.

Promiscuity. Both males and females have sexual relations with multiple partners.

Pronate, pronation. A rotational movement of the forearm or foot. The forearm moves into pronation when the palm of the hand rotates inward from an anterior-facing (thumb pointing outward) to a posterior-facing position, without movement at the shoulder joint. Pronation of the foot causes the sole to face more laterally than in upright stance.

Pronograde. A stance or mode of locomotion in which the body is oriented in parallel with the substrate, as observed in quadrupedal animals.

Quadrupedalism. Posture or locomotion using all four limbs.

Robusticity. A condition in which the massiveness of bones or muscles is high. Gracility is the condition when they are low.

Rugosity. Muscle markings on bones at sites of muscle attachment. Prominent bone rugosities suggest that the muscles were large and powerful.

Scavenging. In the carnivorous sense, this means acquiring food from a dead animal. Confrontational scavenging (also called power or aggressive scavenging) means displacing a predator from prey it has killed and is still eating in order to eat it oneself. Passive scavenging denotes eating parts of carcasses abandoned by predators, or from animals that died of disease, accidents or other natural causes.

Scenario. A loosely defined, informal explanatory structure, often in the form of a story summary, that links several hypotheses in an attempt to account for two or more evolutionary traits (such as bipedal locomotion and size diminution of canine teeth). Scenarios lack the structure or scope of a theory but are more elaborate than a single, theory-free hypothesis.

Science. Science embodies a way of thinking, a process for documenting evidence, and a body of knowledge derived from analysis of the evidence. The process for gathering evidence is based on observation, experimentation, and replication of results verified by other scientists. It strives to obtain objective evidence by use of measuring devices and analytical techniques which exclude insofar as is possible the scientist's wishes and opinions. Major goals of science are the explanation of objective evidence based on the observed laws of nature. Explanation requires human inter- vention: The creativity and insight of the scientific mind must be applied to make sense of the evidence.

Sensu lato. In the broad sense. With regard to a particular taxon, such as a species, this term refers to all of the fossil specimens which might be assigned to that taxon.

Sensu stricto. In the strict sense. This refers to the fossil specimens of a taxon, such as a species, that are distinguished because they strictly exemplify its defining characteristics.

Sexual selection. A subcategory of natural selection characterized by males who fight among themselves for access to females while females choose which males they prefer to mate with.

Sexual size dimorphism. A difference between the male and female genders in body size.

Species. In living organisms, members of a population that can interbreed. In fossils of extinct organisms it is a taxonomic category containing specimens with a degree of similarity judged to be comparable to that within living species.

Supinate, supination. A rotational movement of the forearm or foot. When the forearm moves into supination, the palm faces forward (and the thumb points outward) with the arm at the side of the body. When the foot is supinated, the sole turns inward from the standing position.

Taxon. A single taxonomic unit (plural, taxa). Taxonomy is the science of the classification of plants and animals in a hierarchical scheme. A particular genus (e.g., *Austral-*

GLOSSARY

opithecus) or species (e.g., *afarensis*) constitutes a taxon.

Thenar eminence. The region on the palm of the hand at the base of the thumb overlying a group of muscles. The eminence is larger in the human hand than it is in the hands of apes because the thumb muscles are larger.

Throwing grip. One of the two unique human handgrips (also called a precision grip). It is a fingertip grip of a spheroid.

Valgus. An outward angulation or stress. During a fast throwing motion, the medial elbow musculature, the medialcollateral ligament, the medial joint capsule and the elbow joint itself undergo a valgus stress.

Varus. An inward angulation or stress.

REFERENCES CITED

Adair, R. K. (1990). *The Physics of Baseball*. New York, NY: Harper and Row.

Aiello, L. (1996). Hominine preadaptations for language and cognition. In: Mellars, P. & Gibson, K. (Eds.), *Modeling the Early Human Mind* (89-99). Cambridge, UK: McDonald Institute Monographs.

Aiello, L. & Dean, C. (1990). *An Introduction to Human Evolutionary Anatomy*. New York, NY: Academic Press.

Aiello, L. & Key, C. (2002). Energetic consequences of being a *Homo erectus* female. *Am. J. Hum. Biol., 14,* 551-565.

Aiello, L. & Wheeler, P. (1995). The expensive-tissue hypothesis: The brain and digestive system in human and primate evolution. *Curr. Anthrop., 3,* 199-221.

Alba, D. M., Moyà-Solà, S. & Köhler, M. (2003). Morphological affinities of the *Australopithecus afarensis* hand on the basis of manual proportions and relative thumb length. *J. Hum. Evol., 44,* 225-254.

Alemseged, Z., Spoor, F., Kimbel, W. H., Bobe, R., Geraads, D., Reed, D., & Wynn, J. G. (2006). A juvenile early hominin skeleton from Dikika, Ethiopia. *Nature, 443,* 296-301.

Alméija, S., Moyà-Solà, S. & Alba, D. M. (2010). Early origin for human-like precision grasping: A comparative study of pollical distal phalanges in fossil hominins. *PloS One, 5,* e11727.

Alvard, M. S. (2001). Mutualistic hunting. In: Stanford C. B. & Bunn, H. T. (Eds.), *Meat-Eating and Human Evolution* (261-278). New York, NY: Oxford University Press.

Anderson, F. C. & Pandy, M. G. (2003). Individual muscle contributions to support in normal walking. *Gait and Posture, 17,* 159-169.

Annett, M. (1970). A classification of hand preference by association analysis. *Brit. J. Psychol., 61,* 303-321.

Annett, M. (1973). Handedness in families. *Ann. Hum. Genet., 37,* 93-105.

Antón, S. C. (2003). Natural history of *Homo erectus*. *Yearb. Phys. Anthrop., 46,* 126-170.

REFERENCES CITED

Antón, S. C., Leonard, W. R., & Robertson, M. L. (2002). An ecomorphological model of the initial hominid dispersal from Africa. *J. Hum. Evol., 43,* 773-785.

Asfaw, B., Gilbert, W. H., Beyene, Y., Hart, W. K., Renne, P. R., WoldeGabriel, G., Vrba, E. & White, T. D. (2002). Remains of *Homo erectus* from Bouri, Middle Awash, Ethiopia. *Nature, 416,* 317-320.

Asfaw, B., White, T., Lovejoy, O., Latimer, B., Simpson, S. & Suwa, G. (1999). *Australopithecus garhi:* A new species of early hominid from Ethiopia. *Science, 284,* 629-635.

Ashton, E. H. & Oxnard, C. E. (1964). Functional adaptations in the primate shoulder girdle. *Proc. Zool. Soc. Lond., 142,* 49-66.

Atwater, A. E. (1979). Biomechanics of overarm throwing movements and of throwing injuries. *Exerc. Sport Sci. Rev., 7,* 43-85.

Ayala, F. J. (2010). The difference of being human: Morality. *PNAS 107,* (Suppl. 2) 9015-9022.

Bagesteiro, L. B. & Sainburg. R. L. (2002). Handedness: Dominant arm advantages in control of limb dynamics. *J. Neurophysiol., 88,* 2408-2421.

Barrentine, S. W., Matsuo, T., Escamilla, R. F., Fleisig, G. S. & Andrews, J. R. (1998). Kinematic analysis of the wrist and forearm during baseball pitching. *J. Applied Biomech., 14,* 24-39.

Bartholomew, G. A., & Birdsell, J. B. (1953). Ecology and the protohominids. *Am. Anthrop., 55,* 481-498.

Bartlett, R., Müller, E., Lindinger, S., Brunner, S. & Morriss, C. (1996). Three-dimensional evaluation of the kinematic release parameters for javelin throwers of different skill levels. *J. Applied Biomech., 12,* 58-71.

Baseball.com. (2009). (http://www.baseball-reference.com/awards/mvp-cya.shtml)

Beckner, M. (1959). *The Biological Way of Thought.* New York, NY: Columbia University Press.

Begun, D. R. (2004). The earliest hominins—is less more? *Science, 303,* 1478-1480.

Begun, D. R. (2007). Fossil record of Miocene hominoids.In: Henke, W. & Tattersall, I. (Eds.), *Handbook of Paleoanthropology* (921-977).*Vol. III, Part 2.* Berlin, DE: Springer.

Begun, D. & Walker, A. (1993). The endocast. In: Walker, A. B. & Leakey, R. (Eds.), *The Nariokotome Homo Erectus Skeleton* (326-358). Cambridge, MA: Harvard University Press.

Behrensmeyer, A. K., Todd, N. E., Potts, R. & McBrinn, G. E. (1997). Late Pliocene faunal turnover in the Turkana Basin, Kenya and Ethiopia. *Science, 278,* 1589-1594.

Bell, C. (1837/2004). *The Hand. Its Mechanism and Vital Endowments as Evincing Design.* Elibron Classics Replica Edition, www.elibron.com.

Bénéfice, E., Fouére, T. & Malina, R. M. (1999). Early nutritional history and motor performance of Senegalese children, 4-6 years of age. *Ann. Hum. Biol., 26,* 443-455.

Bénéfice, E. & Malina, R. (1996). Body size, body composition and motor performances of mild-to-moderately undernourished Senegalese children. *Ann. Hum. Biol., 23,* 307-321.

Ben-Itzhak, S., Smith, P. & Bloom, R. A. (1988). Radiographic study of the humerus in Neanderthals and *Homo sapiens. Am. J. Phys Anthrop., 77,* 231-242.

Bennett, M. R., Harris, J. W. K., Richmond, B. G., Braun, D. R., Mbua, E., Kiura, P. Olago, D., Kibunjia, M., Omuombo, C., Behrensmeyer, A. K., Huddart, D. & Gonzalez, S. (2009). Early hominin foot morphology based on 1.5-million-year-old footprints from Ileret, Kenya. *Science, 323,* 1197-1201.

Benton, M. J. & Donoghue, P. C. J. (2007). Paleontological evidence to date the tree of life. *Mol. Biol. Evol., 24,* 26-53.

Berger, L. R. (2000). *In the Footsteps of Eve.* Washington, DC: National Geographic Adventure Press.

Berger, L. R., de Ruiter, D. J., Churchill, S. E., Schmid, P., Carlson, K. J., Dirks, P. H. G. M. & Kibii, J. M. (2010a). *Australopithecus sediba*: A new species of *Homo*-like Australopith from South Africa. *Science, 328,* 195-204.

Berger, L. R., de Ruiter, D. J., Churchill, S. E., Schmid, P., Carlson, K. J., Dirks, P. H. G. M., Kibii, J. M. (2010b). *Australopithecus sediba*: A new species of *Homo*-like Australopith from South Africa. *Science, 328,* Supporting online material, 1-25.

Berna, F., Goldberg, P., Horwitz, L. K., Brink, J., Holt, S.,

Bamford, M. & Chazan, M. [2012]. Microstratigraphic evidence of in situ fire in the Acheulean strata of Wonderwerk Cave, Northern Cape Province, South Africa. *PNAS, 109,* E1215-E1220.

Binford, L. R. (1981). *Bones: Ancient Men and Modern Myths.* New York, NY: Academic Press.

Binford, L. R. (1987). Were there elephant hunters at Torralba? In: Nitecki, M. H. & Nitecki, D. V. (Eds.), *The Evolution of Human Hunting* (47-105). New York, NY: Plenum Press.

Bingham, P. M. (1999). Human uniqueness: A general theory. *Q. Rev. Biol., 74,* 133-169.

Bloom, P. (2010). How do morals change? *Nature, 464,* 490.

Boehm, C. (1999). *Hierarchy in the Forest.* Cambridge, MA: Harvard University Press.

Boesch, C. & Boesch, H. (1989). Hunting behavior of wild chimpanzees in the Taï National Park. *Am. J. Phys. Anthrop. 78,* 547-573.

Boesch, C. & Boesch-Achermnn, H. (2000). *The Chimpanzees of the Taï Forest.* Oxford, UK: Oxford University Press.

Bogin, B. (1999). *Patterns of Human Growth* (2nd Edit.). Cambridge, UK: Cambridge University Press.

Bojsen-Møller, F. (1979). Calcaneocuboid joint and stability of the longitudinal arch of the foot at high and low gear push off. *J. Anat., 129,* 165-176.

Bourdin, M., Rambaud, O., Dorel, S., Lacour, J. R., Moyen, B. & Rahmani, A. (2010). Throwing performance is associated with muscular power. *Int. J. Sports Med., 31,* 505-510.

Bowers, E. J. (2006). A new model for the origin of bipedality. *Hum. Evol. 21,* 241-250.

Bowler, P. J. (1986). *Theories of Human Evolution.* Baltimore, MD: The Johns Hopkins University Press.

Braccini, S., Lambeth, S., Schapiro, S., & Fitch, W. T. (2010). Bipedal tool use strengthens chimpanzee hand preferences. *J. Hum. Evol., 58,* 234-241.

Brace, C. L. (1967). Environment, tooth form, and size in the Pleistocene. *J. Dent. Res., 46,* 809-816.

Bradley, B. J. (2008). Reconstructing phylogenies and

phenotypes: a molecular view of human evolution. *J. Anat., 212,* 337-353.

Brain, C. K. (1981). *The Hunters or the Hunted?* Chicago, IL: University of Chicago Press.

Brain, C. K., Churcher, C. S., Clark, J. D., Grine, F. E., Shipman, P., Susman, R. L. & Turner, A. W. (1988). New evidence of early hominids, their culture and environment from the Swartkrans Cave, South Africa. *S. Afr. J. Sci., 84,* 828-835.

Bramble, D. M., Lieberman, D. E. (2004). Endurance running and the evolution of *Homo. Nature, 432,* 345-352.

Braun, D. R. (2010). Palaeoanthropology: Australopithecine butchers. *Nature, 466,* 828.

Braun, D. R., Tactikos, J. C., Ferraro, J. V. & Harris, J. W. K. (2005). Flake recovery rates and inferences of Oldowan hominin behavior: a response to Kimura, 1999 and Kimura, 2002. *J. Hum. Evol. 48,* 525-531.

Brewer, S. (1978). *The Chimps of Mt. Asserik.* New York, NY: Alfred A. Knopf.

Brockman, J. (2013). *This Explains Everything.* New York, NY: Harper Perennial.

Broom, R. (1951). *Finding the Missing Link.* London, UK: Watts & Co.

Brown, F. H. & McDougall, I. (1993). Geologic setting and age. In: Walker, A. & Leakey, R. (eds.), *The Nariokotome Homo erectus Skeleton* (9-20). Cambridge, MA: Harvard University Press,

Brown, J. H. & Lasiewski, R. C. (1972). Metabolism of weasels: The cost of being long and thin. *Ecology, 523,* 939-943.

Brues, A. (1959). The spearman and the archer—An essay on selection in body build. *Am. Anthrop. 61,* 457-469.

Brunet, M., Guy, F., Pilbeam, D., Lieberman, D. E., Likius, A., Mackaye, H. T., Ponce de Leon, M. S., Zollikofer, C. P. E. & Vignaud, P. (2005). New material of the earliest hominid from the Upper Miocene of Chad. *Nature, 434,* 752-755.

Brunet, M., Guy, F., Pilbeam, D., Mackaye, H. T., Likius, A., Ahounta, D., Beauvilain, A., Blondel, C., Bocherens, H., Boisserie, J-P., De Bonis, L., Coppens, Y., Dejax, J., Denys, C., *et 24 al.* (2002). A new hominid from the Upper Miocene of Chad, Central Africa. *Nature, 418,* 145-151.

Bunn, H. T. (1981). Archaeological evidence for meat-eating by Plio-Pleistocene hominids from Koobi Fora and Olduvai Gorge. *Nature 291,* 574-577

Bunn, H. T. (1994). Early Pleistocene hominid foraging strategies along the ancestral Omo River at Koobi Fora, Kenya. *J. Hum. Evol. 27,* 247-266.

Bunn, H. T. (1997). The bone assemblages from the excavated sites. In: Isaac, G. L. & Isaac, B. (Eds.), *Koobi Fora Resarch Project, Vol. 5. Plio-Pleistocene Archeology* (402-444). Oxford, UK: Clarendon Press.

Bunn, H. T. (2001). Hunting, power scavenging, and butchering by Hadza foragers and by Plio-Pleistocene *Homo*. In: Stanford, C. B. & Bunn, H. T. (Eds.), *Meat-Eating and Human Evolution* (199-218) New York, NY: Oxford University Press.

Bunn, H. T. & Ezzo, J. A. (1993). Hunting and scavenging by Plio-Pleistocene hominids: Nutritional constraints, archaeological patterns, and behavioural implications. *J. Archaeol. Sci., 20,* 365-398.

Bunn, H. T., Harris, J. W. K., Isaac, G., Kaufulu, Z., Kroll, E., Schick, K., Toth, N. & Behrensmeyer, A. K. (1980). FxJj50: an Early Pleistocene site in northern Kenya. *World Archaeol., 12,* 109-139.

Bunn, H. T. & Kroll, E. M. (1986). Systematic butchery by Plio/Pliocene hominids at Olduvai Gorge, Tanzania. *Curr. Anthrop., 27,* 431-452.

Bunn, H. T. & Stanford, C. B. (2001). Conclusions. In: Stanford, C. B. & Bunn, H. T. (Eds.), *Meat-Eating and Human Evolution* (350-359). New York, NY: Oxford University Press.

Burton, R. F. (1884/1987). *The Book of the Sword.* New York, NY: Dover Publications.

Bush, M. E., Lovejoy, C. O., Johanson, D. C. & Coppens, Y. (1982). Hominid carpal, metacarpal and phalangeal bones recovered from the Hadar formation: 1974-1977 collections. *Am. J. Phys. Anthrop., 57,* 651-677.

Butterfield, S. A. & Loovis, E. M. (1993). Influence of age, sex, balance, and sport participation on the development of throwing by children in grades K-8. *Percept. Motor Skills, 76,* 459-464.

Butterfield, S. A. & Loovis, E. M. (1998). Kicking, catching, throwing and striking development by children in grades K-8: preliminary findings. *J. Hum. Movement Studies, 34*, 67-81.

Butterfield, S. A., Loovis, E. M. & Lee, J. (2003). Throwing development by children in grades K-8: A multi-cohort longitudinal study. *J. Hum. Movement Studies, 45*, 31-47.

Calvin, W. H. (1983). *The Throwing Madonna*. New York, NY: McGraw-Hill Book Co.

Cannell, A. (2002). Throwing behaviour and the mass distribution of geological hand samples, hand grenades and Olduvian manuports. *J. Archaeol. Sci., 29*, 335-339.

Carmody, R. N., Weintraub, G. S. & Wrangham, R. W. (2011). Energetic consequences of thermal and nonthermal food processing. *PNAS, 108*, 19199-19203.

Carmody, R. N. & Wrangham, R. W. (2009a). The energetic significance of cooking. *J. Hum. Evol., 57*, 379-391.

Carmody, R. N. & Wrangham, R. W. (2009b). Cooking and the human commitment to a high-quality diet. *Cold Spring Harbor Symp. Quant. Biol., 74*, 427-434.

Carrier, D. (2004). The running-fighting dichotomy and the evolution of aggression in hominids. In: Meldrum, D. J. & Hilton, C. E. (Eds.), *From Biped to Strider: The Emergence of Modern Human Walking, Running, and Resource Transport* (135-162). New York, NY: Kluwer Academic/Plenum Publishers.

Cartmill, M. (1983). Four legs good, two legs bad: Man's place (if any) in nature. *Natural Hist., 92*, 64-79.

Cartmill, M. (1990). Human uniqueness and theoretical content in paleoanthropology. *Int. J. Primatol., 11*, 173-192.

Cartmill, M., Pilbeam, D. & Isaac, G. (1986). One hundred years of paleoanthropology. *Am. Scientist, 74*, 410-420.

Cartmill, M. & Smith, F. H. (2009). *The Human Lineage*. New York, NY: Wiley-Blackwell.

Carvalho, W., Biro, D., Cunha, E., Hockings, K., McGrew, W. C., Richmond, B. G. & Matsuzawa, T. (2012). Chimpanzee carrying behaviour and the origins of human bipedality. *Curr. Biol. 22*, R180-R181.

Cerling, T. E., Wynn, J. G., Andanje, S. A., Bird, M. I., Korir, D. K., Levin, N. E., Mace, W., Macharia, A. N., Quade,

J. & Remien, C. H. (2011). Woody cover and hominin environments in the past 6 million years. *Nature, 476*, 51-56.

Chapelain, A. S. & Hogervorst, E. (2009). Hand preferences for bimanual coordination in 29 bonobos (*Pan paniscus*). *Behav. Brain Res., 196*, 15-29.

Chelly, M. S., Hermassi, S., & Shephard, R. J. (2010). Relationships between power and strength of the upper and lower limb muscles and throwing velocity in male handball players. *J. Strength Cond. Res., 24*, 1480-1487.

Chen, F.-C. & Li, W.-H. (2001). Genomic divergences between humans and other hominoids and the effective population size of the common ancestor of humans and chimpanzees. *Am. J. Hum. Genet., 68*, 444-456.

Christel, M. L., Kitzel, S. & Niemitz, C. (1998). How precisely do bonobos (*Pan paniscus*) grasp small objects? *Int. J. Primatol., 19*, 165-194.

Chu, Y., Fleisig, G. S., Simpson, K. J. & Andrews, J. R. (2009). Bio-mechanical comparison between elite female and male baseball pitchers. *J. Appl. Biomech., 25*, 22-31.

Churchill, S. E. (1993). Weapon technology, prey size selection, and hunting methods in modern hunter-gatherers: Implications for hunting in the Palaeolithic and Mesolithic. In: Peterkin, G. L., Bricker, H. M. & Mellars, P. (Eds.), *Hunting and Animal Exploitation in the LaterPalaeolithic Mesolithic of Eurasia* (11-24). Arlington, VA: Am. Anthrop. Assoc. Archeol. Papers., Number 4.

Churchill, S. E., Holliday, T. W., Carlson, K. J., Jashashvili, T., Macias, M. E., Mathews, S., Sparling, T. L., Schmid, P., de Ruiter, D. J., & Berger, L. R (2013). The upper limb of *Australopithecus sediba*. *Science, 340*, 12 April, DOI: 10.1126/science .1233477.

Churchill, S. E. & Schmitt, D. (2000). Biomechanics in palaeoanthropology: engineering and experimental approaches to the study of behavioural evolution in the genus *Homo*. In: Harcourt, C. & Crompton, R. (Eds.), *New Perspectives in Primate Evolution and Behaviour* (59-89). London, UK: Linnean Society.

Churchill, S. E., Weaver, A. H. & Niewoehner, W. A. (1996).

Late Pleistocene human technological and subsistence behavior: Functional interpretations of upper limb morphology. *Quaternaria Nova, VI,* 413-447.

Clarke, R. J. (1999). Discovery of complete arm and hand of the 3.3 million-year-old *Australopithecus* skeleton from Sterkfontein. *S. Afr. J. Sci., 95,* 477-480.

Clarke, R. J. (2002). Newly revealed information on the Sterkfontein Member 2 *Australopithecus* skeleton. *S. Afr. J. Sci., 98,* 523-526.

Clarke, R. J. & Tobias, P. V. (1995). Sterkfontein Member 2 foot bones of the oldest South African hominid. *Science, 269,* 521-524.

Cooley, D., Oakman, R., McNaughton, L. & Ryska, T. (1997). Fundamental movement patterns in Tasmanian primary school Children. *Percept. Motor Skills, 84,* 307-316.

Coon, C. S. (1971). *The Hunting Peoples.* Boston, MA: Little, Brown and Company.

Coppens Y. (1991). L'originalité anatomique et fonctionelle de la première bipédie. *Bull. Acad. Natle. Mèd. 175,* 977-993.

Corballis, M. C. (1991). *The Lopsided Ape.* New York, NY: Oxford University Press.

Coren, S. (1993). *The Left-Hander Syndrome.* New York, NY: Vintage Books.

Coren, S. & Halpern, D. F. (1991). Left-handedness: A marker for decreased survival fitness. *Psychol. Bull., 109,* 90-106.

Corp, N. & Byrne, R. W. (2004). Sex difference in chimpanzee handedness. *Am. J. Phys Anthrop., 128,* 62-68.

Cosgarea, A. J., Campbell, K. R., Hagood, S. S., McFarland, E. G. & Silberstein, C. E. (1993). Coupling of reach and grasp movements. *Exp. Brain Res., 146,* 511-522.

Crompton, R. H. & Günther, M. (2004). Humans and other bipeds: the evolution of bipedality. *J. Anat., 204,* 317-319.

Crompton, R. H., Sellers, W. I., & Thorpe, S. K. S. (2010). Arboreality, terrestriality and bipedalism. *Phil. Trans. Roy. Soc. B, 365,* 3301-3314.

Crompton, R. H. & Thorpe, S. K. S. (2007). Response to comment on "origin of human bipedalism as an adaptation for locomotion on flexible branches". *Science, 316,* 1066.

Crompton, R. H., Thorpe, S. K. S., Weijie, W., Yu, L., Payne, R., Savage, R., Carey, T., Aerts, P., van Elsacker, L., Hofstetter, A., Günther, M., & Richardson, J. (2003). The biomechanical evolution of erect bipedality. *Cour. Forsch.-Inst. Senckenberg, 243,* 135-146.

Crompton, R. H., Vereecke, E. E. & Thorpe, S. K. S. (2008). Locomotion and posture from the common hominoid ancestor to fully modern hominins, with special reference to the last common panin/hominin ancestor. *J. Anat., 212,* 501-543.

Crompton, R. H., Yu, L., Weijie, W., Günther, M. & Savage, R. (1998). The mechanical effectiveness of erect and "bent-hip, bent-knee" bipedal walking in *Australopithecus afarensis. J. Hum. Evol., 35,* 55-74.

Cross, R. (2004). Physics of overarm throwing. *Am. J. Phys., 72,* 305-312

D'aout, K., Aaerts, P., de Clercq, D., De meester, K. & van Elsacker, L. (2002). Segment and joint angles of hind limb during bipedal and quadrupedal walking of the bonobo (*Pan paniscus*). *Am. J. Phys Anthrop., 119,* 37-51.

Darlington, P. J. Jr (1975). Group selection, altruism, reinforcement, and throwing in human evolution. *PNAS, 72,* 3748-3752.

Dart, R. A. (1925). *Australopithecus africanus*: the man-ape of South Africa. *Nature, 115,* 195-199.

Dart, R. A. (1925a). The round stone culture of South Africa. *S. Afr. J. Sci., 22,* 437-440.

Dart, R. A. (1949a). The predatory implemental technique of *Australopithecus. Am. J. Phys. Anthrop., 7,* 1-38.

Dart, R. A. (1949b). Innominate fragments of *Australopithecus prometheus. Am. J. Phys.Anthrop., 7,* 301-333.

Dart, R. A. (1953).The predatory transition from ape to man. *Internat. Anthrop.Linguistic Rev., 1,* 201-218.

Dart, R. A. & Craig, D. (1959). *Adventures with the Missing Link*. New York, NY: Harper and Brothers.

Darwin, C. (1839/1989). *Voyage of the Beagle*. London, UK: Penguin Books.

Darwin, C. (1859/1979). *The Origin of Species*. New York, NY: Random House Value Publishing.

Darwin, C. (1871/1981). *The Descent of Man, and Selection in Relation to Sex*. Princeton, NJ: Princeton University Press.

Dawkins, R. (1976/1989). *The Selfish Gene*. New York, NY: Oxford University Press.

Dawkins, R. (1996). *The Blind Watchmaker*. New York, NY: W. W. Norton.

Dawkins, R. (2004). *The Ancestor's Tale*. New York, NY: Houghton Mifflin Company.

Debicki, D. B., Gribble, P. L., Watts, S. & Hore J. (2004). Kinematics of wrist joint flexion in overarm throws made by skilled subjects. *Exp. Brain Res., 154,* 382-394.

Debicki, D. B., Gribble, P. L., Watts, S. & Hore, J. (2011). Wrist muscle activation, interaction torque and mechanical properties in unskilled throws of different speeds. *Exp. Brain Res., 208,* 115-125.

Debicki, D. B., Watts, S., Gribble, P. L. & Hore, J. (2010). A novel shoulder-elbow mechanism for increasing speed in a multijoint arm movement. *Exp. Brain Res., 203,* 601-613.

de Heinzelin, J. D., Clark, J. D., White, T., Hart, W., Renne, P., Woldegabriel, G., Beyene, Y. & Vrba, E. (1999). Environment and behavior of 2.5-million-year old Bouri hominids. *Science, 284,* 625-629.

deMenocal, P. B. (1995). Plio-Pleistocene African climate. *Science, 270,* 53-59.

DeSilva, J. M., Holt, K. G., Churchill, S. E., Carlson, K. J., Walker, C. S., Zipfel, B. & Berger, L. R. (2013). The lower limb and mechanics of walking in *Australopithecus sediba*. *Science, 340,* 12 April, DOI: 10.1126 / science 1232999.

DeVore, I. (1964). The evolution of social life. In: Tax, S. (Ed.), *Horizons in Anthropology* (25-36). New York, NY: Aldine Publishing Co.

DeVore, I. & Washburn, S. L. (1963). Baboon ecology and human evolution. In: Howell, F. C. & Bourliere, F. (Eds.), *African Ecology and Human Evolution* (335-367). New Brunswick, NJ : Transaction Publishers.

De Waal, F. (1989). *Chimpanzee Politics*. Baltimore, MD: Johns Hopkins University Press.

Diamond, J. M. (1982). Man the exterminator. *Nature, 298,* 787-789.

Dillman, C. J., Fleisig, G. S. & Andrews, J. R. (1993). Biomechanics of pitching with emphasis upon shoulder kinematics. *J.Orthop. Sports Phys. Therapy, 18,* 402-408.

Dobzhansky, T. (1973). Nothing in biology makes sense except in the light of evolution. *Am. Biol. Teacher, 35,* 125-129.

Domínguez-Rodrigo, M. (2002). Hunting and scavenging by early humans: The state of the debate. *J. World Prehistory, 16,* 1-54.

Domínguez-Rodrigo, M. (2003). Bone surface modifications, power scavenging and the "display" model at early archaeological sites: a critical review. *J. Hum. Evol., 45,* 411-415.

Domínguez-Rodrigo, M. & Barba, R. (2006). New estimates of tooth mark and percussion mark frequencies at the FLK Zinj site: the carnivore-hominid-carnivore hypothesis falsified. *J. Hum. Evol., 50,* 170-194.

Domínguez-Rodrigo, M., de la Torre, I., de Luque, L., Alcalá, L., Mora, R., Serrallonga, J. & Medina, V. (2002). The ST site complex at Peninj, West Lake Natron, Tanzania: Implications for early hominid behavioural models. *J. Archaeol. Sci., 29,* 639-665.

Domínguez-Rodrigo, M., Mabulla, A., Bunn, H. T., Barba, R., Diez-Martin, F., Egeland, C. P., Espilez, E., Egeland, A., Yravedra, J. & Sanchez, P. (2009). Unraveling hominin behavior at another anthropogenic site from Olduvai Gorge (Tanzania): new archaeological and taphonomic research at BK, Upper Bed II. *J. Hum. Evol., 57,* 260-283.

Domínguez-Rodrigo, M. & Pickering, T. R. (2003). Early hominid hunting and scavenging: A zooarcheological review. *Evol. Anthrop., 12,* 275-282.

Domínguez-Rodrigo, M., Pickering, T. R. & Bunn, H. T. (2010). Configurational approach to identifying the earliest hominin butchers. *PNAS, 107,* 20929-20934.

Domínguez-Rodrigo, M., Pickering, T. R., Semaw, S. & Rogers, M. J. (2005). Cutmarked bones from Pliocene archaeological sites at Gona, Afar, Ethiopia: implications for the function of the world's oldest stone tools. *J. Hum. Evol., 48,* 109-121.

Domínguez-Rodrigo, M., Serrallonga, J., Juan-Tresserras, L., Alcalá, L. & de Luque, L. (2001). Woodworking activities by early humans: a plant residue analysis on Acheulian stone tools from Peninj (Tanzania). *J. Hum. Evol., 40*, 289-299.

Doran, D. M. (1992). Comparison of instantaneous and locomotor bout sampling methods. *Am. J. Phys. Anthrop., 89*, 85-99.

Drapeau, M. S. M.. & Ward, C. V. (2007). Forelimb segment length proportions in extant hominoids and *Australopithecus afarensis*. *Am. J. Phys. Anthrop., 132*, 327-343.

Drapeau, M. S. M., Ward, C. V., Kimbel, W. H., Johanson, D. C & Rak, Y. (2005). Associated cranial and forelimb remains attributed to *Australopithecus afarensis* from Hadar, Ethiopia. *J. Hum. Evol., 48*, 593-642.

Du Brul, E. L. (1962). The general phenomenon of bipedalism. *Am. Zoologist, 2*, 205-208.

Dunsworth, H. J., Challis, H. & Walker, A. (2003). Throwing and bipedalism: a new look at an old idea. *Cour. Forsch.-Inst. Senckenberg, 243*, 105-110.

Dusenberry, L. M. (1952). A study of the effects of training in ball throwing by children ages three to seven. *Res. Quart., 23*, 9-20.

Eaton, S. B. & Konner, M. (1985). Paleolithic nutrition. *New Engl. J. Med., 312*, 283-289.

Ehl, T., Roberton, M. A. & Langendorfer, S. J. (2005). Does the throwing "gender gap" occur in Germany? *Res. Quart. Exerc. Sport, 76*, 488-493.

Eimerl, S. & DeVore, I. (1965). *The Primates*. New York, NY: Time Inc.

Elftman, H. (1939). The function of the arms in walking. *Hum. Biol., 11*, 529-535.

Elftman, H. (1944). The bipedal walking of the chimpanzee. *J. Mammal., 25*, 67-71.

Elftman, H. & Manter, J. (1935). Chimpanzee and human feet in bipedal walking. *Am. J. Phys. Anthrop., 20*, 69-79.

Elliott, B., Fleisig, G., Nicholls R. & Escamilla, R. F. (2003). Technique effects on upper limb loading in the tennis serve. *J. Sci. Med. Sport, 6*, 76-87.

Ellison, P. T. (2001). *On Fertile Ground*. Cambridge, MA:

Harvard University Press.

Ellison, P. T. (2003). Energetics and reproductive effort. *Am. J. Hum. Biol., 15,* 342-351.

Escamilla, R. F. & Andrews, J. R. (2009). Shoulder muscle recruitment patterns and related biomechanics during upper extremity sports. *Sports Med., 39,* 569-590.

Escamilla, R., Fleisig, G. S., Barrentine, S., Andrews, J. & Moormann, C. (2002.). Kinematic and kinetic comparisons between American and Korean professional baseball pitchers. *Sports Biomech., 1,* 213-228.

Escamilla, R. F., Fleisig, G. S., Barrentine, S. W., Zheng, N. & Andrews, J. R. (1998). Kinematic comparisons of throwing different types of baseball pitches. *J. Appl. Biomech, 14,* 1- 23.

Escamilla, R. F., Fleisig, G. S., DeRenne, C., Taylor, M. K., Moorman, C. T. III, Imamura, R., Barakatt, E. & Andrews, J. R. (2009). A comparison of age level on baseball hitting kinematics. *J. Appl. Biomech., 25,* 210-218.

Escamilla, R. F., Fleisig, G. S., Yamashiro, K., Mikia, T., Dunning, R., Paulos, L. & Andrews, J. R. (2010). Effects of a 4-week youth baseball conditioning program on throwing velocity. *J. Strength Cond. Res., 24,* 3247-3254.

Escamilla, R. F., Fleisig, G. S., Zheng, N., Barrentine, S. W. & Andrews, J. R. (2001). Kinematic comparisons of 1996 Olympic baseball pitchers. *J. Sports Sci., 19,* 665-675.

Espenschade, A. (1960). Motor development. In: Johnson, W. R. (Ed.), *Science and Medicine of Exercise and Sports* (419-439). New York, NY: Harper and Brothers.

Etherington, J., Harris, P. A., Nandra, D., Hart, D. J., Wolman, R. L., Doyle, D. V. & Spector, T. D. (1996). The effect of weight-bearing exercise on bone mineral density: a study of female ex-elite athletes and the general population. *J. Bone & Mineral Res. 11,* 1333-1338.

Fagot, J., & Vauclair, J. (1991). Manual laterality in nonhuman primates: A distinction between handedness and manual specialization. *Psychol. Bull., 109,* 76-89.

Falk, D. (1980). Language, handedness, and primate brains: Did the australopithecines sign? *Am. Anthrop., 82,* 72-78.

Faurie, C. & Raymond, M. (2005). Handedness, homicide and negative frequency-dependent selection. *Proc. Roy. Soc. B, 272,* 25-28.

Faurie, C., Schiefenhövel, W., Le Bomin, S., Billiard, S. & Raymond, M. (2005). Variation in the frequency of left-handedness in traditional societies. *Curr. Anthrop., 46,* 142-147.

Fedigan, L. M. (1986).The changing role of women in models of human evolution. *Ann. Rev. Anthrop., 15,* 25-66.

Feltner, M. E. & Dapena, J. (1986). Dynamics of the shoulder and elbow joints of the throwing arm during a baseball pitch. *Int. J. Sport Biomech., 2,* 235-259.

Ferring, R., Oms, O., Agusti, J., Berna, F., Nioradze, M., Shelia, T., Tappen, M., Vekua, A., Zhvania, D. & Lordkipanidze, D. (2011). Earliest human occupations at Dmanisi (Georgian Caucasus) dated to 1.85-1.78 Ma. *PNAS, 108,* 14332-10436.

Fifer, F. C. (1987). The adoption of bipedalism by the hominids: A new hypothesis. *Hum. Evol., 2,* 135-147.

Filler, A. G. (2007). *The Upright Ape. A New Origin of the Species.* Franklin Lakes, NJ: The Career Press.

Fisher, H. (1992). *Anatomy of Love.* New York, NY: Random House.

Fleagle, J. G. (1999). *Primate Adaptation and Evolution* (2nd Edit.). New York, NY: Academic Press.

Fleagle, J. G., Stern, J. T. Jr., Jungers, W. L., Susman, R. L., Vangor, A. K. & Wells, J. P. (1981). Climbing: A biomechanical link with brachiation and with bipedalism. *Symp. Zool. Soc. Lond. 48,* 359-375.

Fleisig, G. S., Andrews, J. R., Dillman, C. J. & Escamilla, R. F. (1995). Kinetics of baseball pitching with implications about injury mechanisms. *Am. J. Sports Med., 23,* 233-239.

Fleisig, G. S., Barrentine, S. W., Escamilla, R. F. & Andrews, J. R. (1996). Biomechanics of overhand throwing with implications for injuries. *Sports Med., 6,* 421-437.

Fleisig, G. S., Barrentine, S. W., Zheng, N., Escamilla, R. F. & Andrews, J. R. (1999). Kinematic and kinetic comparison of baseball pitching among various levels of development. *J. Biomech., 32,* 1371-1375.

Fleisig, G. S., Dillman, C. J. & Andrews, J. R. (1989). Proper mechanics for baseball pitching. *Clin. Sports Med., 1,* 151-170.

Fleisig, G. S., Nicholls, R., Elliott, B. & Escamilla, R. F. (2003). Kinematics used by world class tennis players to produce high-velocity serves. *Sports Biomech., 2,* 51-64.

Fleisig, G. S., Phillips, R., Shatley, A., Loftice, J., Dun, S., Drake, S., Farris, J. W. & Andrews, J. R. (2006). Kinematics and kinetics of youth baseball pitching with standard and lightweight balls. *Sports Engin., 9,* 155-163.

Fletcher, A. W. & Weghorst, J. A. (2005). Laterality of hand function in naturalistically housed chimpanzees *(Pan troglodytes). Laterality, 10,* 219-242.

Foley, R. (1987). *Another Unique Species.* New York, NY: Longman, Scientific & Technical.

Foley, R. (2001). The evolutionary consequences of increased carnivory in hominids. In: Stanford, C. B. & Bunn, H. T. (Eds.), *Meat-Eating and Human Evolution,* (305-331). New York, NY: Oxford University Press.

Ford, S. M. (1994). Evolution of sexual dimorphism in body weight in platyrrhines. *Am. J. Primatol., 34,* 221-244.

Forrester, G.S., Leavens, D. A., Quaresmini, C. & Vallortigara, G. (2011). Target animacy influences gorilla handedness. *Anim. Cogn. 14,* 903-907.

Forrester, G. S., Quaresmini, C., Leavens, D. A., Mareschal, D., & Thomas, M. S. (2013). Human handedness: An inherited evolutionary trait. *Behav. Brain Res., 237,* 200-206.

Forrester, G. S., Quaresmini, C., Leavens, D. A., Spiezio, C. & Vallortigara, G. (2012). Target animacy influences chimpanzee handedness. *Anim. Cogn. 15,* 1121-1127.

Fouts, R. (1997). *Next of Kin.* New York, NY: William Morrow.

Frayer, D. W. (1997). Ofnet: Evidence for a Mesolithic massacre. In: Martin, D. L. & Frayer, D. W. (Eds.), *Troubled Times: Violence and Warfare in the Past* (181-216). Amsterdam, NL: Gordon and Breach.

Frayer, D. W., Lozano, M., Bermúdez de Castro, J. M., Carbonell, E., Arsuaga, J. L., Radovcic, J., Fiore, I. & Bondioli, L. (2011). More than 500,000 years of right-handedness in Europe. *Laterality, 14,* 1-19.

Frost, G. T. (1980). Tool behavior and the origins of

laterality. *J. Hum. Evol., 9,* 447-459.
Frost, H. M. (1997). Why do marathon runners have less bone than weight lifters? A vital-biomechanical view and explanation. *Bone, 20,* 183-189.
Fuentes, A. (1998). Re-evaluating primate monogamy. *Am. Anthrop., 100,* 890-907.
Gabunia, L., Antón, S. C., Lordkipanidze, D., Vekua, A., Justus, A. & Swisher, C. C. III (2001). Dmanisi and dispersal. *Evol. Anthrop., 10,* 158-170.
Gabunia, L., de Lumley, M.-A., Vekua, A., Lordkipanidze, D. & de Lumley, H. (2002). Découverte d'un nouvel hominidé à Dmanissi (Transcaucasie, Géorgie). *C. R. Palevol., 1,* 243-253.
Gainor, B. J., Piotrowski, G., Puhjl, J., Allen, W. C. & Hagen, R. (1980). The throw: biomechanics and acute injury. *Am. J. Sports Med., 8,* 114-118.
Galik, K., B. Senut, B., Pickford, M. Gommery, D., Treil, J., Kuperavage, A. J. & Eckhardt, R. B. (2004). External and internal morphology of the BAR 1002'00 *Orrorin tugenensis* femur. *Science, 305,* 1450-1453.
Gaudzinski, S. (2004). Subsistence patterns of Early Pleistocene hominids in the Levant—taphonomic evidence from the 'Ubeidiya Formation (Israel). *J. Archaeol. Sci., 31,* 65-75.
Gaulin, S. J. C. & Sailor, D. (1984). Sexual dimorphism in weight among the primates: The relative impact of allometry and sexual selection. *Int. J. Primatol., 5,* 515-535.
Gesell, A. & Ames, L. B. (1947). The development of handedness. *J. Genet. Psychol., 70,* 155-175.
Gilby, I. C., Thompson, M. E., Ruane, J. D. & Wrangham, R. (2010). No evidence of short-term exchange of meat for sex among chimpanzees. *J. Hum. Evol., 59,* 44-53.
Gomes, C. M., Boesch, C. (2009). Wild chimpanzees exchange meat for sex on a long-term basis. *PloS One 4,* e5116.
Gommery, D. & Senut, B. (2006). La phalange distale du pouce d'*Orrorin tugenensis* (Miocène supérieur du Kenya). *Geobios, 39,* 372-384.
Goodall, J. (1964). Tool-using and aimed throwing in a

community of free-living chimpanzees. *Nature, 201,* 1264-1266.

Goodall, J. (1984). *Among the Wild Chimpanzees.* Washington, DC: National Geographic society Video.

Goodall, J. (1986). *The Chimpanzees of Gombe. Patterns of behavior.* Cambridge, MA: Harvard University Press.

Goodall, J. (1988). *In the Shadow of Man* (Rev. Edit.), Boston, MA: Houghton Mifflin.

Goodall, J. (1990). *Through a Window. My Thirty Years with with the Chimpanzees of Gombe.* Boston, MA: Houghton Mifflin.

Goodall, J. (1992). *The Chimpanzee. The Living Link between "Man" and "Beast".* Edinburgh, UK: Edinburgh University Press.

Gordon, A. D., Green, D. J. & Richmond, B. G. (2008). Strong postcranial size dimorphism in *Australopithecus afarensis*: Results from two new resampling methods for multivariate data sets with missing data. *Am. J. Phys. Anthrop., 135,* 311-328.

Goudge, T. A. (1961). *The Ascent of Life.* London, UK: George Allen & Unwin. Gould, S. J. (1982). Darwinism and the expansion of evolutionary theory. *Science, 216,* 380-387.

Gould, J. L. & Gould, C. G. (1989). *Sexual Selection.* New York, NY: Scientific American Library.

Gould, S. J. (1996). *Full House.* New York, NY: Harmony Books.

Gray, S., Watts, S., Debicki, D. & Hore, J. (2006). Comparison of kinematics in skilled and unskilled arms of the same recreational baseball players. *J. Sports Sci., 24,* 1183-1194.

Green, D. J. & Alemseged, Z. (2012). *Australopithpithecus afarensis* scapular ontogeny, function, and the role of climbing in human evolution. *Science, 338,* 514-517.

Green, D. J. & Gordon, A. D. (2008). Metacarpal proportions in *Australopithecus africanus. J. Hum. Evol., 54,* 705-719.

Greenfield, L. O. (1992). Origin of the human canine: A new solution to an old enigma. *Yearb. Phys. Anthrop., 35,* 153-185.

Greenfield, L. O. & Washburn, A. (1991). Polymorphic aspects of male anthropoid canines. *Am. J. Phys. Anthrop., 84,* 17-34.

Grouios, G., Tsorbatzoudis, H., Alexandris, K. & Barkoukis V. (2000). Do left-handed competitors have an innate superiority in sports? *Percep. Motor Skills, 90,* 1273-1282.

Haapasalo, H., Sievanen, H., Kannus, P., Heinonen, A., Oja, P. & Vuori, I. (1996). Dimensions and estimated mechanical characteristics of the humerus after long-term tennis loading. *J. Bone & Mineral Res., 11,* 864-872.

Haile-Selassie, Y. (2001). Late Miocene hominids from the Middle Awash, Ethiopia. *Nature, 412,* 178-181.

Haile-Selassie, Y. (2010). Phylogeny of early *Australopithecus:* new fossil evidence from the Woranso-Mille (central Afar, Ethiopia). *Phil. Trans. Roy. Soc. B, 365,* 3323-3331.

Haile-Selassie, Y., Latimer, B. M., Alene, M, Deino, A. L., Gilbert, L., Melillo, S. M., Saylor, B. Z., Scott, G. R. & Lovejoy, C. O. (2010). An early *Australopithecus afarensis* postcranium from Woranso-Mille, Ethiopia. *PNAS, 107,* 12121-12126.

Haile-Selassie, Y., Saylor, B. Z., Deino, A., Levin, N. E., Alene, M. & Latimer, B. M. [2012]. A new hominin foot from Ethiopia shows multiple Pliocene bipedal adaptations. *Nature, 483,* 565-569.

Haile-Selassie, Y., Suwa, G. & White, T. D. (2004). Late Miocene teeth from Middle Awash, Ethiopia, and early hominid dental evolution. *Science, 303,* 1503-1505.

Halverson, H. M. (1940). Motor Development. In: Gesell, A., Thompson, H. M., Ilg, F. L., Castner, B. M., Ames, L. B., & Amatruda, C. S. (Eds.), *The First Five Years of Life* (65-107). New York, NY: Harper and Brothers.

Halverson, L. E. & Roberton, M. A. (1979). The effects of instruction on overhand throwing development in children. In: Newell, K., & Roberts, G. (Eds.), *Psychology of Motor Behavior and Sport–1978* (258-264). Champaign, IL: Human Kinetics.

Halverson, L., Roberton, M. A. & Harper, C. J. (1973). Current research in motor development. *J. Res. Develop. Edu., 6,* 56-70.

Halverson, L. E., Roberton, M. A. & Langendorfer, S. (1982).

Development of the overarm throw: Movement and ball velocity changes by seventh grade. *Res.Quart. Exerc. Sport, 53,* 198-205.

Halverson, L. E., Roberton, M. A., Safrit, M. J. & Thomas, W. R. (1977). Effect of guided practice on overhand-throw ball velocities of kindergarten children. *Res. Quart. Exerc. Sport, 48,* 311-318.

Hamrick, M. W., Churchill, S. E., Schmitt, D. & Hylander, W. L. (1998). EMG of the human flexor pollicis longus muscle: Implications for the evolution of hominid tool use. *J. Hum. Evol., 34,* 123-136.

Harcourt-Smith, W. E. H. & Aiello, L. C. (2004). Fossils, feet and the evolution of human bipedal locomotion. *J. Anat., 204,* 403-416.

Harman, E. H. (2006). Size and shape variation in *Australopithecus afarensis* proximal femora. *J. Hum. Evol., 51,* 217-227.

Harris, H. (1981). Rationality in science. In: Heath, A. F. (Ed.), *Scientific Explanation* (36-52). Oxford, UK: Clarendon Press.

Harrison, R. M. & Nystrom, P. (2010). Handedness in captive gorillas *(Gorilla gorilla). Primates, 51,* 251-261.

Harrison, T. (1991). The implications of *Oreopithecus bambolii* for the origins of bipedalism. In: Y. Coppens, & B. Senut (Eds.), *Origine(s) de la Bipédie chez les Hominidés* (235-244). Paris, FR: Cahiers de Paléoanthropologie, Editions du CNRS.

Harrison, T. (2010). Apes among the tangled branches of human origins. *Science, 327,* 532-534.

Hart, D. & Sussman, R. W. (2005). *Man the Hunted.* New York, NY: Westview Press.

Hawkes, K. (2001). Is meat the hunter's property? In: Stanford, C. B. & Bunn, H. T. (Eds.), *Meat-Eating and Human Evolution* (219-236). New York, NY: Oxford University Press.

Hay, R. L. (1976). *Geology of the Olduvai Gorge: A Study of Sedimentation in a Semiarid Basin.* Berkeley, CA: University of California Press.

Hazelton, F. T., Smidt, G. L., Flatt, A. E. & Stephens, R. I. (1975). The influence of wrist position on the force produced by the finger flexors. *J. Biomech., 8,* 301-306.

Hempel, C. G. (1965a). Studies in the logic of explanation. In: Hempel, C. G. (Ed.), *Aspects of Scientific Explanation and Other Essays in the Philosophy of Science* (245-290). New York, NY: The Free Press.

Hempel, C. G. (1965b). Aspects of scientific explanation. In: Hempel, C. G. (Ed.), *Aspects of Scientific Explanation and Other Essays in the Philosophy of Science* (331-489). New York, NY: The Free Press.

Hempel, C. G. (1966). *Philosophy of Natural Science.* Englewood Cliffs, NJ: Prentice-Hall.

Hempel, C. G. (1970). On the "standard conception" of scientific theories. In: Radner, M., Winokur, S. (Eds.), *Minnesota Studies on the Philosophy of Science, Vol. IV* (142-163). Minneapolis, MN:University of Minnesota Press.

Hewes, G. W. (1961). Food transport and the origin of hominid bipedalism. *Am. Anthrop. 63,* 687-710.

Hewes, G. W. (1973). *The Origin of Man.* Minneapolis, MN: Burgess.

Hey, J. (2010). The divergence of chimpanzee species and subspecies as revealed in multipopulation isolation-with-migration analyses. *Mol. Biol., Evol., 27,* 921-933.

Hicks, J. A. (1958). The acquisition of motor skill in young children. *Child Develop., 1,* 90-105.

Hicks, J. H. (1954). The mechanics of the foot. II. The plantar aponeurosis and the arch. *J. Anat., 88,* 25-31.

Hill, K. (1982). Hunting and human evolution. *J. Hum. Evol., 11,* 521-544.

Hilton, C. E. & Meldrum, D. J. (2004). Striders, runners, and transporters. In: Meldrum, D. J.& Hilton, C. E. (Eds.), *From Biped to Strider: The Emergence of Modern Human Walking, Running, and Resource Transport* (1-8). New York, NY: Kluwer Academic/Plenum Publishers.

Hinrichs, R. N. (1999). Upper extremity function in distance running. In: P. R. Cavanagh (Ed.), *Biomechanics of Distance Running* (107-133). Champaign, IL: Human Kinetics Books.

Hirashima, M., Kadota, H., Sakurai, S., Kudo, K. & Ohtsuki, T. (2002). Sequential muscle activity and its functional role in the upper extremity and trunk during overarm throwing. *J. Sports Sci., 20,* 301-310.

Hirashima, M., Kudo, K. & Ohtsuki, T. (2003). Utilization and compensation of interaction torques during ball-throwing movements. *J. Neurophysiol., 89,* 1784-1796.

Hirashima, M., Kudo, K., Watarai, K. & Ohtsuki, T. (2007). Control of 3D limb dynamics in unconstrained overarm throws of different speeds performed by skilled baseball players. *J. Neurophysiol., 97,* 680-691.

Hirashima, M. & Ohtsuki, T. (2008). Exploring the mechanism of skilled overarm throwing. *Exerc. Sport Sci. Rev., 36,* 205-211.

Hirashima, M., Yamane, K., Nakamura, Y. & Ohtsuki, T. 2008).Kinetic chain of overarm throwing in terms of joint rotations revealed by induced acceleration analysis. *J. Biomech., 41,* 2874-2883.

Holloway, R. L. (1967). Tools and Teeth: Some speculations regarding canine reduction. *Am. Anthrop., 69,* 63-67.

Holton, G. (1981). Thematic presuppositions and the direction of scientific advance. In: Heath, A. F. (Ed.), *Scientific Explanation* (1-27). Oxford, UK: Clarendon Press.

Hopkins, W. D. (2006). Comparative and familial analysis of handedness in great apes. *Psychol. Bull., 132,* 538-559.

Hopkins, W. D. & Cantalupo, C. (2005). Individual and setting differences in the hand preferences of chimpanzees *(Pan troglodytes)*: A critical analysis and some alternative explanations. *Laterality, 10,* 65-80.

Hopkins, W. D., Phillips, K. A., Bania, A., Calcutt, S. E., Gardner, M., Russell, J., Schaeffer, J., Lonsdorf, E. V., Ross, S. R. & Schapiro, S. J. (2011). Hand preferences for coordinated biman-ual actions in 777 great apes: Implications for the evolution of handedness in Hominins. *J. Hum. Evol., 60,* 605-611.

Hopkins, W. D., Russell, J. L. & Cantalupo, C. (2007). Neuroanatomical correlates of handedness for tool use in chimpanzees (*Pan troglodytes*). *Psychol. Sci., 18,* 971-977.

Hopkins, W. D, Russell, J. L., Cantalupo, C., Freeman, H. & Schapiro, S. J. (2005). Factors influencing the prevalence and handedness for throwing in captive chimpanzees *(Pan troglodytes)*. *J. Comp. Psychol.*, 119, 363-370.

Hopkins, W. D., Russell, J. L. & Schaeffer, J. A. (2012). The neural and cognitive correlates of aimed throwing in chimpanzees: a magnetic resonance image and behavioural study on a unique form of social tool use. *Phil. Trans. R. Soc. B, 367,* 37-47.

Hopkins, W. D., Wesley, M. J., Izard, M. K., Hook, M. & Schapiro, S. J. (2004). Chimpanzees (*Pan troglodytes*) are predominantly right-handed: Replication in three populations of apes. *Behav. Neurosci., 118,* 659-663.

Hore, J., Debicki, D. B., Gribble, P. L. & Watts, S. (2011). Deliberate utilization of interaction torques brakes elbow Extension in a fast throwing motion. *Exp. Brain Res., 211,* 63-72.

Hore, J., Debicki, D. B. & Watts, S. (2005a). Braking of elbow extension in fast overarm throws made by skilled and unskilled subjects. *Exp. Brain Res., 164,* 365-375.

Hore, J., O'Brien, M. & Watts, S. (2005b). Control of joint rotations in overarm throws of different speeds made by dominant and nondominant arms. *J. Neurophysiol., 95,* 3975-3986.

Hore, J., Ritchie, R. & Watts, S. (1999a). Finger opening in an overarm throw. *Exp. Brain Res., 125* 302-312.

Hore, J., Timmann, D. & Watts, S. (2002). Disorders in timing and force of finger opening in overarm throws made by cerebellar subjects *Ann. N. Y. Acad. Sci., 978,* 1-15.

Hore, J. & Watts, S. (2005).Timing finger opening in overarm throwing based on a spatial representation of hand path. *J. Neurophysiol., 93,* 3189-3199.

Hore, J. & Watts, S. (2011). Skilled throwers use physics to time ball release to the nearest millisecond. *J. Neurophysiol., 106,* 2024-2033.

Hore, J., Watts, S., Leschuk, M. & MacDougall, A. (2001). Control of finger grip forces in overarm throws made by skilled throwers. *J. Neurophysiol., 86,* 2678-2689.

Hore, J., Watts, S., Martin J. & Miller, B. (1995). Timing of finger opening and ball release in fast and accurate overarm throws. *Exper. Brain Res., 103,* 277-286.

Hore, J., Watts, S. & Tweed, D. (1996a). Errors in the control of joint rotations associated with inaccuracies in overarm throws. *J. Neurophysiol., 75,* 1013-1025.

Hore, J., Watts, S. & Tweed, D. (1999b). Prediction and compensation by an internal model for back forces during finger opening in an overarm throw. *J. Neurophysiol., 82,* 1187-1197.

Hore, J., Watts, S., Tweed, D. & Miller, B. (1996b). Overarm throws with the nondominant arm: Kinematics of accuracy. *J. Neurophysiol., 76,* 3693-3704.

Howell, F. C. (1965). *Early Man.* New York, NY: Time-Life.

Hughes, R. E., Johnson, M. E., O'Driscoll, S. W & An, K-N. (1999). Age-related changes in normal isometric shoulder strength. *Am. J. Sports Med. 27,* 651-657.

Hull, D. L. (1974). *Philosophy of Biological Science.* Englewood Cliffs, NJ: Prentice-Hall.

Humle, T. & Matsuzawa, T. (2009). Laterality in hand use across four tool-use behaviors among the wild chimpanzees of Bossou, Guinea, West Africa. *Am. J. Primatol., 71,* 40-48.

Hunt, K. D. (1991). Mechanical implications of chimpanzee positional behavior. *Am. J. Phys. Anthrop., 86,* 521-536.

Hunt, K. D. (1992). Positional behaviour of *Pan troglodytes* in the Mahale Mountains and Gombe Stream National Parks, Tanzania. *Am. J. Phys. Anthrop., 87,* 83-105.

Hunt, K. D. (1994). The evolution of human bipedality: ecology and functional morphology. *J. Hum. Evol., 26,* 183-202.

Hunt, K. D. (1996). The postural feeding hypothesis: An ecological model for the evolution of bipedalism. *S. Afr. J. Sci., 92,* 77-90.

Huxley, T. H. (1863/2003). *Man's Place in Nature.* Mineola, NY: Dover Publications.

Huxley, T. H. (1894/1947). Evolution and ethics. In: Huxley, T. H. & Huxley, J. (Eds.), *Evolution and Ethics* (33-102). London, UK: The Pilot Press LTD.

Isaac, B. (1987). Throwing and human evolution. *Afr. Archaeol. Rev., 5,* 3-17.

Isaac, B. (1992). Throwing. In: Jones, S., Martin, R. & Pilbeam, D. (Eds.), *The Cambridge Encyclopedia of Human Evolution* (358). Cambridge, UK: Cambridge University

Press.

Isaac, G. L. (1977). *Olorgesailie*. Chicago, IL: University of Chicago Press.

Isaac, G. L. (1978). The food-sharing behavior of protohuman hominids. *Sci. Amer., 238,* 90-107.

Isaac, G. L. (1984). The archaeology of human origins: Studies of the Lower Pleistocene in East Africa 1971-1981. *Adv. World Archaeol., 3,* 1-87.

Isaac, G. L., Harris, J. K. & Kroll, E. M. (1997). The stone artefact assemblages: A comparative study. In: Isaac, G. L. & Isaac, B. (Eds.), *Koobi Fora Research Project, Vol. 5. Plio-Pleistocene Archeology* (262-299). Oxford, UK: Clarendon Press.

Ishida, K. & Hirano, Y. (2004). Effects of non-throwing arm on trunk and throwing arm movements in baseball pitching. *Int. J. Sport Health Sci., 2,* 119-128.

Isler, K., Payne, R. C., Günther, M. M., Thorpe, S. K. S., Li, Y., Savage, R. & Crompton, R. H. (2006). Inertial properties of hominoid limb segments. *J. Anat., 209,* 201-218.

Jablonski, N. G. & Chaplin G. (1993). Origin of habitual terrestrial bipedalism in the ancestor of the *Hominidae. J. Hum. Evol. 24,* 259-280.

Jegede, E., Watts, S., Stitt, L. & Hore, J. (2005). Timing of ball release in overarm throws affects ball speed in unskilled but not skilled individuals. *J. Sports Sci., 23,* 805-816.

Jelinek, A. J. (1977). The Lower Paleolithic: Current evidence and interpretations. *Ann. Rev. Anthrop. 6,* 11-32.

Jenkins, F.A. (1972). Chimpanzee bipedalism: Cineradiographic analysis and implications for the evolution of gait. *Science, 178,* 877-879.

Johanson, D., Johanson, L. & Edgar, B. (1994). *Ancestors. In Search of Human Origins.* New York, NY: Villard Books.

Johanson, D. C., Masao, F. T., Eck, G. G., White, T. D., Walter, R. C., Kimbel, W. H., Asfaw, B., Manega, P., Ndessoka, P. & Suwa, G. (1987). New partial skeleton of *Homo habilis* from Olduvia Gorge, Tanzania. *Nature, 327,* 205-209.

Johanson, D. C. & Taieb, M. (1976). Plio-Pleistocene hominid discoveries in Hadar, Ethiopia. *Nature, 260, 293-297.*

Johanson, D. C. & White, T. D. (1979). A systematic

assessment of early African hominids. *Science, 203,* 321-330.

Jolly C. J. (1970). The seed-eaters: a new model of hominid differentiation based on a baboon analogy. *Man 5,* 5-26.

Jones, F. W. (1946). *The Principles of Anatomy as Seen in the Hand* (2nd Edit.). London, UK: Ballière, Tindall and Cox.

Jones, H. H., Priest, J. D., Hayes, W. C., Tichenor, C. C. & Nagel, D. A. (1977). Humeral hypertrophy in response to exercise. *J. Bone & Joint Surg. 59,* 204-208.

Jones, P. R. (1995). Results of experimental work in relation to the stone industries of Olduvai Gorge. In: Leakey, M. D. (Ed.), *Olduvai Gorge. Vol. 5. Excavations in Beds III, IV and the Masek Beds, 1968-1971* (254-298). Cambridge, UK: Cambridge University Press.

Jorgensen, T. P. (1994). *The Physics of Golf.* New York, NY: Springer-Verlag.

Jouffroy, F. K. & Lessertisseur, J. (1960). Les specialisations anatomique de la main chez les Singes a progression suspendue. *Mammalia, 24,* 93-51.

Jungers, W. L. (1982). Lucy's limbs: skeletal allometry and locomotion in *Australopithecus afarensis. Nature, 297,* 676-679.

Kaplan, H. & Hill, K. (1992). The evolutionary ecology of food acquisition. In: Smith, E. A. & Winterhalder, B. (Eds.), *Evolutionary Ecology and Human Behavior* (167-201). New York, NY: Aldine de Gruyter.

Keeley, L. H. (1996). *War Before Civilization.* New York, NY: Oxford University Press.

Keith, A. (1949). *A New Theory of Human Evolution.* New York, NY: Philosophical Library.

Kennedy, G. E. (1985). Bone thickness in *Homo erectus. J. Hum. Evol., 14,* 699-708.

Keogh, J. (1965). *Motor Performance of Elementary School Children.* Los Angeles, CA: University of California, Department of Physical Education.

Kepple, T. M., Siegel, K. L. & Stanhope, S. J. (1997). Relative contributions of the lower extremity joint moments to forward progression and support during gait. *Gait & Posture, 6,* 1-8.

Khosla, T. (1978). Standards on age, height and weight in Olympic running events for men. *Brit. J. Sports Med., 12,* 97-101.

Kibler, W. B. (1995). Biomechanical analysis of the shoulder during tennis activities. *Clinics Sports Med., 14,*79-85.

Kibler, W. B. & McMullen, J. (2003). Scapular dyskinesis and its relation to shoulder pain. *J. Am. Acad. Orthop. Surg., 11,*142-151.

Kimbel, W. H. & Delezene, L. K. (2009). "Lucy" redux: A review of research on *Australopithecus afarensis. Yearb. Phys. Anthrop., 52,* 2-48.

Kimbel, W. H., Johanson, D. C. & Rak, Y. (1997). Systematic assessment of a maxilla of *Homo* from Hadar, Ethiopia. *Am. J. Phys. Anthrop., 103,* 235-262.

Kimbel, W. H., Lockwood, C. A., Ward, C. V., Leakey, M. G., Rak, Y. & Johanson, D. C. (2006). Was *Australopithecus anamensis* ancestral to *A. afarensis*? A case of anagenesis in the hominin fossil record. *J. Hum. Evol., 51,* 134-152.

Kimura, T. (1996). Centre of gravity of the body during the ontogeny of chimpanzee bipedal walking. *Folia Primatol., 66,* 126-136.

Kimura, T., Okada, M. & Ishida, H. (1977). Dynamics of primate bipedal walking as viewed from the force of foot. *Primates 18,* 137-147.

Kimura, T., Okada, M., Yamazaki, N. & Ishida, H. (1983). Speed of the bipedal gaits of man and nonhuman primates. *Ann.Sci. Naturelles, Zool., 5,* 145-158.

King, J. W., Brelsford, H. J. & Tullos, H. S. (1969). Analysis of the pitching arm of the professional baseball pitcher. *Clin. Orthop., 67,* 116-123.

Kingdon, J. (2003). *Lowly Origin. Where, When, and Why Our Ancestors First Stood Up.* Princeton, NJ: Princeton University Press.

Kirschmann, E. (1999). *Das Zeitalter der Werfer.* Eduard Kirschmann, Hannover, Germany, Grünlinde 4, 30459. (Available from the author [webmaster@werfer.de] in an English translation).

Kivell, T. L., Kibii, J. M., Churchill, S. E., Schmid, P. & Berger, L. R. (2011a). *Australopithecus sediba* hand demonstrates mosaic evolution of locomotor and manipulative abilities. *Science, 333,* 1411-1417.

Kivell, T. L., Kibii, J. M., Churchill, S. E., Schmid, P., & Berger, L. R. (2011b). *Australopithecus sediba* hand demonstrates mosaic evolution of locomotor and manipulative abilities. *Science, 333,* Supporting online material, 1-57.

Klein, R. G. (1987). Reconstructing how early people exploited animals: Problems and Prospects. In: Nitecki, M. H. & Nitecki, D. V. (Eds.), *The Evolution of Human Hunting* (11-45). New York, NY: Plenum.

Klein, R. G. (1999). *The Human Career* (2nd Edit.). Chicago, IL: University of Chicago Press.

Klein, R. G. & Edgar, B. (2002). *The Dawn of Human Culture.* New York, NY: John Wiley & Sons.

Knüsel, C. J. (1992).The throwing hypothesis and hominid origins. *Hum. Evol.,7,* 1-7.

Koch, C. (2012). *Consciousness.* Cambridge, MA: MIT Press.

Köhler, W. (1927). *The Mentality of Apes* (2nd Edit.). New York, NY: Harcourt, Brace.

Kontulainen, S., Kannus, P., Haapasalo, H., Heinonen, A., Sievanen, H., Oja, P. & Vuori, I. (1999). Changes in bone mineral content with decreased training in competitive young adult tennis players and controls: a prospective 4-yr follow up. *Med. & Sci. in Sports & Exerc., 31,* 646-652.

Kortlandt, A. (1972). *New Perspectives on Ape and Human Evolution.* Amsterdam, NL: Stichting Voor Psycho-biologie.

Kortlandt, A. (1980). How might early hominids have defended themselves against large predators and food competitors? *J. Hum. Evol., 9,* 79-112.

Kortlandt, A. (1986). The use of stone tools by wild-living chimpanzees and earliest hominids. *J. Hum. Evol.* 15, 77-132.

Kramer, P. A. & Eck, G. G. (2000). Locomotor energetics and leg length in hominid bipedality. *J. Hum. Evol., 38,* 651-666.

Kuhn, T. S. (1962). *The Structure of Scientific Revolutions.* Chicago, IL: University of Chicago Press.

Kumar, S., Filipski, A., Swarna, V., Walker, A. & Hedges, S. B. (2005). Placing confidence limits on the molecular age of the human-chimpanzee divergence. *PNAS, 102,* 18842-18847.

Lambert, P. M. (1997). Patterns of violence in prehistoric hunter-gatherer societies of coastal Southern California. In: Martin, D. L. & Frayer, D. W. (Eds.), *Troubled Times: Violence and Warfare in the Past* (77-109). Amsterdam, NL: Gordon and Breach.

Lancaster, J. B. (1978). Carrying and sharing in human evolution. *Hum. Nature, 1,* 82-89.

Landau, M. (1991). *Narratives of Human Evolution.* New Haven, CT: Yale University Press.

Langdon, J. H. (1985). Fossils and the origin of bipedalism. *J. Hum. Evol. 14,* 615-635.

Langdon, J. H., Bruckner, J., & Baker, H. H. (1991). Pedal mechanics and bipedalism in early hominids. In: Coppens, Y. & Senut, B. (Eds.), *Origine(s) de la Bipédie chez les Hominidés* (159-167). Paris, FR: Cahiers de Paléoanthropogie, Editions du CNRS.

Langendorfer, S. (1987). Prelongitudinal screening of overarm striking development performed under two environmental conditions. In: Clarke, J. E. & J. H. Humphrey, J. H. (Eds.), *Advances in Motor Development Research* (17-47). New York, NY: AMS Press.

Langendorfer, S. J. & Roberton, M. A. (2002). Individual pathways in the development of forceful throwing. *Res. Quart. Exerc. Sport, 73,* 245-256.

Larson, C. S. (2003). Equality for the sexes in human evolution? Early hominid sexual dimorphism and implications for mating systems and social behavior. *PNAS, 100,* 9103-9104.

Larson, S. G. (2007). Evolutionary transformation of the hominin shoulder. *Evol. Anthrop., 16,* 172-187.

Latimer, B. (1991). Locomotor adaptations in *Australopithecus afarensis:* The issue of arboreality. In: Coppens, Y. & Senut, B. (Eds.), *Origine(s) de la Bipédie chez les Hominidés* (169-176). Paris, FR: Cahiers de Paléoanthro- pologie, Editions de CNRS.

Latimer, B. & Lovejoy, C. O. (1989). The calcaneus of *Australopithecus afarensis* and its implications for the evolution of bipedality. *Am. J. Phys. Anthrop.,78,* 369-

386.

Latimer, B., Lovejoy, C. O., Johanson, D. C, & Coppens, Y. (1982). Hominid tarsal, metatarsal and phalangeal bones recovered from the Hadar Formation:1974-1977 collections. *Am. J. Phys. Anthrop., 57,* 701-719.

Latimer, B., Ohman, J. C. & Lovejoy, C. O. (1987). Talocrural joint in African hominoids. *Am. J. Phys. Anthrop., 74,* 155-175.

Latimer, B. & Ward, C. V. (1993). The thoracic and lumbar vertebrae. In: Walker. A. & Leakey, R. (Eds.), *The Nariokotome Homo erectus Skeleton* (266-293). Cambridge, MA: Harvard University Press.

Laughlin, W. S. (1968). Hunting: An integrating biobehavior system and its evolutionary importance. In: Lee, R. B. & DeVore, I. (Eds.), *Man the Hunter* (304-320). Chicago, IL: Aldine Publishing Company.

Le Gros Clark, W. E. (1950). Hominid characters of the Australopithecine dentition. *J. Roy. Anthrop. Inst. of Great Britain and Ireland, 80,* 37-54.

Le Gros Clark, W. E. (1964). *The Fossil Evidence for Human Evolution* (2nd Edit.), Chicago, IL: University of Chicago Press.

Leakey, L. S. B., Tobias, P. V. & Napier, J. R. (1964). A new species of the genus *Homo* from Olduvai Gorge. *Nature, 202,* 7-9.

Leakey, M. D. (1970). Stone artefacts from Swartkrans. *Nature 225,* 1222-1225.

Leakey, M. D. (1971). *Olduvai Gorge. Vol. 3. Excavations in Beds I and II,* (960-1963). Cambridge, UK: Cambridge University Press.

Leakey, M. G., Feibel, C. S., McDougall, I., Ward, C. & Walker, A. (1998). New specimens and confirmation of an early age for *Australopithecus anamensis. Nature 393,* 62-66.

Leakey, M.G. & Walker, A. (1997). Early hominid fossils from Africa. *Sci. Am., 276 (6),* 60-65.

Leakey, R. & Lewin, R. (1992). *Origins Reconsidered.* New York, NY: Doubleday.

LeBlanc, S. A. (2003). *Constant Battles.* New York, NY: St. Martin's Press.

Leca, J.-B., Nahaliage, C. A. D., Gunst, N. & Huffman, M. A. (2008). Stone-throwing by Japanese macaques: form and functional aspects of a group-specific behavioral tradition. *J. Hum. Evol., 55,* 989-998.

Lee, R. B. (1968). What hunters do for a living, or, how to make out on scarce resources. In: Lee, R. B. & DeVore, I. (Eds.), *Man the Hunter* (30-48). Chicago, IL: Aldine Publishing Company.

Lee, R. B. (1979). *The !Kung San.* Cambridge, UK: Cambridge University Press.

Leigh, S. R. & Shea, B. T. (1995). Ontogeny and the evolution of adult body size dimorphism in apes. *Am. J. Primatol., 36,* 37-60.

Leme, S. A. & Shambes, G. M. (1978). Immature throwing patterns in normal adult women. *J. Hum. Movement Studies, 4,* 85-93.

Leonard W. R. & Robertson, M. L (1994). Evolutionary perspectives on human nutrition: The influence of brain and body size on diet and metabolism. *Am. J. Hum. Biol. 6,* 77-88.

Leonard, W. R. & Roberston, M. L. (1997). Comparative primate energetics and hominid evolution. *Am. J. Phys. Anthrop., 102,* 265-281.

Leonard, W. R., Robertson, M. L. & Snodgrass, J. J. (2007). Energetic modes of human nutritional evolution. In: P. S. Ungar (Ed.), *Evolution of the Human Diet* (344-359). New York, NY: Oxford University Press.

Lepre, C. J., Roche, H., Kent, D. V., Harmand, S., Quinn, R. L., Brugal, J.-P., Texier, P.-J., Lenoble, A. & Feibel, C. S. (2011). An earlier origin for the Acheulian. *Nature, 477,* 82-85.

Lewin, R. (1987). *Bones of Contention.* Chicago, IL: University of Chicago Press.

Lewis, O. J. (1977). Joint remodeling and the evolution of the human hand. *J. Anat., 123,* 157-201.

Lewis, O. J. (1989). *Functional Morphology of the Evolving Hand and Foot.* Oxford, UK: Clarendon Press.

Lieberman, D. E. (2012a). Human evolution: Those feet in ancient times. *Nature, 483,* 550-551.

Lieberman, D. E. (2012b). *The Evolution of the Human*

Head. Cambridge, MA: Belknap Press of Harvard University Press.

Lieberman, D. E., Bramble, D. M., Raichlen, D. A. & Shea, J. J. (2009). Brains brawn and the evolution of human endurance running capabilities. In: Grine, F. E., Fleagle, J. G. & Leakey, R. W. (Eds.). *The First Humans: Origin and Early Evolution of the Genus Homo* (77-92). Dordrecht, NE: Springer.

Linton, S. (1971). Woman the gatherer: male bias in anthropology. In: Jacobs, S.-E. (Ed.), *Women in Cross-Cultural Perspective*. Champaign, IL: University of Illinois Press. Reprinted as Slocum, S. (1975). In: Reiter, R. R. (Ed.), *Toward an Anthropology of Women*, (36-50) New York, NY: Monthly Review Press.

Llaurens, V., Raymond, M., Faurie, C. (2009). Why are some people left-handed? An evolutionary perspective. *Phil. Trans. Roy. Soc. B., 364*. 881-894.

Lockwood, C. A., Richmond, B. G., Jungers, W. L. & Kimbel, W. H. (1996). Randomization procedures and sexual dimorphism in *Australopithecus afarensis*. *J. Hum. Evol., 31*, 537-548.

Loftice, J., Fleisig, G. S., Zheng, N. & Andrews, J. R. (2004). Biomechanics of the elbow in longitudinal investigation. *Res. Quart. Exerc. Sport, 69*, 1-10.

Lonsdorf, E. V. & Hopkins, W. D. (2005). Wild chimpanzees show population-level handedness for tool use. *PNAS, 102*, 12634-12638.

Loovis, E. M.. & Butterfield, S. A. (1995). Influence of age, sex, balance, and sport participation on development of sidearm striking by children grades K-8. *Percept. Motor Skills, 81*, 595-600.

Lordkipanidze, D., Jashashvili, T., Vekua, A., Ponce de León, M. S., Zollikofer, C. P. E., Rightmire, G. P., Pontzer, H., Ferring, Oms, O., Tappen, M., Bukhsianidze, M., Agusti, J., Kahlke, R., R., Kiladze, G., Martinez-Navarro, B., Mouskhelishvili, A., Nioradze, M., & Rook, L. (2007). Postcranial evidence from early *Homo* from Dmanisi, Georgia. *Nature, 449*, 305-310.

Lordkipanidze, D., Ponce de León, M. S., Margvelashvili, A., Rak, Y., Rightmire, G. P., Vekua, A., & Zollikofer, C. P. E.

(2013). *Science, 343,* 326-331.

Lovejoy, C. O. (1981). The origin of man. *Science, 211,* 341-350.

Lovejoy, C. O. (1988).The evolution of human walking. *Sci. Am., 259 (5),* 118-125.

Lovejoy, C. O. (1993). Modeling human origins: Are we sexy because we're smart, or smart because we're sexy? In: Rasmussen, D. T. (Ed.), *The Origin and Evolution of Humans and Humanness* (1-28). Sudbury, MA: Jones and Bartlett Publishers.

Lovejoy, C. W. (2005). The natural history of human gait and posture. Part 2. Hip and thigh. *Gait & Posture, 21,* 113-124.

Lovejoy, C. O. (2009a). Reexamining human origins in light of *Ardipithecus ramidus* (Author's summary). *Science, 326,* 74.

Lovejoy, C. O. (2009b). Reexamining human origins in light of *Ardipithecus ramidus. Science, 326,* 74e1-74e8.

Lovejoy, C. O., Latimer, B., Suwa, G., Asfaw, B., & White, T. D. (2009i). Combining prehension and propulsion: The foot of *Ardipithecus ramidus* (Authors' summary). *Science, 326,* 72.

Lovejoy, C. O., Latimer, B., Suwa, G., Asfaw, B. & White, T. D. (2009j). Combining prehension and propulsion: The foot of *Ardipithecus ramidus. Science, 326,* 72e1-72e8.

Lovejoy, C. O., Latimer, B., Suwa, G., Asfaw, B., & White, T. D. (2009k). Combining prehension and propulsion: The foot of *Ardipithecus ramidus. Science, 326.* Supporting Online Material, 1-20.

Lovejoy, C. O., & McCollum, M. A. (2010). Spinopelvic pathways to bipedality: why no hominids ever relied on a bent-hip-bent-knee gait. *Phil. Trans. Roy. Soc. B, 365,* 3289-3299.

Lovejoy, C. O., Simpson, S. W., White, T. D., Asfaw, B. & Suwa, G. (2009d). Careful climbing in the Miocene: The forelimbs of *Ardipithecus ramidus* and humans are primitive (Authors' summary). *Science, 326,* 70.

Lovejoy, C. O., Simpson, S. W., White, T. D,, Asfaw, B. & Suwa, G. (2009e). Careful climbing in the Miocene: The forelimbs of *Ardipithecus ramidus* and humans are primitive. *Science, 326,* 70e1-70e8.

Lovejoy, C. O., Simpson, S. W., White, T. D., Asfaw, B. &

Suwa, G. (2009e*). Careful climbing in the Miocene: The forelimbs of *Ardipithecus ramidus* and humans are primitive. *Science, 326,* Supporting Online Material, 1-42.

Lovejoy, C. O., Suwa, G., Simpson, S. W., Matternes, J. H. & White, T. D. (2009a). The great divides: *Ardipithecus ramidus* reveals the postcrania of our last common ancestors with African apes (Authors' summary). *Science, 326,* 73.

Lovejoy, C. O., Suwa, G., Simpson, S. W., Matternes, J. H. & White, T. D. (2009b). The great divides: *Ardipithecus ramidus* reveals the postcrania of our last common ancestors with African apes. *Science, 326,* 100-106.

Lovejoy, C. O., Suwa, G., Simpson, S. W., Matternes, J. H. & White, T. D. (2009c). The great divides: *Ardipithecus ramidus* reveals the postcrania of our last common ancestors with African apes. *Science, 326,* Supporting online material, 1-28.

Lovejoy, C. O., Suwa, G., Spurlock, L., Asfaw, B. & White, T. D. (2009f). The pelvis and femur of *Ardipithecus ramidus :* The emergence of upright walking (Authors' summary). *Science,* 326: 71.

Lovejoy, C. O., Suwa, G., Spurlock, L., Asfaw, B., & White, T. D. (2009g). The pelvis and femur of *Ardipithecus ramidus*: The emergence of upright walking. *Science, 326,* 71e1-71e6.

Lovejoy, C. O., Suwa, G., Spurlock, L., Asfaw, B., & White, T. D. (2009h). The pelvis and femur of *Ardipithecus ramidus*: The emergence of upright walking. *Science, 326,* Supporting Online Material, 1-14.

MacNeilage, P. F., Rogers, L. J. & Vallortigara, G. (2009). Origins of the left and right brain. *Sci. Amer., 301 (1),* 60-67.

MacWilliams, B. A., Choi, T., Perezous, M. K., Chao, E. Y. S. & McFarland, E. G. (1998). Characteristic ground-reaction forces in baseball pitching. *Am. J. Sports Med., 26,* 66-71.

Malina, R. M. & Buschang, P. H. (1985). Growth, strength and motor performance of Zapotec children, Oaxaca, Mexico. *Hum. Biol., 57,* 163-181.

Malina, R. M., Little, B. B., Shoup, R. F. & Buschang, P. H. (1987). Adaptive significance of small body size: Strength and motor performance of school children in Mexico and Papua New Guinea. *Am. J. Phys. Anthrop., 73,* 489-499.

Manson, J. H. & Wrangham, R. W. (1991). Intergroup aggression in chimpanzees and humans. *Curr. Anthrop. 32,* 369-390.

Marchant, L. F. & McGrew, W. C. (1998). Human handedness: an ethological perspective. *Hum. Evol., 13,* 221-228.

Marchant, L. F., McGrew, W. C. & Eibl-Eibesfeldt, I. (1995). Is human handedness universal? Ethological analyses from three traditional cultures. *Ethol., 101,* 239-258.

Marlow, J. R., Lange, C. B., Wefer, G. & Rosell-Melé, A. (2000). Upwelling intensification as part of the Pliocene-Pleistocene climate transition. *Science, 290,* 2288-2291.

Marques-Bruna, P. & Grimshaw, P. (1997). Three-dimensional kinematics of overarm throwing action of children age 15 to 30 months. *Percept. Motor Skills, 84,* 1267-1283.

Marshall, R. N. & Elliott, B. C. (2000). Long-axis rotation: The missing link in proximal-to-distal segmental sequencing. *J. Sports Sci., 18,* 247-254.

Martin, P. S. (2005). *Twilight of the Mammoths.* Los Angeles, CA: University of California Press.

Marzke, M. W. (1983). Joint functions and grips of the *Australopithecus afarensis* hand, with special reference to the region of the capitate. *J. Hum. Evol., 12,* 197-211.

Marzke, M. W. (1992a). Evolutionary development of the human thumb. *Hand Clinics, 8,* 1-8.

Marzke, M. W. (1992b). Evolution of the hand and bipedality. In: Lock, A. & Peters, A. (Eds.), *Handbook of Human Symbolic Evolution* (126-154). Oxford, UK: Oxford University Press.

Marzke, M. W. (1997). Precision grips, hand morphology, and tools. *Am. J. Phys. Anthrop., 102,* 91-110.

Marzke, M. W. (2005). Who made stone tools? In: Roux, V. & Bril, B. (Eds.), *Stone Knapping* (243-255). Cambridge, UK: McDonald Institute Monographs.

Marzke, M. W., Longhill, J. M. & Rasmussen, S. A. (1988). Gluteus maximus muscle function and the origin of hominid bipedality. *Am. J. Phys. Anthrop., 77,* 519-528.

Marzke, M. W. & Marzke, R. F. (1987). The third metacarpal

styloid process in humans: origin and functions. *Am. J. Phys. Anthrop., 73,* 415-431.

Marzke, M. W. & Marzke, R. F. (2000). Evolution of the human hand: approaches to acquiring, analysing and interpreting the anatomical evidence. *J. Anat., 197,* 121-140.

Marzke, M. W., Marzke, R. G., Linsheid, R. L., Smutz, P., Steinberg B., Reece, S. & An, K. N. (1999). Chimpanzee thumb muscle cross sections, moment arms and potential torques, and comparisons with humans. *Am. J. Phys. Anthrop., 110,* 163-178.

Marzke, M. W. & Shackley, M. S. (1986). Hominid hand use in the Pliocene and Pleistocene: Evidence from experimental archeology and comparative morphology. *J. Hum. Evol. 15,* 439-460.

Marzke, M. W., Toth, N., Schick, K., Reece, S., Steinberg, B., Hunt, K., Linscheid R. L. & An, K-N. (1998). EMG study of hand muscle recruitment during hard hammer percussion manufacture of Oldowan tools. *Am. J. Phys. Anthrop., 105,* 315-332.

Marzke, M. W. & Wullstein, K. L. (1996). Chimpanzee and human grips: a new classification with a focus on evolutionary morphology. *Int. J. Primatol., 17,* 117-139.

Marzke, M. W., Wullstein, K. L. & Viegas, S. F. (1992). Evolution of the power ("squeeze") grip and its morphological correlates in hominids. *Am. J. Phys. Anthrop., 89,* 283-298.

Matsuo, T., Escamilla, R. F., Fleisig, G. S., Barrentine, S. W. & Andrews, J. R. (2001). Comparison of kinematic and temporal parameters between different pitch velocity groups. *J. Appl. Biomech., 17,* 1-13.

Mayr, E. (1961). Cause and effect in biology. *Science,134,* 1501-1506.

Mayr, E. (1963). *Populations, Species and Evolution.* Cambridge, MA: Harvard University Press.

Mayr, E. (2004). *What Makes Biology Unique?* New York, NY: Cambridge University Press.

McDowell, M. A., Fryar, C. D., Hirsch, R. & Ogden, C. L. (2005). *Anthropometric reference data for children and adults: U. S. Population, 1999-2002. Advance data from vital and health statistics: no. 361*. Hyattsville, MD: National Center for Health Statistics.

McFarland, R. (1997). Female Primates: Fat or fit? In: Morbeck, M. E., Galloway, A. & Zihlman, A. L. (Eds.), *The Evolving Female* (163-175). Princeton, NJ: Princeton University Press.

McGrew, W. C. (1992). *Chimpanzee Material Culture: Implications for Human Evolution*. New York, NY: Cambridge University Press.

McGrew, W. C. (2010). In search of the last common ancestor: new findings on wild chimpanzees. *Phil. Trans. Roy. Soc. B, 365,* 3267-3276.

McGrew, W. C. & Marchant, L. F. (1997). On the other hand: Current issues in and meta-analysis of the behavioral laterality of hand function in nonhuman primates. *Yearb.Phys. Anthrop., 40,* 201-232.

McHenry, H. M. (1986). The first bipeds: a comparison of the *A. afarensis* and *A. africanus* postcranium and implications for the evolution of bipedalism. *J. Hum. Evol., 15,* 177-191.

McHenry, H. M. (1991a). Petite bodies of the "robust" Australopithecines. *Am. J. Phys. Anthrop., 86,* 445-454.

McHenry, H. M. (1991b). Sexual dimorphism in *Australopithecus afarensis. J. Hum. Evol., 20,* 21-32.

McHenry, H. M. (1992). How big were early hominids? *Evol. Anthrop., 1,* 15-20.

McHenry, H. M. (1994) Behavioral ecological implications of early hominid body size. *J. Hum. Evol., 27,* 77-87.

McHenry, H. M. & Berger, L. R. (1998). Body proportions in *Australopithecus afarensis* and *A. africanus* and the origin of the genus Homo. *J. Hum. Evol., 35,* 1-22.

McHenry H. M. & Coffing, K. (2000). *Australopithecus* to *Homo:* Transformations in body and mind. *Ann. Rev. Anthrop., 29,* 125-146.

McManus, I. C. (1991). The inheritance of left-handedness. *Ciba Found. Symp., 162,* 251-267.

McPherron, S. P., Alemseged, Z., Marean, C. W., Wynn, J. G., Reed, D., Geraads, Bobe, R., & Béarat, H. A. (2010). Evidence for stone-tool-assisted consumption of animal tissues before 3.39 million years ago at Dikika,

Ethiopia. *Nature, 466,* 857-860.

Mero, A., Komi, V., Korjus, T., Navarrow, E. & Gregor, R. J. (1994). Body segment contributions to javelin throwing during final thrust phases. *J. Appl. Biomech., 10,* 166-177.

Miller, G. (2001). *The Mating Mind.* New York, NY: Anchor Books.

Miller, R. A. (1932). Evolution of the pectoral girdle and fore limb in the primates. *Am. J. Phys. Anthrop. 17:* 1-56.

Milton, K. (1999). A hypothesis to explain the role of meat-eating in human evolution. *Evol. Anthrop., 8,* 11-21.

Mitani, J. C. (2009). Cooperation and competition in chimpanzees: Current understanding and future challenges. *Evol. Anthrop.,18,* 215-227.

Mitani, J. C., Gros-Louis, J. & Richards, A. F. (1996). Sexual dimorphism, the operational sex ratio, and the intensity of male competition in polygynous primates. *Am. Naturalist, 147,* 966-980.

Montoye, H. J., Smith, E. L., Fardon, D. E. & Howley, E. T. (1980). Bone mineral in senior tennis players. *Scand. J. Sports Sci., 2,* 26-32.

Morgan, E. (1972). *The Descent of Woman.* New York, NY: Stein and Day.

Morgan, E. (1990). *The Scars of Evolution.* London, UK: Souvenir Press.

Morgan, E. (1995). *The Descent of the Child.* New York, NY: Oxford University Press.

Morgan, E. (1997). *The Aquatic Ape Hypothesis.* London, UK: Souvenir Press.

Morris, A. M., Williams, J. M., Atwater, A. E., & Wilmore, J. H. (1982). Age and sex differences in motor performance of 3 through 6 year old children. *Res. Quart. Exerc. Sport, 53,* 214-221.

Moriss, C. & Bartlett, R. (1996). Biomechanical factors critical for performance in the men's javelin throw. *Sports Med., 21,* 438-446.

Moyà-Solà, S., Köhler, M., Alba, D. M. & Almécija, S. (2008). Taxonomic attribution of the Olduvai hominid 7 manual remains and the functional interpretation of hand morphology in robust Australopithecines. *Folia Primatol, 79,* 215-250.

REFERENCES CITED

Nagel, E. (1979). *The Structure of Science*. Indianapolis, IN: Hackett Publishing Company.

Nakatsukasa, M. (2004). Acquisition of bipedalism: the Miocene hominoid record and modern analogues for bipedal proto- hominids. *J. Anat. 204,* 385-402.

Nakatsukasa, M., Pickford, M., Egi, N. & Senut, B. (2007). Femur length, body mass, and stature estimates of *Orrorin tugenensis*, a 6 Ma hominid from Kenya. *Primates, 48,* 171-178.

Napier, J. R. (1955). The form and function of the carpometacarpaljoint of the thumb. *J. Anat., 89,* 362-369.

Napier, J. R. (1956). The prehensile movements of the human hand. *J. Bone Joint Surg., 38B,* 902-913.

Napier, J. R. (1960). Studies of the hands of living primates. *Proc. Zool. Soc. Lond., 134,* 647-657.

Napier, J. R. (1961). Hands and handles. *New Scientist, 9,* 797-799.

Napier, J. R. (1962). Fossil hand bones from Olduvai Gorge. *Nature, 196,* 409-411.

Napier, J .R. (1965). Evolution of the human hand. *Proc. Roy. Inst. Great Britain, 40,* 544-557.

Napier, J. R. (1970). *The Roots of Mankind*. New York, NY: Harper & Row.

Napier, J. R. (1993). *Hands*. Princeton, NJ: Princeton University Press.

NAS National Academy of Sciences (2008). *Science, Evolution and Creationism*. Washington, DC: The National Academies Press.

Nelson, J. K., Thomas, J. R., Nelson, K. R. & Abraham, P. C. (1986). Gender differences in children's throwing performance: Biology and environment. *Res. Quart. Exerc. Sport, 57,* 280- 287.

Nelson, K. R., Thomas, J. R. & Nelson, J. K. (1991). Longitudinal changes in throwing performance: Gender differences. *Res. Quart. Exerc. Sport, 62,* 105-108.

Niemitz, C. (2002). A theory on the evolution of the habitual orthograde human bipedalism. The "Amphibische Generalistentheorie." *Anthrop Anz., 60,* 3-66.

Niemitz, C. (2006). Indications for an evolutionary correlation of human upright posture and an ecological niche on the shore. In: Niemitz, C. (Ed.), *Focuses and Perspectives of Modern Physical Anthropology* (37-46). Berlin, DE: Mitt. Berl. Ges. Anthrop. Ethnol. Urgeschichte, Beiheft I.

Niemitz, C. (2010). The evolution of the upright posture and gait—a review and a new synthesis. *Naturwissenschaften, 97,* 241-263.

Nishida, T. (1990). A quarter century of research in the Mahale Mountains: An overview. In: Nishida, T. (Ed.), *The Chimp- anzees of the Mahale Mountains* (3-35). Tokyo, JP: University of Tokyo Press.

O'Connell, J. F., Hawkes, K., Lupo, K. D. & Blurton Jones, N. G. (2002). Male strategies and Plio-Pleistocene archaeology. *J. Hum. Evol., 43,* 831-872.

Oakley, K. P. (1959). Tools makyth man. In: Howells, W. (Ed.), *Ideas on Human Evolution* (422-435). New York, NY: Atheneum.

Oakley, K. P. (1964). *Man the Tool-maker.* Chicago, IL: University of Chicago Press.

Ogden, C. L., Fryar, C. D., Caroll, M. D. & Flegal, K. M. (2005). *Mean body weight, height and body mass index, United States 1960-2002. Advance data from vital and health statistics, no. 347.* Hyattsville, MD: National Center for Health Statistics.

O'Higgins, P. & Elton, S. (2007). Walking on trees. *Science, 316,* 1292-1294.

Ohman, J. C., Slanina, M., Baker, G. & Mensforth, R. P. (1995). Thumbs, tools and early humans. *Science, 268,* 587-589.

Okasha, S. (2002). *Philosophy of Science. A Very Short Introduction.* New York, NY: Oxford University Press.

Overall, C. (2012). *Why Have Children?* Cambridge, MA: MIT Press.

Oxnard, C. E. (1963). Locomotor adaptations in the primate forelimb. *Symp. Zool . Soc. Lond. No. 10 (the Primates),* 165-182.

Oxnard, C. E. (1969). Evolution of the human shoulder: Some possible pathways. *Am. J. Phys. Anthrop., 30,* 319-332.

Panger, M. A., Brooks, A. S., Richmond, B. G. & Wood, B. (2002). Older than the Oldowan? Rethinking the emergence of hominin tool use. *Evol. Anthrop., 11,* 235-245.

Parker, S. T. (1987). A sexual selection model for hominid evolution. *Hum. Evol., 2,* 235-253.

Parker, S. T. & Jaffe, K. E. (2008). *Darwin's Legacy. Scenarios in Human Evolution.* New York, NY: AltaMira Press.

Partridge, T. C., Granger, D. E., Caffee, M. W. & Clarke, R. J. (2003). Lower Pliocene hominid pathological gait. *J. Bone & Joint Surg., 35-A,* 543-558.

Patterson, N., Richter, D. J., Guerre, S., Lander, E. S. & Reich, D. (2006). Genetic evidence for complex speciation of humans and chimpanzees. *Nature, 441,* 1103-1108.

Payne, R. C., Crompton, R. H., Isler, K., Savage, R., Vereecke, E. E., Günther, M. M., Thorpe, S. K. S. & D'Août, K. (2006). Morphological analysis of the hindlimb in apes and humans. I. Muscle architecture. *J. Anat., 208,* 709-724.

Petranek, L. J. & Barton, G. V. (2011). The overarm-throwing pattern among U-14 ASA female softball players: A comparative study of gender, culture, and experience. *Res. Quart. Exerc. Sport, 82,* 220-228.

Pianka, E. R. (1978). *Evolutionary Ecology* (2nd Edit.), New York, NY: Harper and Row, Publishers.

Pickering, T. R. (2012). What's new is old: Comments on (more) archaeological evidence of one-million-year-old fire from South Africa. *S. Afr. J. Sci. 108,* Art. #1250, 2 pages.

Pickering, R., Dirks, P. H. G. M., Jinnah, Z., de Ruiter, D. J., Churchill, S. E., Herries, A. I. R., Woodhead, J. D., Hellstrom, J. C. & Berger, L. R. (2011). *Australopithecus sediba* at 1.977 Ma and implications for the origins of the genus *Homo. Science, 333,* 1421-1423.

Pickering, T. R., Bunn, H. T. (2007). The endurance running hypothesis and hunting and scavenging in savanna-woodlands. *J. Hum. Evol., 53,* 434-438.

Pickering, T. R., Domínguez-Rodrigo, M., Egeland, C. P. & Brain, C. K. (2004). New data and ideas on the foraging behaviour of Early Stone Age hominids at Swartkrans Cave, South Africa. *S. Afr. J. Sci., 100,* 215-219.

Pickford, M., Senut, B., Gommery, D. & Treil, J. (2002). Bipedalism in *Orrorin tugenensis* revealed by its femora.

C. R. Paleovol 1, 191-203.

Pilbeam, D. (1970). *The Evolution of Man.* New York, NY: Funk and Wagnalls.

Pinker, S. (2011). *The Better Angels of Our Nature: Why Violence Has Declined.* New York, NY: Penguin Books.

Pitts, M. & Roberts, M. (1998). *Fairweather Eden.* London, UK: Random House.

Plavcan, J. M. (1993). Canine size and shape in male anthropoid primates. *Am J. Phys. Anthrop., 92,* 201-216.

Plavcan, J. M. (2000). Inferring social behavior from sexual dimorphism in the fossil record. *J. Hum. Evol., 39,* 327-344.

Plavcan, J. M., Lockwood, C.A. Kimbel, W. H. Lague, M. R. & Harmon, E. H. (2005). Sexual dimorphism in *Australopithecus afarensis* revisited: How strong is the case for a human-like pattern of dimorphism? *J. Hum. Evol., 48,* 313-320.

Plavcan, J. M. & van Schaik, C. P. (1992). Intrasexual competition and canine dimorphism in anthropoid primates. *Am. J. Phys. Anthrop., 87,* 461-477.

Plavcan, J. M. & van Schaik, C. P. (1997a). Intrasexual competition and body weight dimorphism in anthropoid primates. *Am. J. Phys. Anthrop., 103,* 37-68.

Plavcan, J. M. & van Schaik, C. P. (1997b). Interpreting hominid behavior on the basis of sexual dimorphism. *J. Hum. Evol., 32,* 345-374.

Pobiner, B. L., Rogers, M. J., Monahan, C. M. & Harris, J. W. K. (2008). New evidence for hominin carcass processing strategies at 1.5 Ma, Koobi Fora, Kenya. *J. Hum. Evol., 55,* 103-130.

Pontzer, H., Raichlen, D. A. & Sockol, M. D. (2009). The metabolic cost of walking in humans, chimpanzees, and early hominins. *J. Hum. Evol., 56,* 43-54.

Pontzer, H., Rolian, C., Rightmire, G. P., Jashashvili, T., Ponce de Léon, M. S., Lordkipanidze, D. & Zollikifer, C P. E. (2010). Locomotor anatomy and biomechanics of the Dmanisi hominins. *J. Hum. Evol., 58,* 492-504.

Potts, M. & Hayden, T. (2008). *Sex and War.* Dallas, TX: Benbella Books.

Potts, R. (1983). Foraging for faunal resources by early hominids at Olduvai Gorge, Tanzania. In: Clutton-Brock, J. & Grigson, C. (Eds.), *Animals and Archeology: Vol. 1, Hunters and Their Prey* (51-62). Oxford, UK: B.A.R. Int. Series 163.

Potts, R. (1984). Home bases and early hominids. *Am. Scientist 72*, 338-347.

Potts, R. (1988). *Early Hominid Activities at Olduvai*. New York, NY: Aldine de Gruyter.

Potts, R. (1989). Olorgesailie: new excavations and findings in Early and Middle Pleistocene contexts, southern Kenya rift valley. *J. Hum. Evol., 18*, 477-484.

Potts, R. (1994). Variables versus models of early Pleistocene hominid land use. *J. Hum. Evol. 27*, 7-24.

Potts, R., Behrensmeyer, A. K., Deino, A., Ditchfield, P. & Clark, J. (2004).Small Mid-Pleistocene hominin associated with East African Acheulean technology. *Science, 305*, 75-78.

Potts, R. & Shipman, P. (1981). Cutmarks made by stone tools on bones from Olduvai Gorge, Tanzania. *Nature, 291*, 577-580.

Preuschoft, H. (2004). Mechanisms for the acquisition of habitual bipedality: are there biomechanical reasons for the acquisition of upright bipedal posture? *J. Anat., 204*, 363-384.

Preuschoft, H. & Witte, H. (1991). Biomechanical reasons for the evolution of hominid body shape. In: Coppens, Y. & Senut, B. (Eds.), *Origine(s) de la Bipédie Chez les Hominidés* (59-77). Paris: Cahiers de Paléoanthropologie, Editions du CNRS.

Prost, J. H. (1980). Origin of bipedalism. *Am. J. Phys. Anthrop., 52*, 175-189.

Race, D. E. (1961). A cinematographic and mechanical analysis of the external movements involved in hitting a baseball effectively. *Res. Quart., 32*, 394-404.

Raichlen, D. A., Armstrong, H. & Lieberman, D. E. (2011). Calcaneus length determines running economy: implications for endurance running performance in modern humans and Neandertals. *J. Hum. Evol., 6*, 299-308.

Ralston, H. J. (1976). Energetics of human walking. In:

Herman, R. M. (Ed.), *Neural Control of Locomotion* (77-98). New York, NY: Plenum Press.

Ransdell, L. B. & Wells, C. L. (1999). Sex differences in athletic performance. *Women in Sport & Phys. Activity J., 8,* 55-81.

Rat-Fischer, L., O'Regan, J. K. & Fagard, J. (2012). *Develop. Psychobiol.,* DOI 10,1002/dev.21078,epub.

Ravey M. (1978). Bipedalism: An early warning system for Miocene hominoids. *Science, 199,* 372.

Raymond, M. & Pontier, D. (2004). Is there geographical variation in human handedness? *Laterality, 9,* 35-51.

Raymond, M., Pontier, D., Dufour, A.-B. & Moller, A. P. (1996). Frequency-dependent maintenance of left handedness in humans. *Proc. Roy. Soc. London B, 206,* 1627-1633.

Reeser, L. A., Susman, R. L. & Stern, J. T. (1983). Electromyographic studies of the human remains from Sterkfontein. *Science, 300,* 607-612.

Reno, P. L., Meindl, R. S., McCollum, M. A. & Lovejoy, C. O. (2003). Sexual dimorphism in *Australopithecus afarensis* was similar to that of modern humans. *PNAS, 100,* 9404-9409.

Reno, P. L., Meindl, R. S., McCollum, M. A. & Lovejoy, C. O. (2005). The case is unchanged and remains robust: *Australopithecus afarensis* exhibits only moderate skeletal dimorphism. A reply to Plavcan et al. (2005). *J. Hum. Evol., 49,* 279-288.

Richmond, B. G. (2007). Biomechanics of phalangeal curvature. *J. Hum. Evol., 53,* 678-690.

Richmond, B. G., Aiello, L. C. & Wood, B. A. (2002). Early hominin limb proportions. *J. Hum. Evol., 43,* 529-548.

Richmond, B. G., Begun, D. R & Strait, D. S. (2001). Origin of human bipedalism: The knuckle-walking hypothesis revisited. *Yearb. Phys. Anthrop., 44,* 70-105.

Richmond, B. G. & Jungers, W. L. (1995). Size variation and sexual dimorphism in *Australopithecus afarensis* and living hominoids. *J. Hum. Evol., 29,* 229-245.

Richmond, B. G. & Jungers, W. L. (2008). *Orrorin tugenensis* femoral morphology and the evolution of hominin bipedalism. *Science, 319,* 1662-1664.

Richmond, B. G., & Strait, D. S. (2000). Evidence that humans evolved from a knuckle-walking ancestor. *Nature, 404,* 382-385.

Ricklan, D. E. (1987). Functional anatomy of the hand of *Australopithecus africanus. J. Hum. Evol., 16,* 643-664.

Ricklan, D. E. (1990). The precision grip in *Australopithecus africanus*: Anatomical and behavioral cor-correlates. In: Sperber, G. H. (Ed.), *From Apes to Angels: Essays in Anthropology in honor of Phillip V. Tobias* (171-183). Wilmington, DE: Wiley-Liss.

Rightmire, G. P. (2004). Brain size and encephalization in Early to Mid-Pleistocene *Homo. Am. J. Phys. Anthrop., 124,* 109-123.

Rightmire, G. P., Lordkipanidze, D. & Vekua, A. (2006). Anatomical descriptions, comparative studies and evolutionary significance of the hominin skulls from Dmanisi, Republic of Georgia. *J. Hum. Evol., 50,* 115-141.

Rippee, N. E., Pangrazi, R. P., Corbin, C. B., Borsdorf, L., Petersen, G. & Pangrazi, D. (1990). Throwing profiles of first and fourth grade boys and girls. *Phys. Educator, 47,* 180-185.

Roberton, M. A. (1977). Stability of stage categorizations across trials: Implications for the "stage theory" of overarm throw development. *J. Hum. Movement Studies, 3,* 49-59.

Roberton, M. A., Halverson, L. E., Langendorfer, S. J. & Williams, K. (1979). Longitudinal changes in children's overarm throw ball velocities. *Res. Quart., 50,* 256-264.

Roberton, M. A. & Konczak, J. (2001). Predicting children's over-arm throw ball velocities from their developmental levels in throwing. *Res. Quart. Exerc. Sport, 72,* 91-103.

Roberton, M. A. & Langendorfer, S. (1980). Testing motor development sequences across 9-14 years. In: Nadeau, C. W., Halliwell, W., Newell, K. & Roberts, G. (Eds.), *Psychology of Motor Behavior and Sport-1979* (269-279). Champaign, IL : Human Kinetics.

Roberts, M. B., Stringer, C. B. & Parfitt, S. A. (1994). A hominid tibia from Middle Pleistocene sediments at Boxgrove, UK. *Nature, 369,* 311-313.

Robinson, J. T. (1963). Adaptive radiation in the Australopithecines and the origin of man. In: Howell, F. C. & Bourli, F. (Eds.), *African Ecology and Human Evolution* (385-416). New Brunswick, NJ: Transaction Publishers.

Robinson, J. T. (1972). *Early Hominid Posture and Locomotion.* Chicago, IL: University of Chicago Press.

Rodman, P. S. & McHenry, H. M. (1980). Bioenergetics and the origin of hominid bipedalism. *Am. J. Phys. Anthrop. 52,* 103-106.

Roebroeks, W. (2001). Hominid behaviour and the earliest occupation of Europe: an exploration. *J. Hum. Evol. 41,* 437-461.

Roebroeks, W. & Villa, P. (2011). On the earliest evidence for habitual use of fire in Europe. *PNAS, 108,* 5209-5214.

Rolian, C., Lieberman, D. E. & Hallgrimsson, B. (2010). The coevolution of human hands and feet. *Evol., 64,* 1558-1568.

Rook, L., Bondioli, L., Köhler, M., Moyà-Solà, S. & Macchiarelli, R.(1999). Oreopithecus was a bipedal ape after all: Evidence from the iliac cancellous architecture. *PNAS, 96,* 8795-8799.

Rose, L. M. (2001). Meat and the early human diet. In: Stanford, C. B. & Bunn, H. T. (Eds.), *Meat-Eating and Human Evolution,* (141-159). New York, NY: Oxford University Press.

Rose, L. M., & Marshall, F. (1996). Meat eating, hominid sociality, and home bases revisited. *Curr. Anthrop. 37,* 307-338.

Rose, M. D. (1984). Food acquisition and the evolution of positional behaviour: the case of bipedalism. In: Chivers, D. J., Wood B. A. & Bilsborough, A. (Eds.), *Food Acquisition and Processing in Primates* (509-523). New York, NY: Plenum Press.

Rose, M. D. (1991). The process of bipedalization in hominids. In: Coppens, Y. & Senut, B. (Eds.), *Origine(s) de la Bipédie chez les Hominidés,* (37-48). Paris, FR: Cahiers de Paléoanthropologie, Editions du CNRS.

Rosenberg, A. & McShea, D. W. (2007). *Philosophy of Biology. A Contemporary Introduction.* New York, NY: Routledge.

Round, J. M., Jones, D. A., Honour, J. W. & Nevill, A. M. (1999). Hormonal factors in the development of differences in strength between boys and girls during adolescence: a longitudinal study. *Ann. Hum. Biol., 26,* 49-62.

Ruff, C. B. (2009). Relative limb strength and locomotion in *Homo habilis. Am. J. Phys. Anthrop., 38,* 90-100.

Ruff, C. B., Trinkaus, E. & Holliday, T. W. (1997). Body mass and encephalization in Pleistocene *Homo. Nature, 387,* 173-176.

Ruff, C. B., Trinkaus, E., Walker, A. & Larsen, C. S. (1993). Post- cranial robusticity in *Homo*. I: temporal trends and mechanical interpretation. *Am. J. Phys. Anthrop. 91,* 21-53.

Ruff, C. B. & Walker, A.(1993). Body size and body shape. In: Walker, A. & Leakey, R. (Eds.), *The Nariokotome Homo erectus Skeleton,* (234-265). Cambridge, MA: Harvard University Press.

Ruff, C. B., Walker, A. & Trinkaus, E. (1994). Postcranial robusticity in *Homo*. III: Ontogeny. *Am. J. Phys. Anthrop., 93,* 35-54.

Runion, B., Roberton, M. A. & Langendorfer, S. J. (2003). Forceful overarm throwing: a comparison of two cohorts measured 20 years apart. *Res. Quart. Exerc. Sport, 74,* 324-330.

Ruxton, G. D. & Wilkinson, D. M. (2011). Thermoregulation and endurance running in hominins: Wheeler's models revisited. *J. Hum. Evol., 61,* 169-175.

Ryan, C. & Jethá, C. (2010). *Sex at Dawn: The Prehistoric Origins of Modern Sexuality.* New York, NY: Harper Collins.

Sahnouni, M. & de Heinzelin, J. (1998). The site of Ain Hanech revisited: New investigations at this Lower Pleistocene site in northern Algeria. *J. Archaeol. Sci. 25,* 1083-1101.

Sahnouni, M., Hadjoulis, D., van der Made, J., Derradji, A., Canals, A., Medig, M., Belahrech, H., Harichene, Z. & Rabhi, M. (2002). Further research at the Oldowan site of Ain Hanech, North-Eastern Algeria. *J. Hum. Evol., 43,* 925-937.

Sahnouni, M., Hadjoulis, D., van der Made, J., Derradji, A., Canals, A., Medig, M., Belahrech, H., Harichene, Z. & Rabhi, M. (2004). On the earliest human occupation in North Africa: a response to Geraads, et al. *J. Hum. Evol.,*

46, 763-775.

Sahnouni, M., Schick, K. & Toth, N. (1997). An experimental investigation into the nature of faceted limestone "spheroids" in the Early Paleolithic. *J. Archaeol. Sci. 24*, 701-713.

Sainburg, R. L. (2002). Evidence for a dynamic-dominance hypothesis of handedness. *Exper. Brain Res., 142*, 241-258.

Sainburg, R. L. (2005). Handedness: Differential specializations for control of trajectory and position. *Exerc. Sport Sci. Rev., 33*, 206-213.

Sainburg, R. L. & Kalakanis, D. (2000). Differences in control of limb dynamics during dominant and nondominant arm reaching. *J. Neurophysiol., 83*, 2661-2675.

Sakurai, S. & Miyashita, M. (1983). Development aspects of overarm throwing related to age and sex. *Hum. Movement Sci., 2*, 67-76.

Salmon, W. C. (1990). *Four Decades of Scientific Explanation*. Pittsburgh, PA: University of Pittsburgh Press.

Sarlegna, F. R. & Sainburg, R. I. (2009). The roles of vision and proprioception in the planning of reaching movements. *Adv. Exp. Med. Biol., 629,* 317-335.

Saunders, J. B., Inman, V. T. & Everhart, H. D. (1953). The major determinants in normal and pathological gait. *J. Bone & Joint Surg., 35-A, 543-558.*

Saunders, J. A. & Knill, D. C. (2004). Visual feedback control of hand movements. *J. Neurosci., 24:* 3223-3234.

Savage-Rumbaugh, S. & Lewin, R. (1994). *Kanzi. The Ape at the Brink of the Human Mind*. New York, NY: John Wiley & Sons.

Schick, K. D. (1987). Modeling the formation of early stone age artifact concentrations. *J. Hum. Evol., 16,* 789-807.

Schick, K. D. & Toth, N. (1993). *Making Silent Stones Speak*. London, UK: Weidenfeld and Nicolson.

Schick, K. D. & Toth, N. (1994). Early Stone Age technology in Africa: A review and case study into the nature and function of spheroids and subspheroids. In: Coruccini, R. S. & Ciochon, R. L. (Eds.), *Integrative Paths to the Past: Paleoanthropological Advances in Honor of F. Clark*

Howell (429-449). Englewood Cliffs, NJ: Prentice Hall.

Schmid, P. (1991). The trunk of the Australopithecines. In: Coppens, Y. & Senut B. (Eds.), *Origine(s) de la Bipédie chez les Hominidés* (225-234). Paris, FR: Cahiers de Paléoanthropologie, Editions du CNRS.

Schmid, P. (2004). Functional interpretation of the Laetoli footprints. In: Meldrum, D. J. & Hilton, C. E. (Eds.), *From Biped to Strider: The Emergence of Modern Human Walking, Running, and Resource Transport* (49-62). New York, NY: Kluwer Academic/Plenum Publishers.

Schmid, P., Churchill, S. E., Nalla, S., Weissen, Carlson, K. J., de Ruiter, D. J. & Berger, L. R. (2013). Mosaic morphology in the thorax of *Australopithecus sediba*. *Science, 340,* 12 April, DOI: 10.1126/science 1234598.

Schmitt, D. (2003). Insights into the evolution of human bipedalism from experimental studies of humans and other primates. *J. Exp. Biol., 206,* 1437-1448.

Schmitt, D., Churchill, S. E. & Hylander, W. L. (2003). Experimental evidence concerning spear use in Neandertals and early modern humans. *J. Archaeol. Sci., 30,* 103-114.

Schoeninger, M. J., Bunn, H. T., Murray, S., Pickering, T. & Moore, J. (2001). Meat-eating by the fourth African Ape. In: Stanford, C. B. & Bunn, H. T. (Eds.), *Meat-Eating and Human Evolution* (179-195). New York, NY: Oxford University Press.

Schultz, A. (1969).*The Life of Primates.* New York, NY: Universe Books.

Seefeldt, V. & Haubenstricker, J. (1982). Patterns, phases, or stages: An analytical model for the study of developmental movement. In: Kelso, J. A. S. & Clark, J. E. (Eds.), *The Development of Movement Control and Coordination* (309-318). New York, NY: Wiley.

Semaw, S. (2000). The world's oldest stone artefacts from Gona, Ethiopia: Their implications for understanding stone technology and patterns of human evolution between 2.6-1.5 million years ago. *J. Archaeol. Sci., 27,* 1197-1214.

Semaw, S., Rogers, M. J. Quade, J., Renne, P. R., Butler, R. F., Domínguez-Rodrigo, M., Stout, D., Hart, W. S., Pickering, T. & Simpson, S. W. (2003). 2.6-million-year-old stone tools and associated bones from OGS-6 and GS-7, Gona, Afar, Ethiopia. *J. Hum. Evol., 45*, 169-177.

Semaw, S., Simpson, S. W., Quade, J., Renne, P. R., Butler, R. F., McIntosh, W. C., Levin, N., Domínguez-Rodrigo, M. & Rogers, M. J. (2005). Early Pliocene hominids from Gona, Ethiopia. *Nature, 433,* 301-305.

Senut, B. (1991). Origine(s) de la bipédie humaine: Approche paléontologique. In: Coppens, Y. & Senut. B. (Eds.), *Origine(s) de la Bipédie chez les Hominidés* (245-257). Paris, FR: Cahiers de Paléoanthropologie, Editions du CNRS.

Senut, B. (2006a). Arboreal origin of bipedalism. In: Ishida, H. Tuttle, R., Pickford, M., Ogihara, N., & Nakatsukasa, M. (Eds.), *Human Origins and Environmental Backgrounds. Developments in Primatology: Progress and prospects* (199-208). New York, NY: Springer.

Senut, B. (2006b). Bipédie et climat. *C. R. Paleovol, 5,* 89-98.

Senut, B. (2007). The earliest putative hominids. In: Henke, W. & Tattersall, I. (Eds.), *Handbook of Paleoanthropology, Vol. III: Phylogeny of Hominines* (1519-1537). New York, NY: Springer.

Senut, B., Pickford, M., Gommery, D., Mein, P., Cheboi, K. & Coppens, Y, (2001). First hominid from the Miocene (Lukeino Formation, Kenya). *C. R. Acad. Sci. Paris, Sci. de la Terre et des Planètes, 332,* 137-144.

Shipman, P. (1986). Studies of hominid-faunal interactions at Olduvai Gorge. *J. Hum. Evol. 15,* 691-706.

Shipman, P. & Walker, A. (1989). The costs of becoming a predator. *J. Hum. Evol., 18,* 373-392.

Shreeve, J. (2010). The evolutionary road. *Nat. Geographic, 218,* 34-67.

Shrewsbury, M. M. & Johnson, R. K. (1983). Form, function, and evolution of the distal phalanx. *J. Hand Surg., 8,* 475-479.

Shrewsbury, M. M., Johnson, R. K. & Ousterhout, D. K. (1972). The palmaris brevis—a reconsideration of its anatomy and possible function. *J. Bone & Joint Surg., 54-A,* 344-348.

Shrewsbury, M. M., Marzke, M. W., Linscheid, R. L. & Reece, S. P. (2003). Comparative morphology of the pollical distal phalanx. *Am. J. Phys. Anthrop., 121,* 30-47.

Sigmon, B. A. (1971). Bipedal behavior and the emergence of erect posture in man. *Am. J. Phys. Anthrop., 34,* 55-60.

Sinclair, A. R. E., Leakey, M. D. & Norton-Griffiths, M. (1986). Migration and hominid bipedalism. *Nature, 324,* 307-308.

Sing, R F. (1984). *The Dynamics of the Javelin Throw.* Cherry Hill, NJ: Reynolds.

Smith, R. J. & Jungers, W. L. (1997). Body mass in comparative primatology. *J. Hum. Evol., 3,* 523-559.

Sober, E. (1984). *The Nature of Selection.* Cambridge, MA: The MIT Press.

Sockol, M. D., Raichlen, D. A. & Pontzer, H. (2007). Chimpanzee locomotor energetics and the origin of human bipedalism. *PNAS, 104,* 12265-12269.

Sollas, W. J. (1924). *Ancient Hunters and their Modern Repre-sentatives* (3rd Edit.), New York, NY: Macmillan.

Spenst, L. F., Martin, A. D. & Donald, T. (1993). Muscle mass of competitive male athletes. *J. Sports Sci., 11,* 3-8.

Spoor, F., Leakey, M. G., Gathogo, P. N., Brown, F. H., Antón, S. C., McDougall, I., Kiarie, C., Manthi, F. K. & Leakey, L. N. (2007). Implications of new early *Homo* fossils from Ileret, east of Lake Turkana, Kenya. *Nature, 448,* 688-691.

Stanford, C. B. (1999). *The Hunting Apes.* Princeton, NJ: Princeton University Press.

Stanford, C. B. (2001). A comparison of social meat-foraging by chimpanzees and human foragers. In: Stanford, C. B. & Bunn H. T. (Eds.), *Meat-Eating and Human Evolution* (122-140). New York, NY: Oxford University Press.

Stanford, C. B. (2003). *Upright. The Evolutionary Key to Becoming Human.* New York, NY: Houghton Mifflin.

Stanford, C. B., Wallis, J., Matama, H. & Goodall, J. (1994). Patterns of predation by chimpanzees on red colobus monkeys in Gombe National Park, Tanzania, 1982-1991. *Am. J. Phys. Anthrop., 94,* 213-229.

Steele, J. (2000). Handedness in past human populations: Skeletal markers. *Laterality, 5,* 193-220.

Steele, J. & Uomini, N. (2005). Humans, tools and handedness. In: Roux, V. & Bril, B. (Eds.), *Stone Knapping* (217-239). Cambridge, UK: McDonald

Institute Mono- graphs.

Steiper, M. E. & Young, N. M. (2006). Primate molecular divergence dates. *Molec. Phylogenet. Evol., 41,* 384-394.

Stern, J. T. Jr. (1972). Anatomical and functional specializations of the human gluteus maximus. *Am. J. Phys. Anthrop., 36,* 315-340.

Stern, J. T. Jr. (1977). Before bipedality. *Yearb. Phys. Anthrop., 19,* 59-68.

Stern, J. T. Jr. (2000). Climbing to the top: A personal memoir of *Australopithecus afarensis*. *Evol. Anthrop. 9,* 113-133.

Stern, J. T. Jr. & Susman, R. L. (1983). The locomotor anatomy of *Australopithecus afarensis*. *Am. J. Phys. Anthrop., 60,* 279-317.

Steudel, K. (1996). Limb morphology, bipedal gait, and the energetics of hominid locomotion. *Am. J. Phys Anthrop., 99,* 345-355.

Steudel-Numbers, K. L. (2001). Role of locomotor economy in the origin of bipedal posture and gait. *Am J. Phys. Anthrop., 116,* 171-173.

Steudel-Numbers, K. L. (2003). The energetic cost of locomotion: humans and primates compared to generalized endotherms. *J. Hum. Evol., 44,* 255-262.

Steudel-Numbers, K. L. (2006). Energetics in *Homo erectus* and other early hominins: The consequences of increased lower-limb length. *J. Hum. Evol., 51,* 445-453.

Steudel-Numbers, K. L. & Tilkens, M. J. (2004). The effect of lower limb length on the energetic cost of locomotion: implications for fossil hominins. *J. Hum. Evol., 47,* 95-109.

Steudel-Numbers, K. L. & Wall-Scheffler, C. M. (2009). Optimal running speed and the evolution of hominin hunting strategies. *J. Hum. Evol., 56,* 355-360.

Steudel-Numbers, K. L., Weaver, T. D. & Wall-Scheffler, C. M. (2007). The evolution of human running: Effects of changes in lower-limb length onlocomotor economy. *J. Hum. Evol., 53,* 191-196.

Stodden, D. F., Fleisig, G. S., McLean, S. P. & Andrews, J. R. (2005). Relationship of biomechanical factors to baseball pitching velocity: within pitcher variation. *J. Appl. Biomech., 21,* 44-56.

Stodden, D. F., Fleisig, G. S., McLean, S. P., Lyman, S. L. & Andrews, J. R. (2001). Relation of pelvis and upper torso kinematics to pitched baseball velocity. *J Applied Biomech 17,* 164-172.

Stodden, D. F., Langendorfer, S. J., Fleisig, G. S.,& Andrews, J. R. (2006a). Kinematic constraints associated with the acquisition of overarm throwing. Part I: step and trunk actions. *Res. Quart. Exerc. Sport, 77,* 417-427.

Stodden, D. F., Langendorfer, S. J., Fleisig, G. S. & Andrews, J. R. (2006b). Kinematic constraints associated with the acquisition of overarm throwing. Part II: upper extremity actions. *Res. Quart. Exerc. Sport, 77,* 428-436.

Straus, L. G. (1993). Upper Paleolithic hunting tactics and weapons in Western Europe. In: Peterkin, G. L., Bricker, H. M. & Mellers, P. (Eds.), *Hunting and Animal Exploitation in the Later Paleolithic and Mesolithic of Eurasia* (83-93). Archeol. Papers Am. Anthrop. Assoc., No. 4.

Straus, W. L. (1962). Fossil evidence of the evolution of the erect, bipedal posture. *Clin. Orthop. Related Res., 25,* 9-19.

Stretch, R. A., Buys, F. J., & Viljoen, G. (1995). The kinetics of the drive off the front foot in cricket batting: Hand grip force. *J. Res. Sport, Phys. Ed., Rec., 18,* 83-93.

Stringer, C. B. (2012). *Lone Survivors: How We Came to be the Only Humans on Earth.* New York, NY: Times Books.

Stringer, C. B. & Gamble, C. (1993). *In Search of the Neanderthals.* New York, NY: Thames and Hudson.

Stringer, C. B. & McKie, R. (1996). *African Exodus.The origins of Modern Humanity.* New York, NY: Henry Holt.

Stringer, C. B., Trinkaus, E., Roberts, M. B., Parfitt, S. A. & Macphail, R. I. (1998). The Middle Pleistocene human tibia from Boxgrove. *J. Hum. Evol., 34,* 509-547.

Suppe, F. (1977). *The Structure of Scientific Theories* (2nd Edit.). Chicago, IL: University of Illinois Press.

Susman, R. L. (1979). Comparative and functional morphology of hominoid fingers. *Am. J. Phys. Anthrop., 50,* 215-236.

Susman, R. L. (1983). Evolution of the human foot: Evidence from Plio-Pleistocene hominids. *Foot & Ankle, 3,* 365-376.

Susman, R. L. (1988a). New postcranial remains from Swartkrans and their bearing on the functional morphology and behavior of *Paranthropus robustus*. In: Grine, F. W. (Ed.), *Evolutionary History of the Robust Australo- pithecines* (149-172). New York, NY: Aldine de Gruyter.

Susman, R. L. (1988b). Hand of *Paranthropus robustus* from Member 1, Swartkrans: Fossil evidence for tool behavior. *Science, 240,* 781-784.

Susman, R. L. (1991). Who made the Oldowan tools? Fossil evidence for tool behavior in Plio-Pleistocene hominids. *J. Anthrop. Res., 47,* 129-151.

Susman, R. L. (1993). Hominid postcranial remains from Swartkrans. In: Brain, C. K. (Ed.), *Swartkrans. A Cave's Chronicle of Early Man* (117-136). Pretoria, SA: Transvaal Museum.

Susman, R. L. (1994). Fossil evidence for early hominid tool use. *Science, 265,* 1570-1573.

Susman, R. L. (1998). Hand function and tool behavior in early hominids. *J. Hum. Evol., 35,* 23-46.

Susman, R. L. (2008). Brief Communication: Evidence bearing on the status of *Homo habilis* at Olduvai Gorge. *Am. J. Phys. Anthrop., 137,* 356-361.

Susman, R. L. & Brain, T. M. (1988). The first metatarsal (SKX 5017) from Swartkrans and the gait of *Paranthropus robustus. Am. J. Phys. Anthrop., 77,* 7-15.

Susman, R. L. & Creel, N. (1979). Functional and morphological affinities of the subadult hand (O.H. 7) from Olduvai Gorge. *Am. J. Phys. Anthrop. 51,* 311-332.

Susman, R. L. & de Ruiter, D. J. (2004). New hominin first metatarsal (SK 1813) from Swartkrans. *J. Hum Evol., 47,* 171-181.

Susman, R. L., de Ruiter, D. & Brain, C. K. (2001). Recently identified postcranial remains of *Paranthropus* and Early *Homo* from Swartkrans Cave, South Africa. *J. Hum. Evol., 41,* 607-629.

Susman, R. L. & Stern, J. T. Jr. (1982). Functional morphology of *Homo habilis. Science, 217,* 931-934.

Susman, R. L. & Stern, J. T. Jr. (1991). Locomotor behavior of early hominids: Epistemology and fossil evidence. In:

Coppens, Y. & Senut, B. (Eds.), *Origine(s) de la bipédie chez les hominidés* (121-131). Paris: Cahiers de Paléoanthropologie, Editions du CNRS.

Susman, R. L., Stern, J. T. Jr. & Jungers, W. L. (1984). Arboreality and bipedality in the Hadar hominids. *Folia Primatol., 43,* 113-156.

Suwa, G., Asfaw, B., Haile-Selassie, Y., White, T., Katoh, S., WoldeGabriel, G., Hart, W. K., Nakaya, H. & Beyene, Y. (2007a). Early Pleistocene *Homo erectus* fossils from Konso, southern Ethiopia. *Anthrop. Sci., 115,* 133-151.

Suwa, G., Kono, R. T., Katoh, S., Asfaw, B. & Beyene, Y. (2007b). A new species of great ape from the late Miocene epoch in Ethiopia. *Nature, 448,* 921-924.

Suwa, G., Kono, R. T., Simpson, S. W., Asfaw, B., Lovejoy, C. O. & White, T. D. (2009a). Paleolobiological implications of the *Ardipithecus ramidus* dentition. *Science, 326,* 94-99.

Suwa, G., Kono, R. T., Simpson, S. W., Asfaw, B., Lovejoy, C. O. & White, T. D. (2009b). Paleobiological implications of the *Ardipithecus ramidus* dentition, *Science, 326,* Supporting online material, 1-73.

Swindler, D. R., & Wood, C. D. (1982). *An Atlas of Primate Gross Anatomy*. Malabar, FL: Robert E. Krieger Publishing Company.

Sylvester, A. D. & Kramer, P. A. (2008). Brief Communication: Stand and shuffle: When does it make energetic sense? *Am. J. Phys. Anthrop., 135,* 484-488.

Symons, D. (1979). *The Evolution of Human Sexuality*. New York, NY: Oxford University Press.

Szalay, F. S. (1975). Hunting-scavenging protohominids: A model for human origins. *Man, 10,* 420-429.

Szymanski, D. J., Szymanski, J. M., Schade, R. L., Bradford, T. J., McIntyre, J. S., DeRenne, C. & Madsen, N. H. (2010). The relation between anthropometric and physiological variables and bat velocity of high-school baseball players before and after 12 weeks of training. *J. Strength. Cond. Res., 24,* 2933-2943.

T & F News Panel (2006). 2005 World Rankings. *Track & Field News, 59,* 20-24.

Tanner, J. M. (1978). *Fetus into Man*. Cambridge, MA: Harvard University Press.

Tanner, N. & Zihlman, A. (1976). Women in Evolution. Part I: Innovation and selection in human origins. *Signs: J. Women in Culture and Society, 1,* 585-608.

Tardieu, C. (1992). Location of the body center of gravity in primates and other mammals: Implications for the evolution of hominid body shape and bipedalism. In: Matano, S., Tuttle, R. H., Ishida, H. & Goodman, M. (Eds.), *Topics in Primatology, Vol. 3* (191-208). Tokyo, JP: University of Tokyo Press.

Tardieu, C., Aurengo, A. & Tardieu, B. (1993). New method of three-dimensional analysis of bipedal locomotion for the study of displacements of the body and body-parts centers of mass in man and non-human primates: evolutionary framework. *Am. J. Phys. Anthrop., 90,* 455-476.

Tattersall, I. (1995). *The Fossil Trail.* New York, NY: Oxford University Press.

Tattersall, I. (1998). *Becoming Human.* New York, NY: Harcourt Brace.

Tattersall, I. (2012). *Masters of the Planet.* New York, NY: Palgrave Macmillan.

Teaford, M. F. & Ungar, P. S. (2000). Diet and the evolution of the earliest ancestors. *PNAS, 97,* 13506-13511.

Teixeira, L. A. & Gasparetto, E. R. (2002). Lateral asymmetries in the development of the overarm throw. *J. Motor Behav., 34,* 151-161.

Teleki, G. (1973). *The Predatory Behavior of Wild Chimpanzees.* Lewisburg, PA: Bucknell University Press.

Thieme, H. (1997). Lower Paleolithic hunting spears from Germany. *Nature, 385,* 807-810.

Thieme, H. (2005). The Lower Paleolithic art of hunting. In: Gamble, C. & Porr, M. (Eds.), *The Hominid Individual in Context,* (115-132). New York, NY: Routledge.

Thomas, J. R. (2000). Children's control, learning, and performance of motor skills. *Res. Quart. Exerc. Sport, 71,* 1-9.

Thomas, J. R., Alderson, J. A., Thomas, K. T., Campbell, A. C. & Elliott, B. C. (2010). Developmental gender differences for overhand throwing in Aboriginal Australian children. *Res. Quart. Exerc. Sport, 81,* 432-441.

Thomas, J. R., & French, K. E. (1985). Gender differences across age in motor performance: A meta-analysis. *Psychol. Bull., 98*, 260-282.

Thomas, J. R., & Marzke, M. W. (1992). The development of gender differences in throwing: Is human evolution a factor? In: *Enhancing Human Performance in Sport: New Concepts and Developments* (60-76). Am. Acad. Phys. Ed. Papers, 25. Champaign, IL: Human Kinetics.

Thomas, J. R., Michael, D. & Gallagher, J. D. (1994). Effects of training on gender differences in overhand throwing: a brief quantitative literature analysis. *Res. Quart. Exerc. Sport, 65,* 67-71.

Thomas, J. R. & Thomas, K. T. (1988). Development of gender differences in physical activity. *Quest, 40,* 219-229.

Thomas, K. T., Gallagher, J. D. & Thomas, J. R. (2001). Motor development and skill acquisition during childhood and adolescence. In: Singer, R. N. (Ed.), *Handbook of Sport Psychology* (20-52). New York, NY: Wiley.

Thompson, M. E. & Wrangham, R. W. (2008). Diet and reproductive function in wild female chimpanzees (*Pan troglodytes schweinfurthii*) at Kibale National Park, Uganda. *Am. J. Phys. Anthrop., 135,* 171-181.

Thorpe, S. K. S. & Crompton, R. H. (2006). Orangutan positional behavior and the nature of arboreal locomotion in Hominoidea. *Am. J. Phys. Anthrop., 131,* 384-401.

Thorpe, S. K. S., Crompton R. H., Günther, M. M., Ker, R. F. & Alexander, R. M. (1999). Dimensions and moment arms of the hind- and forelimb muscles of common chimpanzees (*Pan troglodytes*). *Am. J. Phys. Anthrop., 110,* 179-199.

Thorpe, S. K. S., Holder, R. L. & Crompton, R. H. (2007a). Origin of human bipedalism as an adaptation for locomotion on flexible branches. *Science, 316,* 1328-1332.

Thorpe, S. K. S., Holder, R. L., & Crompton, R. H. (2007b). Response to Schwartz. Letters. *Science, 318,* 1065.

Thorpe, S. K. S., Holder, R. L., & Crompton, R. H. (2009). Orangutans employ unique strategies to control branch flexibility. *PNAS, 106,* 12646-12651.

Timmann, D., Citron, R., Watts, S. & Hore, J. (2001).

Increased variability in finger position occurs throughout overarm throws made by cerebellar and unskilled subjects. *J. Neurophysiol., 86,* 2690-2702.

Timmann, D., Lee, P., Watts, S. & Hore, J. (2008). Kinematics of arm joint rotations in cerebellar and unskilled subjects associated with the inability to throw fast. *Cerebellum, 7,* 366-378.

Tobias, P. V. (1964). The Olduvai Bed I hominine with special reference to its cranial capacity. *Nature, 202,* 3-4.

Tobias, P. V. (1983). Hominid evolution in Africa. *Can. J. Anthrop. 3,* 163-185.

Tocheri, M. W. (2007). Three-dimensional riddles of the radial wrist: Derived carpal and carpometacarpal joint morphology in the genus *Homo* and the implications for understanding the evolution of stone tool-related behaviors in hominins. *Ph.D. Diss.,* Tempe, AZ: Arizona State University.

Tocheri, M. W., Marzke, M. W., Liu, D., Bae, M., Jones, G. P., Williams, R. C. & Razdan, A. (2003). Functional capabilities of modern and fossil hominid hands: Three-dimensional analysis of trapezia. *Am. J. Phys. Anthrop., 122:* 101-112.

Tocheri, M. W., Orr, C. M., Jacofsky, M. C. & Marzke, M. W. (2008). The evolutionary history of the hominin hand since the last common ancestor of *Pan* and *Homo. J. Anat., 212,* 544-562.

Tooby, J. & DeVore, I. (1986). The reconstruction of hominid evolution through strategic modeling. In: Kinzey, W. G. (Ed.), *Evolution of Human Behavior: Primate Models* (183-237). New York, NY: SUNY Press.

Toriola, A. L. & Igbokwe, N. U. (1986). Age and sex differences in motor performance of pre-school Nigerian children. *J. Sports Sci., 4,* 219-227.

Toth, N. (1985). Archaeological evidence for preferential right handedness in the Lower and Middle Pleistocene, and its possible implications. *J. Hum. Evol., 14,* 607-614.

Toth, N. (1987). Behavioral inferences from early stone artifact assemblages: An experimental model. *J. Hum. Evol. 16,* 763-787.

Toth, N. (1997). The artefact assemblages in the light of

experimental studies. In: Isaac, G. L. & Isaac, B. (Eds.), *Koobi Fora Resarch Project, Vol. 5. Plio-Pleistocene Archeology* (363-388). Oxford, UK: Clarendon Press.

Toyoshima, S., Hoshikawa, T., Miyashita, M. & Oguri, T. (1974). Contribution of the body parts to throwing performance. In: Nelson, R. C. & Morehouse, C. (Eds.), *Biomechanics IV; Proceedings* (169-174). Baltimore, MD: University Park Press.

Toyoshima, S. & Miyashita, M. (1973). Force-velocity relation in throwing. *Res. Quart.*, 44, 86-95.

Trinkaus, E. (1977). A functional interpretation of the axillary border of the Neandertal scapula. *J. Hum. Evol.*, 6, 231-234.

Trinkaus, E. (1983). *The Shanidar Neandertals*. New York, NY: Academic Press.

Trinkaus, E. (1984). Does KNM-ER 1481 establish *Homo erectus* at 2.0 myr BP? *Am. J. Phys. Anthrop.*, 64, 137-139.

Trinkaus, E. (1986). The Neandertals and modern human origins. *Ann. Rev. Anthrop.*, 15, 193- 218.

Trinkaus, E. (1997). Appendicular robusticity and the paleobiology of modern human emergence. *PNAS*, 94, 11367-13373.

Trinkaus, E. & Churchill, S. E. (1999). Diaphyseal cross-sectional geometry of Near Eastern Middle Paleolithic Humans: The humerus. *J. Archaeol. Sci.* 26, 173-184.

Trinkaus, E., Churchill, S. E. & Ruff, C. B. (1994). Postcranial robusticity in *Homo*. II: Humeral bilateral asymmetry and bone plasticity. *Am. J. Phys. Anthrop.*, 93, 1-34.

Trinkaus, E. & Hilton, C. E. (1996). Neandertal pedal proximal phalanges: diaphyseal loading patterns. *J. Hum. Evol.*, 30, 399-425.

Trinkaus, E. & Shipman, P. (1993). *The Neandertals*. New York, NY: Alfred A. Knopf.

Tullos, H. S. & King, J. W. (1973). Throwing mechanism in sports. *Orthop. Clinics North Am.*, 4, 709-720.

Tuttle, R. H. (1967). Knuckle-walking and the evolution of hominoid hands. *Am. J. Phys. Anthrop.*, 26, 171-206.

Tuttle, R. H. (1974). Darwin's apes, dental apes, and the descent of man: Normal science in evolutionary anthropology. *Curr. Anthrop.*, 15, 389-426.

Tuttle, R. H. (1981). Evolution of hominid bipedalism and prehensile capabilities. *Phil. Trans. Roy. Soc. Lond. B, 292,* 89-94.

Tuttle, R. H., Basmajian, J. V. & Ishida, H. (1979). Activities of pongid thigh muscles during bipedal behavior. *Am. J. Phys. Anthrop., 50,* 123-136.

Tuttle, R. H., Webb, D. M., & Tuttle, N. I. (1991). Laetoli footprint trails and the evolution of hominid bipedalism. In: Coppens, Y. & Senut, B. (Eds.), *Origine(s) de la Bipédie Bipédie chez les Hominidés* (187-198). Paris, FR: Cahiers de Paléoanthropologie, Editions du CNRS.

Ulrich, B. D. & Ulrich, D. A. (1985). The role of balancing ability in performance of fundamental motor skills in 3,- 4,- and 5-year-old children. In: Clark, J. E. & Humphrey, J. H. (Eds.), *Motor Development: Current Selected Research, Vol. 1* (87-97). Princeton, NJ: Princeton Book. Co.

Ungar, P. S. & Sponheimer, M. (2011). The diets of early hominins. *Science, 334,* 190-193.

Vallortigara, G. & Rogers, L. J. (2005). Survival with an asymmetrical brain: Advantages and disadvantages of cerebral lateralization. *Behav. Brain Sci., 28,* 575-633.

van den Tillar, R. & Ettema, G. (2004). A force-velocity relationship and coordination patterns in overarm throwing. *J. Sports Sci. Med., 3,* 211-219.

Van Lawick, H. (1991). *People of the Forest: The Chimps of Gombe.* Bethesda, MD: The Discovery Channel Video Library.

Van Lawick-Goodall, J. (1968). The behaviour of free-living chimpanzees in the Gombe Stream Reserve. *Animal Behav. Monogr. 1,* 161-311.

Van Lawick-Goodall, J. (1975). The Behaviour of the Chimpanzee. In: Kurth, G. & Eibl-Eibesfeldt, I. (Eds.), *Hominisation and Behavior* (74-136). Stuttgart, DE: Gustav Fischer.

Van Valkenburgh, B. (2001). The dog-eat-dog world of carnivores. In: Stanford, C. B. & Bunn, H. T. (Eds.), *Meat-Eating and Human Evolution* (101-121). NewYork, NY: Oxford University Press.

Vasey, N., & Walker, A. (2001). Neonate body size and

hominid carnivory. In: Stanford, C. B. & Bunn, H. T. (Eds.), *Meat-Eating and Human Evolution* (332-349). New York, NY: Oxford University Press.

Vekua, A., Lordkipanidze, D., Rightmire, G. P., Agusti, J., Ferring, R., Maisuradze, G., Mouskhelishvili, A., Nioradze, M., Ponce de Leon, M., Tappen, M., Tvaichrelidze, M. & Zollikofer, C. (2002). A new skull of early *Homo* from Dmanisi, Georgia. *Science, 297,* 85-89.

Videan, E. N. & McGrew, W. C. (2001). Are bonobos (*Pan paniscus*) really more bipedal than chimpanzees (*Pan troglodytes*)? *Am. J. Primatol., 54,* 233-239.

Vlcek, E. (1973). Postcranial skeleton of a Neandertal child from Kiik-Koba, U.S.S.R. *J. Hum. Evol., 2,* 537-544.

Waldstein, D. (2012). Built for the gridiron, at home on the mound. *N.Y.Times, 27 Feb,* D1, D4.

Walker, A. (1993). Perspectives on the Nariokotome discovery. In: Walker, A. & Leakey, R. (Eds.), *The Nariokotome Homo Erectus Skeleton* (411-430). Cambridge, MA: Harvard University Press.

Walker, A. & Leakey, R. (1993). The postcranial bones. In: Walker, A. & Leakey, R. (Eds.), *The Nariokotome Homo Erectus Skeleton,* (95-160). Cambridge, MA: Harvard University Press.

Walker, A. & Shipman, P. (1997). T*he Wisdom of the Bones*. New York, NY: Vintage Books.

Walker, A. & Shipman, P. (2005). *The Ape in the Tree.* Cambridge, MA: Harvard University Press.

Walker, J., Cliff, R. A. & Latham, A. G. (2006). U-Pb isotopic age of the StW 573 hominid from Sterkfontein, South Africa. *Science, 314,* 1592-1594.

Ward, C. V. (2002). Interpreting the posture and locomotion of *Australopithecus afarensis*: Where do we stand? *Yearb. Phys. Anthrop. 45,* 185-215.

Ward, C. V. (2007). Postcranial and locomotor adaptations of hominoids. In: Henke, W. & Tattersall, I. (Eds.), *Handbook of Paleoanthropology, Vol. III, Part 2* (1011-1030). Berlin, DE: Springer.

Ward, C. V., Kimbel, W. H., & Johanson, D. C. (2011). Complete fourth metatarsal and arches in the foot of *Australopithecus afarensis. Science, 331,* 750-753.

Ward, C. V., Leakey, M. G., Brown, B., Brown, F., Harris, J. & Walker, A. (1999a). South Turkwel: A new Pliocene hominid site in Kenya. *J. Hum. Evol., 36,* 69-95.

Ward, C. V., Leakey, M. G. & Walker, A. (1999b). The new hominid species *Australopithecus anamensis. Evol. Anthrop., 7,* 197-205.

Ward, C. V., Leakey, M. G. & Walker, A. (2001). Morphology of *Australopithecus anamensis* from Kanapoi and Allia Bay, Kenya. *J. Hum. Evol., 41,* 255-368.

Ward, C. V., Plavcan, J. M. & Manthi, F. K. (2010). Anterior dental evolution in the *Australopithecus anamensis-afarensis* lineage. *Phil. Trans. Roy. Soc. B, 365,* 3333-3344.

Washburn, S. L. (1950). The analysis of primate evolution with particular reference to the origin of man. *Cold Spring Harbor Symp. Quant. Biol., 15,* 67-77.

Washburn, S. L. (1967). Behaviour and the origin of man. *Proc. Roy. Anthrop. Inst. Great Britain & Ireland for 1967,* 21-27.

Washburn, S. L. & Howell, F. C. (1960). Human evolution and culture. In Tax. S. (Ed.), *After Darwin, Vol. 2* (33-56). Chicago, IL: University of Chicago Press.

Washburn, S. L. & Lancaster, C. S. (1968). The evolution of hunting. In: Lee, B., DeVore, I. (Eds.), *Man the Hunter* (293-303). Chicago, IL: Aldine Publishing Company.

Washburn, S. L. & Moore, R. (1974). *Ape Into Man.* Boston, MA: Little Brown & Company.

Watanabe, H. (1968). Subsistence and ecology of northern food gatherers with special reference to the Ainu. In: Lee, R. B. & DeVore, I. (Eds.), *Man the Hunter* (69-77). Chicago, IL: Aldine Publishing Company.

Watkins, R. G., Dennis, S., Dillin, W. H., Schnebel, B., Schneiderman, G., Jobe, F., Farfan, H., Perry, J. & Pink, M. (1989). Dynamic EMG analysis of torque transfer in professional baseball pitchers. *Spine, 14,* 404-408.

Watts, S., Pessotto, I. & Hore, J. (2004). A simple rule for controlling overarm throws to different targets. *Exp. Brain Res., 59,* 329-339.

Weidenreich, F. (1939). The duration of life of fossil man in China and the pathological lesions found in his skeleton.

Chinese Med. J., 55, 34-44.
Weidenreich, F. (1943). *The Skull of Sinanthropus Pekinensis.* New York, NY: Hafner.
Weidenreich, F. (1947). The trend of human evolution. *Evolution, 1,* 221-236.
Welch, C. .M, Banks, S. A., Cook, F. F. & Draovitch, P. (1995). Hitting a baseball: a biomechanical description. *J. Orthop. Sports Phys. Therapy, 22,* 193-201.
Wells, C. L. (1991). *Women, Sport, & Performance* (2nd Edit.). Champaign, IL: Human Kinetics Books.
Werner, S. L., Fleisig, G. S., Dillman, C. J. & Andrews, J. R. (1993). Biomechanics of the elbow during baseball pitching. *J. Orthop. Sports Phys. Therapy, 17,* 274-278.
Werner, S. L., Gill, T. J., Murray, T A., Cook, T. D. & Hawkins, R. J. (2001). Relationships between throwing mechanics and shoulder distraction in professional baseball pitchers. *Am. J. Sports Med., 29:* 354-358.
Weyand, P. G. & Davis, J. A. (2005). Running performance has a structural basis. *J. Exp. Biol., 14,* 2625-2636.
Wheeler, P. E. (1993). The influence of stature and body form on hominid energy and water budgets: a comparison of *Australo pithecus* and early *Homo* physiques. *J. Hum. Evol., 24,* 13-28.
White, T. D. (1980). Evolutionary implications of Pliocene hominid footprints. *Science, 208,* 175-176.
White, T. D., Asfaw, B., Beyene, Y., Haile-Selassie, Y., Lovejoy, C. O., Suwa, G. & WoldeGabriel, G. (2009). *Ardipithecus ramidus* and the paleobiology of early hominids. *Science, 326,* 75-85.
White, T. D., Johanson, D. C. & Kimbel, W. H. (1981). *Australopithecus africanus*: its phyletic position reconsidered. *S. Afr. J. Sci., 77,* 445-470.
White, T. D., Suwa, G. & Asfaw, B. (1994). *Australopithecus ramidus,* a new species of early hominid from Aramis, Ethiopia. *Nature, 371, 306-312.* Corrigendum (1995). *371,* 88.
White, T. D., Suwa, G., Hart, W. K., Walter, R. C., Woldegabriel, G., de Heinzelin, J., Clark, J. D., Asfaw, B., & Vrba, E. S. (1993). New discoveries of *Australopithecus* at Maka in Ethiopia. *Nature, 366,* 261-265.

REFERENCES CITED

White, T. D., WoldeGabriel, G., Asfaw, B., Ambrose, S., Beyene, Y., Bernor, R. L., Boisserie, J.-R., Currie, B., Gilbert, H., Haile-Selassie, Y., Hart, W. K., Hlusko, L. J., Howell, F. C., Kono, R. T., Lehmann, T., Louchart, A., Lovejoy, C. O., Renne, P. R., Saegusa, H.,Vrba, E. S., Wesselman, H. & Suwa , G. (2006). Asa Issie, Aramis and the origin of *Australopithecus. Nature, 440,* 883-889.

Whittaker, J. C. (1994). *Flintknapping.* Austin, TX: University of Texas Press.

Wickstrom, R. L. (1977). *Fundamental Motor Patterns.* Philadelphia, PA: Lea & Febiger.

Wild, M. R. (1938). The behavior pattern of throwing and some observations concerning its course of development in children. *Res. Quart. 9,* 20-24.

Wile, I. S. (1934). *Handedness: Right and Left.* Boston, MA: Lothrop, Lee and Shepard.

Wilk, K. E., Meister, K., Fleisig, G. S. & Andrews, J. R. (2000). Biomechanics of the overhead throwing motion. *Sports Med. Arthroscopy Rev., 8,* 124-134.

Wilkins, J., Schoville, B. J., Brown, K. S. & Chazan, M. (2012). Evidence for early hafted hunting technology. *Science, 338,* 942-946.

Williams, E. M., Gordon, A. D. & Richmond, B. G. (2010). Upper limb kinematics and the role of the wrist during stone tool production. *Am. J. Phys. Anthrop, 143,* 134-145.

Williams, E. M., Gordon, A. D. & Richmond, B. G. (2012). Hand pressure distribution during Oldowan stone tool production. *J. Hum. Evol., 62,* 520-532.

Williams, K., Haywood, K. & VanSant, A. (1998). Changes in throwing by older adults: a longitudinal investigation. *Res. Quart. Exerc. Sport, 69,* 1-10.

Willoughby, P. R. (1985). Spheroids and battered stones in the African Early Stone Age. *World Archaeol. 17,* 44-60.

Willoughby, P. R. (1987). *Spheroids and Battered Stones in the African Early and Middle Stone Age.* Oxford, UK: Cambridge Monographs in African Archaeology 17. BAR International Series 321.

Wolfe, S. W., Crisco, J. J., Orr, C. M. & Marzke, M. W. (2006). The dart-throwing motion of the wrist: Is it unique to humans? *J. Hand Surg., 31,* 1429-1437.

Wolpoff, M. H. (1976). Some aspects of the evolution of early hominid sexual dimorphism. *Curr. Anthrop., 17,* 579-606.

Wood, B. (1992). Origin and evolution of the genus *Homo*. *Nature, 355,* 783-790.

Wood, B. (2010). Reconstructing human evolution: Achievements, challenges and opportunities. *PNAS, 107,* Suppl. 2, 8902-8909.

Wood, B. (2011). Did early *Homo* migrate "out of" or "in to" Africa? *PNAS, 108,* 10375-10376.

Wood, B. & Collard, M. (1999). The human genus. *Science, 28,* 65-71.

Wood, B. & Harrison, T. (2011). The evolutionary context of the first hominins. *Nature, 470,* 347-352.

Wood, B. & Lonergan, N. (2008). The hominin fossil record: taxa, grades and clades. *J. Anat., 212,* 354-376.

Wood, C. J. & Aggleton, J. P. (1989). Handedness in "fast ball" sports: Do left-handers have an innate advantage? *Brit. J. Psychol., 80,* 227-240.

Woodhead, J. D., Hellsltrom, J. C. & Berger, L. R. (2011). *Australopithecus sediba* at 1.977 Ma and implications for the origins of the genus *Homo*. *Science, 333,* 1421-1423.

Wrangham R. W. (1980). Bipedal locomotion as a feeding adaptation in Gelada baboons, and its implications for hominid evolution. *J. Hum. Evol. 9,* 329-331.

Wrangham, R. W. (1999). Evolution of coalitionary killing. *Yearb. Phys. Anthrop., 42,* 1-30.

Wrangham, R. W. (2001). Out of the *Pan*, into the fire. In: De Waal, F. B. M. (Ed.), *Tree of Origin* (119-143). Cambridge, MA: Harvard University Press.

Wrangham, R. (2009). *Catching Fire*. New York, NY: Basic Books.

Wrangham, R. & Peterson, D. (1996). *Demonic Males*. New York, NY: Houghton Mifflin Company.

Yamazaki, N., Ishida, H., Kimura, T. & Okada, M. (1979). Biomechanical analysis of primate bipedal walking by computer simulation. *J. Hum. Evol., 8,* 337-349.

Young, R. W. (2003). Evolution of the human hand: the role of throwing and clubbing. *J. Anat., 202,* 165-174.

Young, R. W. (2009). The ontogeny of throwing and striking. *Hum. Ontogenetics, 3,* 1-13.

Young, R. W. (2010). Hominin evolution: Genetics and be-behavior. In Osborne, M. A. (Ed.), *Advances in Genetics Research, Vol. 1*, (1-77). Hauppauge, NY: Nova Science Publishers.https://www.novapublishers.com/catalog/advanced search result?keywords=Hominin

Zackowski, K. M., Thach, W. T. Jr. & Bastian, A. J. (2002). Cerebellar subjects show impaired coupling of reach and grasp movements. *Exp. Brain Res., 146,* 511-522.

Zheng, N., Barrentine, S. W., Fleisig, G. S. & Andrews, J. R. (2008). Swing kinematics for male and female pro golfers. *Int. J. Sports Med. 29,* 487-493.

Zihlman, A. L. (1978). Interpretations of early hominid locomotion. In: Holly, C. J. (Ed.), *Early Hominids of Africa* (363-377). London, UK: Duckworth.

Zihlman, A. L. (1981). Women as shapers of the human adaptation. In: Dahlberg, F. (Ed.), *Woman the Gatherer* (75-120). New Haven, CT: Yale University Press.

Zihlman, A. L. & Brunker, L. (1979). Hominid bipedalism: Then and now. *Yearb. Phys. Anthrop., 22,* 132-162.

Zihlman, A. L. & Hunter, W. S. (1972). A biomechanical interpretation of the pelvis of *Australopithecus. Folia Primatol., 18,* 1-19.

Zipfel, B., DeSilva, J. M. & Kidd, R. S. (2009). Earliest complete hominin fifth metatarsal—Implications for the evolution of the lateral column of the foot. *Am. J. Phys. Anthrop., 140,* 532-545.

Zollikofer, C. P. E., Ponce de León, M. S., Liegerman, D. E., Guy, F., Pilbeam, D., Likius, A., Mackayer, H. T., Vignaud, P. & Brunet,M. (2005). Virtual cranial reconstruction of *Sahelanthropus tschadensis. Nature, 434,* 755-759.

INDEX

Acceleration distance, 82, 174
Accuracy
  of throwing, 120, 181, 183, 232, 233, 236
  and finger opening, 121
  role of finger opening, 233
Acheulian tools, 58
Adaptation, evolutionary, 9, 110
  for fighting in males, 45
  of human hand, 260
  requirement of reproductive benefits, 256
  to arborealism, 85
  in Orrorin, 123
  lost in early *Homo*, 71
  in pre-*Homo*, 74
  to arborealism in early hominins, 71
  to bipedal weapons use, 55, 94
  to child-bearing in females, 45
  to cooked food, 76
  to gripping missiles and clubs, 112
  to throwing, 243
  to throwing and clubbing, 122, 256, 268
  to throwing, in brain, 230, 235
  to weapons use, 40
Adaptive suite, 244-247
African apes, 12
Aggression, male
  towards strangers, 20
Aggressiveness
  higher in boys than girls, 44

A./H. habilis, 48, 58, 61, 62, 65, 67, 68, 70, 72, 74, 98
  and bipedalism, 213
  bipedal on ground, 70
  arborealism, 129
  definition, 57
  hand bones, 128
  humero/femoral index, 69
  robusticity, 167
  size dimorphism, 48
  small body size, 69
Ain Hanech,32, 34, 35
A.L. 288-1, Lucy, 48, 68
Allia Bay, 125
Ambrona, 105
Anatomy, ancestral
  divergence from, 53
Ancestor, common
  chimpanzees and hominins, 11, 15, 16, 74, 106
  panins and hominins, 12
Androgens, 44
Animal tissues, 81, 84, 86, 98
  acquisition of, 86, 95, 108
  and use of weapons, 87
  and reduction of chewing apparatus, 94
  and transition to *Homo*, 87
Animals, naive, 107
Anthropoids, 51
Aramis, 124, 210
Arboreal heritage, 74
Arborealism, 11, 15, 17, 65, 74, 85
  abandoned by *Homo*, 53, 77, 80, 86, 95
  and curved phalanges, 125
  loss of, 76

INDEX

pre-*Homo*, 54
Archaeological sites, 30, 31
Archeologists, 29
Ardipithecus, 67
Ardipithecus kadabba, 13
   and bipedalism, 210
   canine teeth, 156
   hand bones, 124
Ardipithecus ramidus, 68, 74
   and bipedalism, 210, 211
   arborealism, 125
   canine teeth, 156
   flexor pollicis longus, 124
   foot, 210, 211
   habitat, 85
   hand, 111
   hand bones, 124
   handgrips, 125
   humero/femoral index, 68
   molars, 71
   pelvis, 59, 210
   robusticity, 125, 166
   size dimorphism, 47
   thumb, 124
   vertebral column, 210
Arm-hanging, 66, 74, 113
Atapuerca, 168
Australopithecine(s), 54, 56, 57, 63, 65-67, 71, 84
   and hunting, 103
   food, 84
   robust, 72
Australopithecus, 57, 59, 60, 63, 71, 83, 161
   anatomical conservatism, 54
   and arborealism, 85
   and short lumbar spine, 90
   and small body size, 59
   hip width, 66
   hunting with stones and bones, 104
   robust upper limbs, 70
   size of, 67
   thoracic and lumbar region, 66
Australopithecus afarensis, 40, 47, 48, 50, 56, 58, 62, 68, 64, 66, 71, 74, 99
   and bipedalism, 211, 212
   apical tufts, 127
   canine teeth, 156
   dimorphism paradox, 50
   foot, footprints, 211
   habitat, 85
   hand bonesand grips, 126
   humero/femoral index, 68
   hunting by, 103
   robusticity, 166
   size dimorphism, 47
Australopithecus africanus, 11, 48, 49, 56, 62, 64
   and bipedalism, 212
   arborealism, 127
   canine teeth, 157
   hand bones, 127
   humero/fermoral index, 69
   robusticity, 167
   styloid process, 127
Australopithecus anamensis, 13, 47, 48, 54, 61, 63, 71
   and bipedalism, 211
   canine teeth, 156
   habitat, 85
   hand bones, 125
   robusticity, 166
   tooth enamel, 163
Australopithecus garhi, 56, 70, 71, 98, 100. 103, 157, 212
   pedal phalanx, 212
Australopithecus habilis, 75
Australopithecus robustus (Paranthro-

INDEX

pus), 48, 49, 54, 62, 71, 167
   and bipedalism, 213
   apical tufts, 129
   canine teeth, 157
   hand bones, grips, 129
   size dimorphism, 48

Australopithecus sediba, 56, 65, 66, 128
   and bipedalism, 213
   arborealism, 128
   canine teeth, 157
   hand bones, 128
   robusticity, 167

Back force
   during throwing, 122
   missile on hand, 121
   of club on hand, 174
   of ball on fingers, 232
   of missile on third finger, 121

Balance, upright, 66, 181, 195, 199
   during throwing and striking, 196, 237
   dynamic, 208, 221, 223, 225, 226, 238, 258
   in chimpanzees, 149, 206
   lateral, 209
   static, 207
   static and dynamic, 184

Bauplan, 58
Behavior, bipedal, 54
Bipedal use of weapons theory
   access to food, mates and safe habitats, 257, 258
   acquisition of animal food, 252, 261, 265
   assumptions, 250
   bipedal locomotion, 253, 258, 259
   canine tooth diminution, 253, 270
   causal effects, 251, 260
   change in arm and leg strength, 264
   change in relative forearm length, 265
   chewing apparatus diminution, 253, 265
   closing remarks, 271
   CNS control of throwing and clubbing, 252, 268-270
   congruent with Darwinian theory, 253
   differences between bipedal walking and use of weapons, 259
   evolution of the hand, 251, 260
   explanatory principle, 257
   gender disparity in throwing and clubbing, 253, 256, 268
   gender size difference, 261, 262
   handedness, 251, 266
   hominin origins, 251, 258
   human origins, 258
   increase in *Homo* size, 261
   increase in shoulder mobility, 262, 263
   major strengths, 256
   meets requirements of a scientific theory, 253
   modern human hand anatomy, 260
   modern human handgrips, 260
   modern human throwing and clubbing, 268, 252
   musculoskeletal robusticity, 252, 266
   natural selection, 258, 259
   ontogeny of throwing and clubbing, 251, 267
   origin of waist in *Homo*, 263
   reorientation of shoulder joint, 263

reproductive benefits, 256-258
settling disputes over scarce resources, 256
simple explanation, broad scope, 270
theoretical principles, 250
theoretical scope, 251
thorax change in shape, 262
transition to *Homo*, 252, 261
transition to walking, 258

Bipedalism. *See also* **Locomotion, bipedal** and **Weapons, bipedal use**
dissimularity of walking and bipedal use of weapons, 221
in fossil hominins, 209, 210, 211, 212, 213
origin of, 224, 257
similarity of walking and bipedal use of weapons, 221
simultaneous evolution bipedal locomotion and use of weapons, 224, 227
terrestrial, 57, 85
two forms of, 219
which bipedal behavior came first?, 224

Body mass
calculating in fossils, 47
increased in *H. erectus*, 61

Body size, 39, 53, 63, 64, 74, 76
and brain size in *H. erectus*, 63
and cost of locomotion, 91
and throwing and clubbing, 171-176
physical principles, 174
dimorphism, 47
increase in *Homo*, 48, 59, 61, 67, 69, 74-79, 86, 87, 91, 97, 261, 262

gender difference, 87
increase, male hominins, 45, 87
stable in *Australopithecus*, 54
variation in *habilis* 57

Body size, canine size
dimorphism 49

Body size, large
advantages for fighting, and weapons use, 78
advantages, females, 79
and nutrition, 81
disadvantages for locomotion, 78
energy cost, females, 78
gender effect, 79
need of more food, 78

Body type
and running, throwing, clubbing, 175

Body, upper
strong in males, 45

Bonobos, 46, 67-69
Bouri, 100, 103
Bovids, 100, 101
Boxgrove, 105
Brachial index 70, 71, 94
and throwing, 230, 235
energy requirements, 81
size, 54, 56-59, 62-64, 81

Brain. *See also* **Central nervous system.**

Brain/body weight relationship, 63
Butchery, 99, 101, 103
Calories, 75, 80, 81
Canine teeth, 47, 49-51, 54, 71, 242-244, 246
in anthropoids, 154
and hominin status, 153
change in size and shape, 153, 157, 162
chimpanzee male in figure 10, p. 155

diminution in size, 161
functional honing, 155
gender size dimorphism, 154, 162
in anthropoids, chimpanzees and humans, 154
fossil evidence, 154
Carnivore guild, 107
Carnivore interaction
less at hominin sites, 102
Carnivores, 24, 26, 80, 86, 99, 100, 101, 102, 103, 107
Carnivory, 81, 99, 100, 308, 349
and transition to *Homo*, 97
Caucasus, 58
Cave of Hearths, 33, 35
Central nervous system (CNS).
control of dominant and nondominant arm, 235
control of hand and arm during throwing, 231
control of throwing, 189, 229, 230, 231, 233
complexity of, 237
computational power required, 238
conscious and unconscious mechanisms, 237
output of motor commands, 234, 235
vision and proprioception, 233
internal model of movement dynamics, 235
prediction of time to release missile, 235
torques, predicted and controlled by, 233
Cerebellum
and throwing, 233
precision in timing, 234
Cervids, 101

Chewing
reduction by cooking, 76
Chewing apparatus
diminution in *Homo*, 71, 84, 94
enlarged in australopithecines, 84
Children
nurturing of, 87
Chimpanzees, 7, 20, 46, 48, 49, 60, 66, 69, 70, 103, 106
bipedalism, 206
hand, 111
figure 4, p. 112
hand bones
figure 5, p.114
and muscles, 111
hand phalanges, 111, 124
hand, arboreal adaptation, 111
knuckle-walking, 111
lack of weapons use, 20
large upper bodies, 67, 70
thoracic and lumbar region, 66
thumb, 111, 113
thumb and arborealism, 113
wrist flexion and extension, 111
Chimpanzees, common
Pan troglodytes, 20
Climate, 6, 69, 85
in Africa, 8-3 Mya, 54
Climbing
vertical supports, 113
Clubbing, 22, *See also* **Striking**
overarm or sidearm, 89
Clubbing motion. *See* **Striking motion**
Clubbing, development of. *See* **Striking, development**
Clubhandles

INDEX

smooth, cylindrical, 30
Clubs
  and chimpanzees, 21
  axes, maces, swords, 30
  bones, 30
  characteristics, 30
  cylindrical handles, 30
  cyndroidal handles, 109, 123
Club-swinging, 10, 73, *See also* **Striking**
  bipedal stance, 30
Coalition aggression, 106
Coalitions, male
  in chimpanzees, 16
  in hominins, 24, 27, 95
Cobbles
  naturally weathered, 31
  stream worn, 32
  water-rounded, 31
Common ancestor
  hominins and chimpanzees, 15
Competition
  male-male, 16
Conspecifics, 14
  and biological needs, 24
  belligerant, 88
Cooking, 73, 75, 76, 80-82, 84
  beginning of, 75, 76
  benefits of, 76
  digestive system changes, 77
  reduced chewing apparatus, 94
Cooperation
  in child care, 79
  in hunting, 106, 107
Cores, *36*, 37
  core tools, flakes, 31
  knapping flakes from, 103
Cranial capacity
  and body mass, 64
  in *H. habilis*, 57
  in hominins, 63

Darwinian explanation, 106, 256
Darwinian theory of evolution, 3, 241, 249, 253, 255
Dentition
  changes, 71, 84, 94, 158
Dexterity, manual, 109
Diet, 72, 78, 84, 107
  and chewing, 94
  and tooth structure, 163
  change in, 94, 158
  in *A. robustus*, 54
  in chimpanzees, 21
  requiring heavy chewing, 161
  tough, abrasive, 158
  vegetarian, 98
Dikika, 99, 103
Dimorphism
  sexual (gender) size, 16
Divergence
  hominin, panin, 13, 14
Dmanisi, 32, 58, 102, 168, 213
Dmanisi hominins, 59
  bipedalism, 213
  canine teeth, 157
  early *H. erectus*, 67
  foot, 214
  glenoid fossa, 65
  hallux, 58
  meat acquisition, 100
  robusticity, 168
  small brains, 58
  small size, 67
Dominance hierarchy
  and food access, 20
  and use of weapons, 24
  and access to females, 24
  and access to food, 24
Dominance status, 16
East Africa, 56, 58, 61
Elephants, 101, 108
Encephalization

INDEX

quotient
in *H. Heidelburgensis*, 64
Endurance running, 170
Energy
dietary, 75
transfer to shoulder, 85
in *H. erectus* females, 78, 79
Energy, kinetic, 197
applied to target, 122
generation, transfer to weapon, 93
linkage 196, 232
transmitted by second and third fingers, 232
transmitted to weapon, 89, 90, 93, 94, 121, 122, 222
Enriched habitats
access to, 6, 10
Environment, 5, 6, 13, 26, 29 36, 78, 85, 103
Equids, 100, 101, 105
Estrogen, 45
Ethnographic studies, 30
Evidence, 1, 3, 11, 13, 63
anatomical, 116
biomechanical, 92, 186, 193
chronological, 55
comparative anatomical, 45
comparative anatomy, 11, 111, 154
comparative hand use, 140
developmental, 180, 198
of clubbing, 30
ethnographic, 123
ethological, 139
experimental, 123, 216
fossil, 12, 13, 14, 16, 30, 39, 46, 47, 49, 60
fossil and molecular, 12, 14
from modern humans, 25
lack of archaeological, 29
morphological, 40, 48, 49, 58, 126, 156, 168, 212
neurophysiological, 230
objective, 3
of access to meat, 87
of antemortem trauma, 105
of butchery, 99
of aggressiver scavenging, 99
of controlled fire, 76, 77
of cooking effects, 94
of hominin carnivory, 99
of hunting, 99
of locomotion
bipedal and quadrupedal, 91
of meat-eating, 98
of stone tool-making, 109
of throwing and clubbing, 109
of weapons in environment, 36
paleontological, 55, 123, 154, 165, 166, 209
paleontological, 59
paleontological and archaeological, 81
phytolithic, 104
scientific, 3
transition to *Homo*, 53, 55
Evidence, explanation, 3, 40
Evidence, paleontological
first appearance of *Homo*, 55
Evolution, human, 2, 6, 27, 39
Evolutionary adaptation
to throwing and clubbing, 85
Exaptation,
definition, 133
of hand for stone tool-making, 133
for modern uses, 133

INDEX

of throwing motion
for stone tool-making, 134
Explanation
by adaptation to throwing
and clubbing, 61
by natural selection for
throwing and clubbing, 9
by practice of cooking, 76
of carpometacarpal joint
two close-packed
positions, 121
of central nervous system
control of throwing, 229
evolutionary, 51, 88, 110,
172, 198 ultimate cause,
180
for acquisition of meat, 99,
102-104, 105-108
for canine change in shape,
160
for hominin hand
evolution, 110, 135, 136
for longer, stronger legs in
Homo, 93
for reduction in hand size,
94
for reduction in size of
upper limbs in Homo, 93
for reduction of brachial
index in Homo, 94
for stone tool cut marks,
101
for transition to Homo, 95
for acquisition of meat,
103
hand, stone tool-making,
129
problems with, 130, 131
modern human hand
anatomy, 116
of abandoning
arborealism, 95
of bipedal locomotion,
220, 225, 226
of bipedalism, 224

of body size increase, 86-88
in Homo females, 87
of body size increase in
Homo, 77, 78
of canine diminution, 158,
161
of change in shape of
thorax, 82
of change in thorax shape,
88
of chest and shoulder
anatomy, 82
of early onset of throwing
and striking, 201
of extension of fingers,
wrist, 122
of gender differences in
throwing
and clubbing, 200, 201
of gender size dimorphism,
39, 41
genetic element in
throwing, 199
of glenoid fossa
reorientation, 89
of handedness, 147, 149
of hominin hand evolution,
136
of hominin
musculoskeletal
robusticity, 172, 175
of hominin origins, 19, 23,
24, 27, 251
of shoulder joint mobility,
88
of innate throwing
ontogeny, 198
of innate throwing,
clubbing motions, 199
of lower limb size in Homo,
90, 91
of modern human
handgrips, 116
of modern human

throwing and clubbing
prowess 180, 199, 202
of ontogeny of throwing,
clubbing, 198, 200, 201
of persisting left-
handedness, 150
of rotation of fingers
during flexion, 122
of size change in upper and
lower limbs in Homo, 92
of size dimorphism, 51
diminution in Homo, 79
of spheroid function, 35
of the opposable thumb,
119, 122
of the transition to Homo,
73, 77, 85
of the waist in Homo, 89,
90
of throwing prowess
in modern human females,
87
of two weapons categories
123
of upper limb size in Homo,
92
ontogeny of bipedal
throwing and striking, 200
throwing and clubbing, 109
Explanation, nature of
adaptive suite, 244
adaptive suite and hominin
evolution, 245
adaptive suite, critique of,
246, 247
and assumptions, 250
and formalization, 249
and prediction, 248
biological, 240
causal factors, 240
Darwinian, 240
different forms of, 240
evidence, 239
evolutionary, 239, 247
general rules of, 240, 257

goal of, 240
hypotheses and theories,
96, 241, 247
scenarios, 242
structure of, 240, 246, 248
theory and parsimony,
247, 257
theory, scope of, 248, 257
theory-based hypotheses,
241
ultimate causes, 241
Extinction, 7, 8, 12, 72,
80, 86, 95
Facts
verified evidence, 3
Family, 242
Felids, 101
Femoral/tibial index,
58
Fighting, 42, 87
and canine teeth, 154, 155
and gender dimorphism,
43
and handedness, 143, 150
avoided by females, 23
by chimpanzee males, 20,
22
by males in hominins, 22,
40. 41, 45, 48, 51, 86, 151,
159, 200, 201
cessation of, 159
chimpanzee male
behavior, 20
decrease in, 176
for scarce resources, 78
with carnivores, 86
with enemies, 159
with forelimbs, 221
with weapons, 51, 86

Finger robusticity
shift in, 112
figure 5, p. 114
Fingers
and missile release, 121

curvature, 11, 14, 15, 17, 57, 74
and arborealism, 113
reduction of curvature, 54
rotation when flexed, 115, 122
Fingertip pads, 119
grip of spheroidal objects, 122
Fire, control of, 75-77, 243
Flakes, 36
Flight distance, 107
Food, 80
access to, 6, 10, 19, 20, 24
and tooth anatomy, 163
and vested provisioning, 245
animals, 76
for sex in *Ardipithecus*, 245
for sex in *H. erectus*, 79
need for more in *Homo*, 88
nutritious, 25-27, 80, 86, 95
plants, 75, 80
provisioning, 243
sharing, 21, 242, 243
Foot strike, in throwing, 93
Force, 87, *See also* **Back forces**
applied to missile or club, 174
muscular, 174
of club impact, 122
on missile by hand, 121
Force-velocity curve, 175
Forearm during throwing, 121
at missile release, 89 figure 12, p. 190
in *H. erectus*, 94
in chimpanzees, 92
Forearms
and brachial index, 94 61, 70
length diminished in *H. erectus*, 74

long in *Australopithecus*, 56
long in pre-*Homo* hominins, 71
reduced mass in *Homo*, 92
Forests
gallery, 85
Gender differences
in behavior, 22, 23
Genes, 4, 9, 26, 87, 89, 106
Genome, 5, 8, 69
Genotype, 5
Genus, 9
definitions of, 8, 57
Gestation, 23, 25, 78, 79, 88, 201
Giraffids, 100, 101, 108
Glenoid fossa. *See also* **Shoulder joint**
change in orientation, 83
in Homo, 65
in throwing, 189
orientation, 83
orientation during throwing
figure 12, p. 191
orientation in A. afarensis, 64
orientation in A. sediba, 65
orientation in arboreal apes, 88
orientation in Homo, 88
orientation in Nariokotome boy, 65
Gona, 31, 100, 103, 124
Gorillas, 12, 46, 48, 49, 68, 69
Grasslands, 84, 85, 95, 101
wooded, 55
Gripping
cylinders, 30
cylindrical clubhandles, 30
quadrupedal, 15
spheres and cylinders, 110
spheroidal stones, 37
stones, 30

tree limbs, 15, 113
Grips, 109
    chimpanzee, 113, 116
    clubbing, 175
    cylinder squeeze against the palm, 116
    fingertip pad, 116
    fingertip pad on missile, 110
    firm on club, 110, 175
    for stone tool-making, 131, 132
    forceful precision, 130
    hominin, 30, 54
    hominin, evolution of, 136
    palmar squeeze on a cylinder, 119
    power, 116, 118, 123, 119
    power/clubbing, 113, 116, 119, 122, 123
        figure 7, p. 118
        figure 8, p. 120
    precision, 113, 116
        figure 6, p. 117
        figure 8, p. 120
    precision and power, 115, 130, 120
    precision/throwing, 116, 121, 123
        figure 6, p. 117
        figure 8, p. 120
    sphere held by the fingertip pads, 116
    throwing and clubbing, 94, 136
Grotte de l'Ours, 34
Group aggression
    by males, 106
Habitats, 85
Habitats, open
    dry, 84
    expansion of, 55, 85
    temperate, 100
Hadar, 126, 167, 211
Hammerstones, 33, 36, 37, 130 figure 9, p. 131
Hand
    chimpanzee, 111-114
    in hominin ancestor ape-like, 111
    arboreal traits, 111
Hand, hominin, 30, 54. *See also* **individual hominin taxa**
    exapted for stone tool-making, 133
Hand, hominin ancestor, 110
Hand, modern human, 109, 113
    and throwing, 188
    apical tufts, 113
    ball release sequence, 233
    evolution of, 110
    exaptations, 133
        figure 4, p. 112
    fingertip pads, 113
    flexor pollicis longus, 115, 119, 130
    grasp of weapons, 122
    hand bones
        figure 5, p. 114
    metacarpals, 113
    movement during throwing, clubbing 121, 233
    palm, palmaris brevis, 115
    phalanges, 92. 94, 113
    shift in finger robusticity, 113
    styloid process, 115
    thumb musculoskeletal robusticity, 122
    thumb opposability, 115, 122
    thumb, fingers, palm, 113
    thumb, mobility of, 112
    trajectory during throwing, 233
    wrist extension, 113

INDEX

Handedness
    ancient hominin trait, 147
    and fighting, 151
    and language, 144, 145
    and skill, 146
    and task complexity, 146
    and throwing, striking, 148
    definition, 139
    development of, 140
    explanation of, 149
    hypotheses for, 143
    in chimpanzees, 141
    in *H. neanderthalensis*, 147
    in hominin/panin ancestor142
    in Paleolithic *H. sapiens*, 148
    in modern humans: inherited or learned?, 139
    modern human ratio, 139
    less evidence in apes, 140
    neural control of, 235
    neural control of inter-segmental dynamics, 236
    persistance of left-handers, 140
    positive evidence in apes, 141
    task-specific, in apes, 142
    unique aspect in humans, 140, 143

Hippopotamus, 101, 108

Hominin origins, 7, 11, 12, 23, 29, 86, 175, 255
    and use of weapons, 23

Hominins, early, 26, ,75, 88
    earliest anatomy, 14, 16
    first appearance of, 12
    population variability, 69
    requirements for throwing effectively, 237
    transitional, 55

Hominins, fossil *see also* **individual hominin taxa**
    assigning gender, 46
    size dimorphism, 46

Homo, *see also* **Body Size, increase in *Homo***
    and enhanced access to meat, 81
    and lengthening of lumbar spine, 90
    dedicated bipedal locomotion, 57
    early, 61, 86
    emergence of, 54
    reduced upper limb robusticity, 70
    transition to, 9, 59, 61, 66, 69, 73, 75-77, 83, 98, 161
    and mosaic evolution, 59

Homo antecessor
    clavicle, 65

Homo erectus, 56-59, 61, 70, 71, 74, 75, 78, 81, 98, 102, 168
    abandoned the trees, 85
    and control of fire, 76
    and increase of carnivory, 103
    and use of weapons, 104
    humero/femoral index, 67
    iliac blades, 66
    loss of arboreal adaptation, 75
    mothers need help?, 79
    Nariokotome boy, 65
    robusticity, 168
    shortened forearm in, 71
    size dimorphism, 48, 53, 54,
    small brain size increase, 64
    time of appearance, 55
    transition to, 53, 55, 56, 59,
        figure 3, p. 60, 63, 68, 74, 84-86, 90, 100, 107

***Homo*, early**, 88
    and modern limb

INDEX

proportions, 71
Homo ergaster, 55
Homo habilis, 56, 57,
  See also **A./H. habilis**
  and small body size, 59
  clavicle, 56
  cranial capacity, 56
  hand and foot, 56
  highly diverse, 57
  two species concept, 57
  widely dispersed, 57
Homo heidelbergensis,
  62, 170
  increase in brain size in, 64
  robusticity, 168
Homo neanderthalensis,
  129, 147, 168, 169
  robusticity, 168
  shoulders, clavicles, 65
Homo sapiens, 7, 59,
  62, 68, 72, See also
  Humans, Modern
  only surviving hominins, 8
Human evolution, 2
Human origins, 2
Humans, modern, 7,
  14, 16, 22, 25, 26, 40,
  43, 45-50, 56, 65, 69,
  70, 76, 84, 87, 98, 171-
  173, 180
  and development of
  throwing and striking, 179,
  201
  and maturation of
  throwing and striking, 179
  bipedal locomotion, 207
  central nervous system
  control of throwing, 229
  clavicle, 65
  handedness, 139
  neural control of, 235
  throwing and walking
    figure 15, p. 208
Humero/femoral

index, 56,
  apes, hominins, 67
  Dmanisi hominins, 58
  Table IV, 68
Humerus
  orientation during
  throwing, 89, 189-191
  rotation during throwing,
  89, 187
Hunting, 27, 78, 80, 86,
  99, 100-103, 106, 108,
  242
  and team aggression, 105
  avoided by females, 23
  by male humans, 23
  by males, 201
  cooperative by males, 106
  in chimpanzees, 21
  males in groups, 106
  persistence, 170
  with weapons, 103, 105,
  107
Hypotheses
  for bipedal locomotion
  aquatic, 217
  carrying, 215, 245
  combining hypotheses, 219
  cooking, 75, 81
  fighting, 151
  forelimb fighting, 159
  display, 216
  energetic, 216
  genetic mutation, 218
  orthograde clambering,
  218
  postural feeding, 215
  sentinel, 215
  shore dweller, 217
  tool use, 216
  for canine change in shape
  change in diet, 161
  improved chewing of food,
  160
  for canine diminution
  collaboration and

373

cooperation, 159
hand-held weapons, 159-161
male pacificism, 158, 159
for hand evolution, food gathering, processing, 135
foot-hand pleiotropy, 136
loss of foot grip, 135
manipulation, 135
relaxed selection, 135
selection for climbing, 135
stone tool-making, 129
for handedness, bipedalism, tool use, 144
genetic mutation, 143
language, 144
for hominin robusticity
change in diet, 169
long-distance walking, 170
mechanical loading, 169
resistance to thrown stones, 169
for reduced hominin robusticity, 176
decreased mate competition, 176
improved locomotion, nutrition, 177
glenoid fossa orientation, 89
reach-and-grasp, 234
Ileret, 102, 214
footprints, 214
Ilium
in *H. erectus*, 66
Injury protection factors,
during throwing, 122, 231
Intermembral index
chimpanzee, 69
gorilla, 69
Homo, 70
orangutan, 69
Isimila, 34, 35
Isolation
geographic, 7

reproductive, 8, 12, 55
Joint, carpometacarpal
close-packed positions, 121
Kanapoi, 125, 211
KNM-WT 15000
Nariokotome boy, 55, 64, 66, 67
Koobi Fora, 32, 55, 57, 100, 101
La Quina, 34
Lactation, 23, 25, 78, 79, 88, 201
Laetoli, 126, 211
footprints, 211, 214
Laws, biological, 2
Leg thrusts
and ground reaction forces, 93
and missile velocity, 93
in throwing, 93
in throwing and clubbing, 90
Legacy, hominin, 110
Length, relative
of arms and legs, 67
Lever length, 174
Lever-arm
maximal biomechanical, 89, 120, 189, 191, 231
Limb length, upper
similar in *Homo* and Chimpanzees, 70
Limb proportions
pre and post-*Homo* hominins, 71
Limbs, lower, 61
advantages of length increase, 91
and energetic costs of locomotion, 91
and robusticity in *H. erectus*, 91
in *Homo*, 70
increase in size in *Homo*,

INDEX

90
length and transition to
  Homo, 91
role in throwing and
  clubbing, 93
size and locomotor
  energetic costs, 91
Limbs, upper, 61
in arborealism and bipedal
  walking, 92
in Nariokotome Boy, 70
manipulation, tool use, 92
reduced mass in *Homo*, 92
similarities in apes and
  humans, 92
Limbs, upper, lower
in *Homo,* adaptation to
  throwing and clubbing,
  92, 93
ape and monkey, 12
australopithecine, 85
hominin, 8, 97
Lineage, hominin, 26, 86
beginning, 7, 11
continuity of, 9
origin in Africa, 11
Locomotion, arboreal, 67, 70, 83, 90
Locomotion, bipedal,
  See also **Bipedalism**
  and **Weapons, bipedal**
  use of, 10, 13, 15, 19,
  58, 66, 203, 204, 207,
  243-246,
and weapons, 221
and human origins, 204, 257
and throwing, 221
body type and efficiency, 176
explanation of origin, 220
hypothetical ancestor
  model, 205

in *A. afarensis*, 91
in fossil hominins, 209
lack of a compelling
  explanation, 204
long legs and energetic
  effects, 90
chimpanzee model, 205
modern human and
  chimpanzee, 207
obstacle to explanation of, 219
preadaptation by
  arborealism, 205
the primordial hominin
  behavior?, 204
Locomotion, human
adaptive changes in
  modern humans, 209
Locomotion, quadruped 16
Lower body
rotation, 66
Maka, 211
Makapansgat, 34
Malapa, 56, 128, 167, 213
Malnourishment, 25
Manuports, 31, 58, 100
used as weapons, 31. 104
at Olduvai, 32
cobblestone, 31
Mass, center of, 66, 67, 93
Mastication, heavy, 72
Mating opportunities,
access to, 6, 10, 16 24-26
Meat, 27, 75, 76, 80, 81, 84, 86, 101, 107
acquisition of, 81, 86, 94, 95, 98, 102
animal tissues, 80
sharing in chimpanzees, 21
Meat-eating
in chimpanzees, 98
in modern humans, 98
Meat-sharing for sex
in chimpanzees, 21

Megadontia, 71
   in robust australopithecines, 84

Metacarpal, third
   styloid process, 122

Metacarpals
   in chimpanzees and modern humans figure 5, p. 114

Miocene apes, 13-15, 111

Missiles
   aerodynamic properties, 30
   in rock collections, 31
   spheroidal, 109, 123

Molars, 71

Momentum
   applied to target, 174

Monkeys, 21, 49, 101, 106, 124

Monogamy, 42, 43, 50

Mosaic, evolutionary
   of primitive and derived traits, 59, 69

Musculature, lower limbs
   large in *Homo*, 70

Musculature, upper limbs
   large in Chimpanzees, 70

Nariokotome boy, 65-68

Natural selection, 5, 6, 9, 14, 20, 24, 29, 45, 51, 54, 69, 73, 75, 76, 82, 85, 161, 162, 169, 172, 200, 241, 247, 256, 261, 262, 264, 268
   and arborealism, 95
   and differential reproduction, 26
   and genetic variability, 69
   and glenoid fossa orientation, 89
   and hominin origins, 27
   and net reproductive advantage, 176
   for bipedal throwing and striking, 220, 255
   for bipedal use of weapons, 77, 85, 86, 92, 93, 95, 97, 105, 108, 109, 151
   for endurance running, throwing and clubbing, 176
   for fighting, 41
   for gripping spheres and cylinders, 133
   for endurance locomotion, 76
   for throwing and clubbing behavior, 180
   for throwing, striking, 175
   of behavior, 6

Needs, biological, 26, 75, 86

Nutrition, 5, 21, 25, 26, 78-80, 87, 97

O.H. 62 *Homo habilis*, 61

Oblique muscles
   external and internal, 66

Oldowan tools, 58, 100, 130

Olduvai Gorge, 31, 32, 34, 35, 56, 58, 99, 100-102, 104

Olorgesailie, 32, 34, 105

Ontogeny of striking, 193-195

Ontogeny of throwing, 180-186

Orangutan, 12, 46, 49, 68, 69

Oreopithecus, 12, 13, 15

Origins, hominin, 8, 10, 13, 30, 95

Origins, human, 2, 9, 14, 17, 19, 27

Orrorin tugenensis, 13, 124
   and bipedalism, 209
   canine tooth, 155

flexor pollicis longus, 123
hand bones, 123
humerus, 123
molars, 71
robusticity, 166
thumb phalanx, 111
Outgroups
chimpanzee, 20, 106
hominin, 24
marauding, 26, 95
Ovulatory crypsis, 243, 246
Pair-bonding, 243, 245, 246
Paleodeme, 56, 58
Paleolithic, Middle, 108
Pan
chimpanzees, bonobos, 67
Pan paniscus
bonobo chimpanzees, 14
Pan troglodytes
common chimpanzee, 14
Paranthropus see Australopithecus robustus
Pelvis, 65
rotation of, 66, 83, 90, 93, 187, 193
Phalanges
curvature of, 56
in chimpanzees and modern humans, figure 5, p. 114
Phenotype, 5
Pleistocene, 108
Polygyny, 41, 50
Predation
at Olduvai, 31
Predator guild, 107
Predators, 6, 24, 26, 27, 31-33, 69, 75, 86, 88, 98, 99, 106. 107
Prediction
and future evolution, 249
of hand evolution, 109
Predictions from the theory, 275

accuracy of throwing and striking, 276
bipedal weapons use, 282
CNS aiming, 276
CNS anatomy, 276
CNS control of lower body, 276
CNS stored motions, 276
dynamic vertical balance, 277
elbow joint, 279
forearm muscles, 280
gluteus maximus, 277
H. erectus hand bones, 280
humerus rotation, 278
injury protection mechanisms, 281
lower limb structure, 277
muscles and humerus rotation, 279
ontogeny of striking, 275
pectoralis major, 279
rotation of pelvis, 277
shoulder anatomy, 278
upper limb, 280
Prey, 6, 21-23, 27, 86, 99, 101, 107
Promiscuity, 41, 43
Provisioning
by males, 79, 243, 245
Puberty, 44
Reproductive advantages, 6, 14, 24, 51, 53, 54, 74, 75, 83, 85, 92
and access to food, 26
of large size in *Homo*, 78
of lateral glenoid fossa orientation, 89
of weapons use, 27
Reproductive benefits, 77, 85, 9, 10, 19, 20, 29, 75,86, 88, 95, 106, 107, 177, 180, 247
and dominance in

chimpanzees, 20
of team aggression, 106
Reproductive success, 5, 14, 24, 26, 30, 40, 44, 219
different gender perspectives, 25, 200, 201
fighting, chimpanzees, 20
from weapons use, 110, 225, 255
limited by mating opportunities, 25
Resources
scarce, 26
Robusticity, musculoskeletal, 61, 78, 87, 166, 177
and genetics, 172
correlation with striking prowess, 174
correlation with throwing and striking prowess, 172
definition, 165
high in fossil hominins, 166
high in power throwers, 171
high in elite throwers and strikers, 172, 173
low in endurance runners, 171
of childen, 168
recent decline in, 169
Sahelanthropus tchadensis, 13, 154
and bipedalism, 209
Savannas, 86, 86, 95
Scavenging, 80, 99, 103
aggressive, 27, 99
confrontational, 99, 100, 101, 102, 105, 106, 107
with weapons, 103, 107
definition of, 98
passive, 99, 100
power, 99
Scavenging animals, 32
Scenarios
cooking, 243
man the hunter, 242
man the provisioner, 243
man the thrower, 243
woman the gatherer, 242
Schönigen
and control of fire, 77
Science, 1
Sexual selection, 40, 45, 272
females choose, 25, 40, 243, 245
males fight, 40, 49
Shoulder
muscles, *A. afarensis*, 66
Shoulder joint. *See also* **Glenoid foss**
mobility, 65, 82, 88
orientation, 74
orientation during throwing
figure 12, p. 190
orientation in *A./H. habilis*, 65
orientation in apes, 64
orientation in *Australopithecus*, 56, 64
orientation in *homo,,83*
Shoulder strength
higher in males, 44
Shoulders
Shoulders, 74
breadth, 65, 82, 88
breadth, change in puberty, 44
plane of, 89
Sima de los Huesos, 147
Size dimorphism, 39, 43, 49-51
and behavior, 40
and puberty, 44
and sex hormones, 44
and sexual selection, 40
decrease in *Homo*, 63, 79

INDEX

figure 2, p. 42
gender (sexual), 39
in *A. afarensis*, 40, 48, 49
in *Australopithecus*, 16
in early hominins, 48
in hominins, 86, table II, p. 48
in primates, 45
reduced in *H. erectus*, 61
Size dimorphism paradox
solution to, 50
Size dimorphism, primates
table I, p, 45
Size monomorphism
and monogamy, 41
South Africa, 56
South Turkwel, 127
Spears, 77
oldest known, 30
throwing, 105, 107
thrusting, 105, 123
thrusting or throwing, 104
Species, 5
biological, 7
competition among members of, 6
different, difficult to document, 7
each a special narrative, 6
extinct, 8
fossil, 47
phenotypic differences, 8
H. habilis, one species or two?, 57
hominin, 77
hominin ancestral, 24
origin of, 4, 7, 24, 55, 56, 85
relationships with, 6
reproduction of, 85
variability in, 5
Spheroids
composition, 33
dimensions, 35

function of, 35
geographical range, 34
hominin-made, 33
quantitative analysis, 35
use as missiles, 35
Spheroids, subspheroids
end products of tool-making, 33, 36
figure 1, p. 34
Stance, bipedal, 10, 13, 45, 66, 86, 110, 226, 242, 243
Stature
and locomotion, 78
and thermoregulation, 78
and throwing and clubbing, 172- 174
calculating in fossils, 47
increased in *H. erectus*, 61
Sterkfontein, 34, 35, 127, 167, 212
Stockpiles, stone, 31
Stone Age, Early, 35
Stone Age, Middle, 35
Stone balls, 34
Stone cobbles
throwing and tool-making, 32
Stone tool cut marks, 98-103
Stone tool-making, 32, 33, 36, 103, 129, 130
and reproductive benefits, 133
figure 9, p, 131
hands of first makers, 132
original purpose, 132
Stone tools, 99, 102, 105
no hand made = no weapons, 29, 160, 245
sharp flakes, 132
Stones
for throwing and tool-making, 32
smooth-surfaced, water-rounded, 31
spheroidal, 29

379

suitable for throwing, 30
used as weapons, 104
Stones, throwing
characteristics, 30
Stones, unmodified
suitable for throwing, 31
used as weapons, 32
Stride length
and bipedal locomotion, 90, 91
and throwing, 93
Striking, 42, *See also* **Clubbing** and **Club-swinging**
Striking motion
complexity of, 223
contrast with walking, 223
description of, 194, 195
in adult males, 194
in boy and adult male figure 14, p. 195
overarm figure 13, p. 194
overarm similar to throwing, 193, 194
similar to throwing motion, 194
similarity in children and adults, 196
Striking, development of, 193
gender differences, 193, 194
similarity to throwing development, 193
Suids, 100, 101
Swartkrans, 34, 77, 129, 213
Team aggression, 16, 106
Teeth
diminution in *Homo*, 72
enamel, 71, 84, 94
molars, reduction in size, 94
size decrease in *H. erectus*, 72
Territorial patrol
by male chimpanzees, 20
Territory
access to, 21
Testosterone
and muscle mass, 44
effects at puberty, 45
Thermoregulation, 78
Thorax, 82
barrel-shaped, 66, 82
change in shape, 82
cone-shaped, 65
funnel-shaped, 65, 74, 88
in apes and early hominins, 65
in *H. erectus*, 88
rotation of, 83, 90, 187
Thorax/pelvis immobility
in apes and early hominins, 65
Throwing, 10, 22, 30, 32, 34, 35, *42*, 73, 82, 242, 243
a stone or a sphere, 110
and chest and shoulder anatomy, 82
begins shortly after birth in *H. sapiens*, 181
compared with walking figure 15, p. 208
in chimpanzees, 20-23
in female hominins, 23
in modern humans, 84
injury prevention mechanisms, 191
acceleration phase, 191
at ball release, 191
overarm or sidearm, 89
role of central nervous system, 229
strong genetic element, 199
timing of release, 233
Throwing and clubbing, 87, 88, 90, 97, 104, 105, 107, *See also* **Throwing and striking**
aggressive, 24

INDEX

    development and maturation, 179, 180
    increase in prowess, 95
    Marzke's perspective, 134
    overarm, sidearm, 189
    unique in human ontogeny, 200
Throwing and clubbing motions, 88, 90, 174
    biomechanics of, 93
    importance of genetics, 199
    role of legs, hip and trunk, 93
Throwing and club-swinging, 27, 77
    adaptation to, 61
    advanced in boys compared to girls, 44
Throwing and striking, 24,
    See also **Throwing and clubbing**
    by male chimpanzees, 22
    evolution of gender differences, 200
    gender difference in hominins, 87
    gender effects in adult athletes, 198
    gender similarities and differences in adults, 197
    in chimpanzees, 21, 122
    in hominins reproductive benefits, 22
    kinetic energy linkage mechanism, 196
Throwing motion
    figure 15, 208
    complexity of, 223
    contrast with walking, 223
    description of, 186, 187, 189
    develops in all modern human children, acceleration phase, 89, 121, 122, 189
    adult athletes, 186
    biomechanics of, 92
    central nervous system control, 231
    compared with walking, 182
    elite females and males, 197
    gender difference, 181
    gender difference in adults, 192
    in child and adult, figure 11, p. 188
    precursor of mature form, 186, 189
    kinematics and age, 186
    missile release, 188
    organization of CNS control, 235
    rapidity of, 189, 235
    same in all elite throwers, 189
    standardized, 181
    stride and foot strike, 187
    stride length in adult males, 192
    stride length in children, 192
    three stages, figure 16, p. 222
    windup, 93, 187
Throwing development
    biomechanics, 186
    boys outperform girls, 181
    effects of instruction, 184, 185
    evidence of a genetic basis, 198
    gender difference, 182-185
    in young children, 180
    individually variable, 182
    occurs in all children, 181
    stages in, 180
Thumb bones

chimpanzee and modern humans, figure 5, p. 114
Toes
    curvature of, 11, 15, 57, 74
Tool-making, 31
Tools
    sharp-edged, 33, 104
Tools, spheroidal, 36
Tooth marks, 99, 101
Torques
    gravity, 232
    interaction, 231
    muscle, 231
    predicted and controlled by CNS, 233
    velocity-dependent, 232
    viscoelastic, 231
Torralba, 105
Trunk
    rotation of, 66
Tubers, 76, 80
'Ubeidiya, 101
Ungulates, 86, 99, 107
Upper body, 66, 82, 44, 89
Variation, genetic, 5
Velocity, 87
    internal rotation of humerus, 191
    of ball release, 184, 232
    of flexion during throwing, 121
    of missile or club, 93, 174, 175, 192
    gender difference, 197
    of thrown ball, 181-183, 175, 185, 186, 192, 196, 198, 231
Vertebrae, lumbar, 65,66
Waist, 66, 74, 83
    and rotation of thorax and pelvis, 89
    in apes and early hominins, 65
    in *H. erectus*, 61, 66
Walking. *see* **Bipedalism,** **Locomotion, bipedal**
Weapons,27, 86,*See* also **Bipedalism** and **Locomotion, bipedal**
    and ancient hominins, 97
    and carcass acquisition, 103
    and female choice, 25
    and hominin origins, 10
    and size dimorphism, 39
    available in environment, 29
    available to early hominins, 36
    bipedal stance, 23, 107, 110
    bipedal use,17, 19, 55
    bipedal use and bipedal locomotion, 45, 50, 51, 77, 86, 88, 116, 155, 220. 222, 225, 244, 255
    missiles and clubs, 51
    clubs, stones, knives, spears, 104
    early hominin, 29
    grip of, 109
    hand-held, 10, 41, 42, 86
    and biological needs, 26
    mass, 30, 175
    rocks, clubs, spears, stones, bones, 22, 29, 78, 82, 86, 105, 107
    branches, sticks, 29
    throwing, 32, 35
    unmodified natural objects, 29
    use to defend territory, 24
    use to deter predators, 24
    use to expand territory, 24
    use to obtain meat, 103
    used by throwing and clubbing, 104
    utility of, 27
West Turkana, 58
    weapons, 19. 160, 245

made by hominins, 29
missiles & clubs. 51
no hand-made = no weapons, 29, 160, 245
present in environment, 36
projectile, 105
replacement of canines for fighting, 160, 162

Woodlands, 85, 95, 101

closed-canopy, 85
riverine, 85

Wrist
movement during clubbing, 122
movement during throwing, 121

Zhoukoudian, 105

www.ingramcontent.com/pod-product-compliance
Lightning Source LLC
Chambersburg PA
CBHW051758170526
45167CB00005B/1799